T0137218

Studies in Computational Intelligence

Volume 681

Series editor

Janusz Kacprzyk, Polish Academy of Sciences, Warsaw, Poland
e-mail: kacprzyk@ibspan.waw.pl

About this Series

The series "Studies in Computational Intelligence" (SCI) publishes new developments and advances in the various areas of computational intelligence—quickly and with a high quality. The intent is to cover the theory, applications, and design methods of computational intelligence, as embedded in the fields of engineering, computer science, physics and life sciences, as well as the methodologies behind them. The series contains monographs, lecture notes and edited volumes in computational intelligence spanning the areas of neural networks, connectionist systems, genetic algorithms, evolutionary computation, artificial intelligence, cellular automata, self-organizing systems, soft computing, fuzzy systems, and hybrid intelligent systems. Of particular value to both the contributors and the readership are the short publication timeframe and the worldwide distribution, which enable both wide and rapid dissemination of research output.

More information about this series at http://www.springer.com/series/7092

Krassimir Georgiev · Michail Todorov
Ivan Georgiev
Editors

Advanced Computing in Industrial Mathematics

Revised Selected Papers of the 10th Annual Meeting of the Bulgarian Section of SIAM December 21–22, 2015, Sofia, Bulgaria

Editors
Krassimir Georgiev
Institute of Information and Communication Technologies
Bulgarian Academy of Sciences
Sofia
Bulgaria

Michail Todorov
Faculty of Applied Mathematics and Informatics
Technical University of Sofia
Sofia
Bulgaria

Ivan Georgiev
Institute of Information and Communication Technologies, Institute of Mathematics and Informatics
Bulgarian Academy of Sciences
Sofia
Bulgaria

ISSN 1860-949X ISSN 1860-9503 (electronic)
Studies in Computational Intelligence
ISBN 978-3-319-84185-4 ISBN 978-3-319-49544-6 (eBook)
DOI 10.1007/978-3-319-49544-6

Softcover reprint of the hardcover 1st edition 2017

Printed on acid-free paper

This Springer imprint is published by Springer Nature
The registered company is Springer International Publishing AG
The registered company address is: Gewerbestrasse 11, 6330 Cham, Switzerland

Preface

This book contains peer-reviewed papers presented during the 10th Annual Meeting of the Bulgarian Section of SIAM (BGSIAM), December 21–22, 2015, Sofia, Bulgaria. The conference was hosted by the Institute of Mathematics and Informatics of the Bulgarian Academy of Science (http://www.math.bas.bg/IMIdocs/BGSIAM/).

The conference's topics of interest were as follows:

- Industrial mathematics;
- Scientific computing;
- Numerical methods and algorithms;
- Hierarchical and multilevel methods;
- High performance computing and applications in the industrial mathematics;
- Partial differential equations and their applications;
- Control and uncertain systems;
- Monte Carlo and quasi-Monte Carlo methods;
- Neural networks, metaheuristics, genetic algorithms;
- Financial mathematics.

The list of the plenary invited speakers includes several internationally recognized scienists: *Tzanio Kolev* (Lawrence Livermore National Laboratory, USA); *Johannes Kraus* (Universität Duisburg-Essen, Germany); *Maya Neytcheva* (Uppsala University, Sweden) and *Vladimir Veliov* (Vienna University of Technology, Austria).

The further development of the society is in deep connection with the successful solution of very challenging and extremely difficult real-life problems. Mathematicians (both theoretical and applied), computer scientists, engineers, physicians, chemists, biologists, etc. are developing and using complicated and robust mathematical and computer models in the attempts to resolve successfully such kind of problems which are appearing very often.

The Industrial Mathematics is one of the most prominent examples of an interdisciplinary area involving mathematics, computer science, scientific computations, engineering, physics, chemistry, medicine etc.

The tools of Industrial Mathematics are usually based on mathematical models and corresponding computer codes that are used to perform virtual experiments to obtain new data or to better understand the existing experimental results.

The modern fast supercomputers are one of the main tools to find accurate enough and fast enough solutions of many of the nowadays large and very complicate problems. However, unfortunately, not in all cases and not for all important problems. *Arthur Jaffe* predicted in 1984 (A. Jaffe, "Ordering the universe: The role of mathematics", SIAM Review., Vol. 26 (1984), pp. 475–488) *Although the fastest computers can execute millions of operations in one second, they are always too slow. This may seem a paradox, but the heart of the matter is: the bigger and better computers become, the larger are the problems scientists and engineers want to solve.*

We, the editors of this issue, would like to thank all the referees of the presented papers (also the referees of the not published papers) for preparing in time and for their professional reviews and for the constructive criticism which resulted in considerable improvements of the quality of the accepted papers.

Sofia, Bulgaria

Krassimir Georgiev
Michail Todorov
Ivan Georgiev

Contents

Factorizations in Special Relativity and Quantum Scattering on the Line

Danail S. Brezov, Clementina D. Mladenova and Ivaïlo M. Mladenov

Abstract We extend an old result due to Piña on three-dimensional rotations to the hyperbolic case and utilize it to construct a specific factorization scheme for the isometries in $\mathbb{R}^{2,1}$. Although somewhat restrictive compared to Euler-type decompositions, our setting allows for gimbal lock control and decouples the dependence on the compound transformation's invariant axis and angle (rapidity), which provides convenience from both theoretical and practical point of view. In some particularly symmetric cases the solutions are obtained in a simple way from the natural parameterization of the Poincaré disk. Apart from the obvious relation to 2+1 dimensional relativity, we discuss this approach in the context of the monodromy matrix description of quantum mechanical scattering. In the case of $\mathbb{R}^{3,1}$ isometries, our construction is shown to generalize naturally the well-known Wigner decomposition.

1 Preliminaries

To the knowledge of the authors, Piña [1] was the first to suggest an expression for the generic rotation that links two arbitrarily given points on the unit sphere $\mathscr{R} : \mathbf{u} \to \mathbf{v} \in \mathbb{S}^2$. Adapting his precise argument for the proper Lorentz group $\mathsf{SO}^+(2, 1)$, we obtain the family of pseudo-rotations $\mathscr{P}_\lambda : \mathbf{u} \to \mathbf{v}$, $\lambda \in \mathbb{RP}^1$ for two given vectors $\mathbf{u}, \mathbf{v} \in \mathbb{R}^{2,1}$ with equal norms. Namely, we set $\mathbf{u}^2 = \mathbf{v}^2 = \varepsilon$, where the square is obtained with respect to the flat Lorentz metric $\eta = \mathrm{diag}\,(1, 1, -1)$ and $\varepsilon = \pm 1$ in the

D.S. Brezov
Department of Mathematics, University of Architecture Civil Engineering and Geodesy, 1 Hristo Smirnenski Blvd., 1046 Sofia, Bulgaria
e-mail: danail.brezov@gmail.com

C.D. Mladenova
Institute of Mechanics, Bulgarian Academy of Sciences, Acad. G. Bonchev Str., Bl. 4, 1113 Sofia, Bulgaria
e-mail: clem@imbm.bas.bg

I.M. Mladenov (✉)
Institute of Biophysics, Bulgarian Academy of Sciences, Acad. G. Bonchev Str., Bl. 21, 1113 Sofia, Bulgaria
e-mail: mladenov@bio21.bas.bg

K. Georgiev et al. (eds.), *Advanced Computing in Industrial Mathematics*, Studies in Computational Intelligence 681, DOI 10.1007/978-3-319-49544-6_1

space-like (respectively, time-like) case and $\varepsilon = 0$ in the light-like one, in which the normalization is arbitrary.

We choose to work with the vector-parameter description of $\mathsf{SO}^+(2,1)$ as being most efficient. Let $\zeta = (\zeta_0, \boldsymbol{\zeta})$, $\zeta_0^2 - \boldsymbol{\zeta}^2 = 1$ be a unit split-quaternion given by its real and imaginary parts. Then, it represents an element of $\mathsf{SU}(1,1) \cong \mathsf{SL}(2,\mathbb{R})$ and thus, induces an isometry of $\mathbb{R}^{2,1}$ via the adjoint action on the corresponding Lie algebra identified with the space of three-vectors. Homogeneity allows for projecting and thus obtaining the associated *vector-parameter* $\mathbf{c} = \dfrac{\boldsymbol{\zeta}}{\zeta_0}$ (see [2]). Note, however, that $\mathbf{c}$ is not a vector as it inhibits a projective space, e.g. vector-parameters of half-turns correspond to infinite points. Moreover, if we express the hyperbolic dot and cross products in terms of their Euclidean analogues as

$$\mathbf{c}_2 \cdot \mathbf{c}_1 = (\,\mathbf{c}_2, \eta\,\mathbf{c}_1), \qquad \mathbf{c}_2 \curlywedge \mathbf{c}_1 = \eta\,(\mathbf{c}_2 \times \mathbf{c}_1)$$

we obtain the natural composition of such objects

$$\langle\,\mathbf{c}_2,\ \mathbf{c}_1\rangle = \frac{\mathbf{c}_2 + \mathbf{c}_1 + \mathbf{c}_2 \curlywedge \mathbf{c}_1}{1 + \mathbf{c}_2 \cdot \mathbf{c}_1} \tag{1}$$

inherited from split-quaternion multiplication. The latter reduces to vector summation only in the infinitesimal case or if $\mathbf{c}_{1,2}$ are both mutually parallel and orthogonal null vectors. The above definition also allows for representing the associated $\mathbb{R}^{2,1}$ isometry (pseudo-rotation) in the Cayley form

$$\mathrm{Cay}\colon\ \mathbf{c}\ \in\ \mathbb{RP}^3 \quad\longrightarrow\quad \mathscr{P}(\mathbf{c}) = \frac{\mathscr{I} + \mathbf{c}^{\curlywedge}}{\mathscr{I} - \mathbf{c}^{\curlywedge}}\ \in\ \mathsf{SO}(2,1) \tag{2}$$

where $\mathscr{I}$ stands for the identity in $\mathbb{R}^{2,1}$ and $\mathbf{c}^{\curlywedge} = \eta\,\mathbf{c}^{\times} \in \mathfrak{so}(2,1)$ extends the Hodge duality and thus yields $\mathbf{c}_2^{\curlywedge}\,\mathbf{c}_1 = \mathbf{c}_2 \curlywedge \mathbf{c}_1$ in analogy with $\mathbf{c}_2^{\times}\mathbf{c}_1 = \mathbf{c}_2 \times \mathbf{c}_1$.

Now, let us return to Piña's construction. He argues that the invariant axis of a rotation $\mathscr{R}: \mathbf{u} \to \mathbf{v}$ must lie in the plane of mirror symmetry between $\mathbf{u}$ and $\mathbf{v}$. The same is true in the hyperbolic case as well, so we may write

$$\mathbf{c} = \alpha\,(\mathbf{u} \curlywedge \mathbf{v}) + \beta\,(\mathbf{u} + \mathbf{v}), \qquad \alpha, \beta \in \mathbb{RP}^1$$

and applying formula (2), substitute the above expression into the relation

$$\left(\mathscr{I} + \mathbf{c}^{\curlywedge}\right)\mathbf{u} = \left(\mathscr{I} - \mathbf{c}^{\curlywedge}\right)\mathbf{v}.$$

Under the reasonable assumption $\mathbf{u} \neq \mathbf{v}$, this yields $\alpha = -\dfrac{1}{\varepsilon + \mathbf{u}\cdot\mathbf{v}}$, but fails to determine the second parameter β. Then, the solution can be written in the form

$$\mathbf{c} = -\frac{1}{\varepsilon + \mathbf{u}\cdot\mathbf{v}}\,(\,\mathbf{u} \curlywedge \mathbf{v} + \lambda\,(\mathbf{u}+\mathbf{v})\,) \tag{3}$$

which is more suitable for our further study. Note that the parameter $\lambda \in \mathbb{RP}^1$ may become infinite in the time-like case $\varepsilon = -1$ that corresponds to a half-turn. For space-like vectors, i.e., $\varepsilon = 1$, $\lambda = \infty$ yields a time-reversing boost, which is not an element of the proper Lorentz group $\mathsf{SO}^+(2, 1)$, while in the isotropic case $\varepsilon = 0$ such choice of λ leads to a divergent pseudo-rotation matrix (2) and is thus not allowed. Similar divergence is present for $\mathbf{c}^2 = 1$. Such peculiarities of the hyperbolic setting related to its light cone structure lead to the appearance of a singularity other then the gimbal lock. In [3] we investigate this phenomenon (naturally referred to as *isotropic* or *light cone singularity*) in the context of generalized Euler-type decompositions.

2 The Conjugated Decomposition Technique

Elsewhere (cf. [2, 3]) we have developed techniques for decomposing generic transformations in $\mathsf{SO}(2, 1)$ and $\mathsf{SL}(2, \mathbb{R}) \cong \mathsf{SU}(1, 1)$, which generalize the Euler and Iwasawa decompositions, and discuss possible applications in $2+1$ dimensional special relativity and quantum mechanical scattering. The focus of the present article lies on a more specific type of factorizations, namely

$$\mathscr{P}(\mathbf{c}) = \mathscr{P}_1^{-1}\mathscr{P}_2^{-1}\mathscr{P}_3\mathscr{P}_2\mathscr{P}_1, \qquad \mathscr{P}_i = \mathscr{P}(\mathbf{c}_i) \tag{4}$$

we refer to as "conjugated" due to the following property

$$\mathscr{P}_1^{-1}\mathscr{P}_2^{-1}\mathscr{P}_3\mathscr{P}_2\mathscr{P}_1 = \mathscr{P}\left(\mathscr{P}_1^{-1}\mathscr{P}_2^{-1}\mathbf{c}_3\right) \quad \Longrightarrow \quad \mathbf{c}_3 = \mathscr{P}_2\mathscr{P}_1\mathbf{c} \tag{5}$$

which demands that $\mathbf{c}_3$ has the same norm as $\mathbf{c}$ and we may apply formula (3) to obtain the λ-labelled set of solutions for the vector-parameter $\tilde{\mathbf{c}} = \langle\, \mathbf{c}_2,\, \mathbf{c}_1 \rangle$.

Let us first denote $\mathbf{c} = \tau\mathbf{n}$ and $\mathbf{c}_i = \tau_i\hat{\mathbf{c}}_i$, where $\mathbf{n}$ and $\hat{\mathbf{c}}_i$ are normalized in the sense that $\varepsilon = \mathbf{n}^2 = \pm 1$ in the space-like (respectively time-like) case and $\varepsilon = 0$ in the isotropic one, and similarly for $\varepsilon_i = \hat{\mathbf{c}}_i^2$. Moreover, we make use of the notation (no summation is involved)

$$\upsilon_i = \mathbf{n} \cdot \hat{\mathbf{c}}_i, \qquad \tilde{\upsilon}_k = \varepsilon_{ijk}\, \hat{\mathbf{c}}_i \curlywedge \hat{\mathbf{c}}_j \cdot \mathbf{n}, \qquad \omega = \hat{\mathbf{c}}_1 \curlywedge \hat{\mathbf{c}}_2 \cdot \hat{\mathbf{c}}_3$$

where ε_{ijk} is the Levi-Civita symbol and i, j, k take different values.

Then, applying formula (3) to the transformation $\mathscr{P}_2\mathscr{P}_1\colon\ \mathbf{n} \to \hat{\mathbf{c}}_3$, we obtain

$$\tilde{\mathbf{c}} = \langle\, \mathbf{c}_2,\, \mathbf{c}_1 \rangle = -\frac{1}{\varepsilon + \upsilon_3}\left(\mathbf{n} \curlywedge \hat{\mathbf{c}}_3 + \lambda\,(\mathbf{n} + \hat{\mathbf{c}}_3)\right) \tag{6}$$

and the next step is to express the above vector-parameter as a composition

$$\tilde{\mathbf{c}} = \langle\, \tau_2\, \hat{\mathbf{c}}_2,\, \tau_1\, \hat{\mathbf{c}}_1 \rangle$$

with suitable coefficients $\tau_{1,2} \in \mathbb{RP}^1$. Multiplying both sides of the above equality on the left with $\tilde{\mathbf{c}}^{\curlywedge}$ and taking dot products with $\hat{\mathbf{c}}_{1,2}$ yields

$$\tau_1 = \frac{\tilde{\mathbf{c}} \cdot \hat{\mathbf{c}}_1 \curlywedge \hat{\mathbf{c}}_2}{\tilde{\mathbf{c}} \cdot \hat{\mathbf{c}}_{[2} g_{1]1}}, \qquad \tau_2 = \frac{\tilde{\mathbf{c}} \cdot \hat{\mathbf{c}}_1 \curlywedge \hat{\mathbf{c}}_2}{\tilde{\mathbf{c}} \cdot \hat{\mathbf{c}}_{[1} g_{2]2}} \tag{7}$$

where we use the standard notation $a_{[i}b_{j]} = a_i b_j - a_j b_i$ for anti-symmetrization and the *scalar parameters* $\tau_{1,2}$ are determined by substituting (6) into (7).

Note that the above solutions are valid only on a zero-measure set, since there are two parameters available and the dimension of the group is three. The condition may be derived from the invariant axis theorem asserting that

$$\tilde{\mathscr{P}} = \mathscr{P}_2 \mathscr{P}_1 \quad \Longleftrightarrow \quad \hat{\mathbf{c}}_2 \cdot \tilde{\mathscr{P}} \hat{\mathbf{c}}_1 = \hat{\mathbf{c}}_2 \cdot \hat{\mathbf{c}}_1 \tag{8}$$

which we write, using again the Cayley representation (2), in the form

$$\tilde{\mathbf{c}} \curlywedge \hat{\mathbf{c}}_2 \cdot \sum_{n=0}^{\infty} (\tilde{\mathbf{c}}^{\curlywedge})^n \hat{\mathbf{c}}_1 = \frac{1}{1 - \tilde{\mathbf{c}}^2} \left(\hat{\mathbf{c}}_1 \cdot \tilde{\mathbf{c}} \curlywedge \hat{\mathbf{c}}_2 + \tilde{\mathbf{c}} \curlywedge \hat{\mathbf{c}}_1 \cdot \tilde{\mathbf{c}} \curlywedge \hat{\mathbf{c}}_2 \right) = 0.$$

This ultimately leads to the simple equality

$$\tilde{\mathbf{c}} \cdot \hat{\mathbf{c}}_1 \curlywedge \hat{\mathbf{c}}_2 = \tilde{\mathbf{c}} \curlywedge \hat{\mathbf{c}}_1 \cdot \tilde{\mathbf{c}} \curlywedge \hat{\mathbf{c}}_2 \tag{9}$$

that will be used later to determine the value of λ in (6). Let us first point out, however, that formula (5) demands that the vectors $\mathbf{c}$ and $\mathbf{c}_3$ are of the same type ($\varepsilon_3 = \varepsilon$). This is a very restrictive condition and it is therefore wise to choose the three axes in the decomposition to represent the three types of vectors in $\mathbb{R}^{2,1}$ as in the classical Iwasawa setting, after which, to allow permutations.

Condition (9) leads to a quadratic equation for λ in the form

$$\begin{aligned} &((\upsilon_1 + g_{13})(\upsilon_2 + g_{23}) - 2g_{12}(\upsilon_3 + \varepsilon))\,\lambda^2 - \left(\tilde{\upsilon}_1\tilde{\upsilon}_2 + g_{12}(\upsilon_3^2 - \varepsilon^2) + \upsilon_{[1}g_{2]3}(\upsilon_3 + \varepsilon)\right) \\ &\qquad + (\tilde{\upsilon}_2(\upsilon_2 + g_{23}) - \tilde{\upsilon}_1(\upsilon_1 + g_{13}) + (\tilde{\upsilon}_3 + \omega)(\upsilon_3 + \varepsilon))\,\lambda = 0 \end{aligned}$$

whose discriminant needs to be non-negative, i.e.,

$$\begin{aligned} \Delta = {} & 4\,((\upsilon_1 + g_{13})(\upsilon_2 + g_{23}) - 2g_{12}(\upsilon_3 + \varepsilon))\left(\tilde{\upsilon}_1\tilde{\upsilon}_2 + g_{12}(\upsilon_3^2 - \varepsilon^2) + \upsilon_{[1}g_{2]3}(\upsilon_3 + \varepsilon)\right) \\ & + (\tilde{\upsilon}_2(\upsilon_2 + g_{23}) - \tilde{\upsilon}_1(\upsilon_1 + g_{13}) + (\tilde{\upsilon}_3 + \omega)(\upsilon_3 + \varepsilon))^2 \geq 0 \end{aligned} \tag{10}$$

in order to ensure the existence of real solutions given by the formula

$$\lambda^{\pm} = \frac{\tilde{\upsilon}_1(\upsilon_1 + g_{13}) - \tilde{\upsilon}_2(\upsilon_2 + g_{23}) - (\tilde{\upsilon}_3 + \omega)(\upsilon_3 + \varepsilon) \pm \sqrt{\Delta}}{2\,((\upsilon_1 + g_{13})(\upsilon_2 + g_{23}) - 2g_{12}(\upsilon_3 + \varepsilon))}. \tag{11}$$

Then, the scalar parameters $\tau_{1,2}$ may be expressed as

$$\tau_1^{\pm} = \frac{(\upsilon_3 + \omega)\,\lambda^{\pm} - \upsilon_{[1}g_{2]3}}{(g_{1[1}\upsilon_{2]} + g_{1[1}g_{2]3})\,\lambda^{\pm} - \varepsilon_1\tilde{\upsilon}_1 - g_{12}\tilde{\upsilon}_2}$$
$$\tau_2^{\pm} = \frac{(\upsilon_3 + \omega)\,\lambda^{\pm} - \upsilon_{[1}g_{2]3}}{(g_{2[2}\upsilon_{1]} + g_{2[2}g_{1]3})\,\lambda^{\pm} + g_{12}\tilde{\upsilon}_1 + \varepsilon_2\tilde{\upsilon}_2}. \tag{12}$$

In particular, for $\mathbf{n} = \pm\hat{\mathbf{c}}_1$ one has the condition $\hat{\mathbf{c}}_2 \in \mathrm{Span}\{\mathbf{n} \curlywedge \hat{\mathbf{c}}_3, \mathbf{n} \pm \hat{\mathbf{c}}_3\}$ that yields $g_{23} = \pm\upsilon_3$, which is obviously guaranteed in the Davenport setting. Note that in the case of two axes ($\hat{\mathbf{c}}_3 = \hat{\mathbf{c}}_1$) the equation for λ is reduced to

$$(\upsilon_2 - g_{12})\,\lambda^2 + 2\,\tilde{\upsilon}_3\lambda + \varepsilon\,\upsilon_2 + g_{12}(\varepsilon - 2\,\upsilon_1) = 0. \tag{13}$$

For example, in the ZXZ setting one may express the unit time-like vector

$$\mathbf{n} = (\mathrm{sh}\,\theta\cos\varphi,\ \mathrm{sh}\,\theta\sin\varphi,\ \mathrm{ch}\,\theta)^t$$

in terms of its coordinates θ, φ on the unit hyperboloid and thus to obtain

$$\lambda_{\pm} = \frac{-n_2 \pm \sqrt{n_3^2 - 1}}{n_1} = \frac{1 \mp \sin\varphi}{\pm\cos\varphi}$$

which finally yields for the scalar parameters

$$\tau_1^{\pm} = \frac{n_3\lambda_{\pm} + n_1}{n_1\lambda_{\pm} + n_2} = \mathrm{ch}\,\theta\sec\varphi \pm (\cos\varphi - \mathrm{cth}\,\theta)$$
$$\tau_2^{\pm} = \frac{n_3\lambda_{\pm} + n_1}{(1 + n_3)\lambda_{\pm}} = \frac{1}{2}\left(1 + \mathrm{th}\frac{\theta}{2}\right)^2 \pm \mathrm{th}\frac{\theta}{2}\sin\varphi. \tag{14}$$

More Examples

Similarly, for the XYZ decomposition one has

$$\lambda_{\pm} = \frac{n_2^2 + (1 + n_3)(1 \pm \sqrt{n_1^2 + 1})}{n_1 n_2}$$

and the solutions for $\tau_{1,2}$ are even more complicated, namely

$$\tau_1^{\pm} = \frac{1 - n_3}{n_2 - \lambda_{\pm}^{-1}n_1}, \qquad \tau_2^{\pm} = \frac{1 - n_3}{n_1 + \lambda_{\pm}^{-1}n_2}. \tag{15}$$

This complexity implies that there should be a more appropriate setting for the hyperbolic case. One such setting is given by the classical Iwasawa decomposition of the spin covering group $\mathsf{SL}(2,\mathbb{R})$, which induces in $\mathsf{SO}(2,1)$ configurations like WXZ, ZWY and XYW, where the axis OW is the bisectrix of OX and OZ determined by the isotropic directional vector $\mathbf{w} = (1,0,1)^t$. Some of these configurations, however, involve light cone singularities, as we shall see below. Let us consider the WYZ setting, in which it is not hard to obtain

$$\lambda_+ = \frac{n_1 - n_3 - 1}{n_2}, \qquad \lambda_- = \frac{n_2}{n_3 - n_1 + 1}.$$

Due to the presence of isotropic singularity, we cannot apply (12) directly, but rather use the formulas (60) in [3] instead, which in this case amounts to

$$\tau_1 = \frac{(\hat{\mathbf{c}}_2 \curlywedge \tilde{\mathbf{c}})^\circ}{(\tilde{\mathbf{c}} \cdot \hat{\mathbf{c}}_2)\, \hat{\mathbf{c}}_1^\circ - (\hat{\mathbf{c}}_1 \curlywedge \hat{\mathbf{c}}_2)^\circ}, \qquad \tau_2 = \frac{(\hat{\mathbf{c}}_1 \curlywedge \tilde{\mathbf{c}})^\circ}{(\hat{\mathbf{c}}_1 \curlywedge \hat{\mathbf{c}}_2)^\circ - (\tilde{\mathbf{c}} \cdot \hat{\mathbf{c}}_1)\, \hat{\mathbf{c}}_2^\circ}. \tag{16}$$

Here $\mathbf{x}^\circ = (\mathbf{x}, \mathbf{w})$ stands for the Euclidean dot product of a vector $\mathbf{x} \in \mathbb{R}^{2,1}$ with the isotropic direction $\mathbf{w}$. One may easily verify that the solutions

$$2\,\tau_1^\pm = \frac{(1 + n_3 + n_1)\lambda_\pm + n_2}{n_2\lambda_\pm + 1 + n_3 - n_1}, \qquad \tau_2^\pm = \frac{n_2}{1 + n_3}\lambda_\pm \tag{17}$$

yield an ill-defined transformation (2) for λ_+, so only $\lambda = \lambda_-$ is relevant. Another way to see this is to note that only λ_- yields $\tilde{\mathbf{c}} \perp \mathbf{w}$, which is necessary in the presence of a light cone singularity. Simplifying the so obtained solutions finally yields (note that in the isotropic case we have $\phi = 2\tau$)

$$\begin{aligned} 2\,\tau_1 &= \frac{n_2}{n_3 - n_1} &&\Longrightarrow \quad \phi_1 = \frac{\operatorname{th}\theta \sin\varphi}{1 - \operatorname{th}\theta\cos\varphi} \\ \tau_2 &= \frac{n_2^2}{(1+n_3)(1+n_3-n_1)} &&\Longrightarrow \quad \phi_2 = 2\operatorname{arcth}\left(\frac{\operatorname{th}^2\frac{\theta}{2}\sin^2\varphi}{1 - \operatorname{th}\frac{\theta}{2}\cos\varphi}\right). \end{aligned} \tag{18}$$

These expressions have a nice geometric interpretation. Namely, if $\mathbf{n} \to z \in \mathbb{D}$ is the standard projection from the hyperboloid to the unit Poincaré disk given by the linear-fractional transformation $z = \dfrac{n_1 + in_2}{1 + n_3} = \operatorname{th}\dfrac{\theta}{2}\,\mathrm{e}^{\mathrm{i}\varphi}$, then ϕ_2 is equal to the hyperbolic norm $\left\| \dfrac{\Im^2(z)}{1 + \Re(z)} \right\|_{\mathbb{D}}$ and similarly $\phi_1 = \dfrac{\Im(\tilde{z})}{1 + \Re(\tilde{z})}$, where $\tilde{z} = \dfrac{n_1 + in_2}{n_3} = \operatorname{th}\theta\,\mathrm{e}^{\mathrm{i}\varphi} \in \mathbb{D}$.

Next, we consider the WYX setting for the space-like case $\varepsilon = 1$, in which

$$\Delta = \left((n_1 - n_3 + 1)^2 - n_2^2\right)^2 \geq 0$$

and the solutions for $\lambda_\pm$ are given by the expressions

$$\lambda_+ = \frac{n_1 - n_3 + 1}{n_2}, \qquad \lambda_- = \frac{n_2}{n_1 - n_3 + 1}.$$

Note, however, that the light cone singularity is present here as well, so once more we need to use formula (16), which yields the solutions in the form

$$2\tau_1^\pm = \frac{n_2 + (n_1 + n_3 + 1)\lambda_\pm}{n_2\lambda_\pm - (n_1 - n_3 + 1)}, \qquad \tau_2^\pm = -\frac{n_2\lambda_\pm + n_3}{1 + n_1}.$$

In particular, we see that $\tau_1^+ = \infty$ and $\tau_2^+ = -1$ that are not admissible values for an isotropic (respectively, space-like) direction, hence, $\lambda \equiv \lambda_-$ and

$$2\tau_1 = \frac{n_2}{n_3 - n_1}, \qquad \tau_2 = \frac{n_1 - n_3 - 1}{n_1 - n_3 + 1}. \tag{19}$$

Substituting one possible parameterization of space-like unit vectors, namely

$$\mathbf{n} = (\sin\varphi \operatorname{ch}\theta,\ \cos\varphi,\ \sin\varphi \operatorname{sh}\theta)^t$$

in the above solutions, we easily obtain for the corresponding rapidities

$$\phi_1 = -\mathrm{e}^\theta \cot\varphi, \qquad \phi_2 = -2\operatorname{arcth}\left(\frac{\mathrm{e}^\theta - \sin\varphi}{\mathrm{e}^\theta + \sin\varphi}\right). \tag{20}$$

Finally, consider the case of isotropic invariant vectors

$$\mathbf{n} = \left(u^2 - v^2,\ 2uv,\ u^2 + v^2\right)^t, \qquad u, v \in \mathbb{R}$$

in which it is convenient to decompose in the XYW setting that leads to

$$\Delta = 4n_2^4 \geq 0$$

and respectively

$$\lambda_+ = 0, \qquad \lambda_- = -\frac{2n_2^2}{n_3 - n_1}.$$

Since $\tau_2^+ = 1$ is not an admissible value for a space-like direction, we conclude once more that $\lambda \equiv \lambda_-$ and obtain the solutions

$$\tau_1 = \frac{n_2}{1 - n_3} - \frac{n_2}{n_3 - n_1}, \qquad \tau_2 = 1 - 2\,\frac{n_3 - 1}{n_1 - 1}. \tag{21}$$

With the aid of the alternative parameterization

$$\mathbf{n} = \left(\Re\, z^2,\ \Im\, z^2,\ |z\,|^2\right)^t, \qquad z = u + \mathrm{i}v \in \mathbb{C}$$

the above expressions may be written also in the compact form

$$\tau_1 = \frac{\Im\, z^2}{1 - |z\,|^2} + \arg \mathrm{i}z, \qquad \tau_2 = 1 - 2\,\frac{1 - |z\,|^2}{1 - \Re\, z^2}. \tag{22}$$

3 Applications

A major advantage of the approach presented here, noticeable at first glance, is that it separates the dependence on the orientation of the compound transformation's invariant axis $\mathbf{n}$ (affecting only $\tau_{1,2}$) and the one on the scalar parameter $\tau = \tau_3$. Thus, we obtain a coordinate change $(\theta, \varphi, \phi) \to (\phi_1, \phi_2, \phi_3)$ on $\mathsf{SO}^+(2,1)$, which resembles the cylindrical coordinates in the Euclidean case: namely, it is trivial for the third parameter $\phi_3 = \phi$ and $\phi_{1,2} = \phi_{1,2}(\theta, \varphi)$. This might come quite handy for the study of infinitesimal variations of the compound transformation's vector-parameter $\mathbf{c} \to \mathbf{c} + \delta\mathbf{c}$. The total derivative has a longitudinal and normal component $\dot{\mathbf{c}} = \dot{\tau}\mathbf{n} + \tau\dot{\mathbf{n}}$ with $\dot{\mathbf{n}} \cdot \mathbf{n} = 0$ and for the former we have $\dot{\tau} = \dot{\tau}_3$, while the latter is given as a linear combination of the partial ones $\partial_{\theta,\varphi}\phi_{1,2}$. Consider for example the standard parameterization

$$\mathbf{n} = (\mathrm{sh}\,\theta\,\cos\varphi,\ \mathrm{sh}\,\theta\,\sin\varphi,\ \mathrm{ch}\,\theta)^t$$

of a unit time-like vector $\mathbf{n} \in \mathbb{R}^{2,1}$. Straightforward differentiation yields

$$\dot{\mathbf{n}}(\theta, \varphi) = \dot{\theta}\frac{\partial \mathbf{n}}{\partial \theta} + \dot{\varphi}\frac{\partial \mathbf{n}}{\partial \varphi} = \begin{pmatrix} \dot{\theta}\,\mathrm{ch}\,\theta\,\cos\varphi - \dot{\varphi}\,\mathrm{sh}\,\theta\,\sin\varphi \\ \dot{\theta}\,\mathrm{ch}\,\theta\,\sin\varphi + \dot{\varphi}\,\mathrm{sh}\,\theta\,\cos\varphi \\ \dot{\theta}\,\mathrm{sh}\,\theta \end{pmatrix}.$$

On the other hand, working with the two projections to the unit disc presented above, namely $z = \mathrm{th}\,\frac{\theta}{2}\,\mathrm{e}^{\mathrm{i}\varphi}$ and $\tilde{z} = \mathrm{th}\,\theta\,\mathrm{e}^{\mathrm{i}\varphi}$, for which we have

$$\dot{z} = \Big(\frac{\dot{\theta}}{2}\big(1 - \mathrm{th}^2\frac{\theta}{2}\big) + \mathrm{i}\,\dot{\varphi}\,\mathrm{th}\,\frac{\theta}{2}\Big)\mathrm{e}^{\mathrm{i}\varphi}, \qquad \dot{\tilde{z}} = \big(\dot{\theta}\big(1 - \mathrm{th}^2\theta\big) + \mathrm{i}\,\dot{\varphi}\,\mathrm{th}\,\theta\,\big)\,\mathrm{e}^{\mathrm{i}\varphi}$$

and using the alternative form of the solutions (18), we finally obtain

$$\dot{\tau}_1 = \frac{\Im(\dot{\tilde{z}}) + \dot{\varphi}|\tilde{z}|^2}{2\,(1 + \Re(\tilde{z}))^2}, \qquad \dot{\tau}_2 = \frac{\Im(z)}{(1 + \Re(z))^2}\big(\Im(\dot{z})(2 + \Re(z)) + \dot{\varphi}|z|^2\big). \tag{23}$$

Moreover, we may directly parameterize with $z, \tilde{z} \in \mathbb{D}$, such that $\tilde{z} = \tilde{r}e^{i\varphi}$ has the argument of $z = re^{i\varphi}$ and twice its hyperbolic norm, i.e.,

$$\tilde{r} = \frac{2r}{1+r^2}, \qquad r = \frac{1-\sqrt{1-\tilde{r}^2}}{\tilde{r}}, \qquad r, \tilde{r} \in [\,0, 1\,).$$

Then, working only in terms of r and φ, we easily express

$$\dot{z} = (\dot{r} + \mathrm{i}\, r\dot{\varphi})\, e^{i\varphi}, \qquad \dot{\tilde{z}} = \frac{2\, e^{i\varphi}}{(1+r^2)^2}\left((1-r^2)\,\dot{r} + \mathrm{i}\, r(1+r^2)\dot{\varphi}\right)$$

after which the derivatives of the solutions are obtained from Eq. (23).

Although the time-like case is somewhat special due to its importance for the $2+1$ dimensional relativity,[1] a similar approach may be used also for space-like or isotropic invariant axes $\mathbf{n}$ and for various decomposition settings. Consider for example the variation of a space-like vector-parameter in the WYX decomposition setting discussed above. We obviously have

$$\dot{\mathbf{n}} = \dot{\theta}\frac{\partial \mathbf{n}}{\partial \theta} + \dot{\varphi}\frac{\partial \mathbf{n}}{\partial \varphi}$$

and the scalar parameters may be derived directly from (20) in the form

$$\dot{\tau}_k = \dot{\theta}\frac{\partial \tau_k}{\partial \theta} + \dot{\varphi}\frac{\partial \tau_k}{\partial \varphi}$$

which yields in this particular case

$$\dot{\tau}_1 = \frac{e^{\theta}}{2}\left(\dot{\varphi}\csc^2\varphi - \dot{\theta}\cot\varphi\right), \qquad \dot{\tau}_2 = \frac{2\, e^{\theta}}{(e^{\theta} + \sin\varphi)^2}\left(\dot{\varphi}\cos\varphi - \dot{\theta}\sin\varphi\right).$$

The Monodromy Matrix

Another possible application may be found in the quantum mechanical description of scattering on the line via the so-called *monodromy matrix*

$$\mathscr{M} = \begin{pmatrix} a & b \\ \bar{b} & \bar{a} \end{pmatrix} \in \mathsf{SU}(1,1) \tag{24}$$

that relates left and right free particle asymptotic solutions (see [4] for details). Its entries are expressed in terms of the *transition* and *reflection* coefficients (usually

[1] $\mathbf{n}$ may be interpreted as a generator of the Wigner little group of a massive particle.

denoted by t and r, respectively) as $a = \frac{1}{\bar{t}}$, $b = -\frac{\bar{r}}{\bar{t}}$. The correspondence between split-quaternions and hyperbolic vector-parameters[2]

$$\zeta_\circ^\pm = \pm(1-\mathbf{c}^2)^{-\frac{1}{2}}, \qquad \boldsymbol{\zeta}^\pm = \zeta_\circ^\pm \mathbf{c} \tag{25}$$

allows for applying the construction from the previous section and the compound vector-parameter $\mathbf{c}$ associated with $\mathscr{M}$ can be written in the form

$$\mathbf{c} = \frac{1}{\Re(t)}\big(-\Re(r\bar{t}),\, \Im(r\bar{t}),\, \Im(t)\big)^t. \tag{26}$$

Therefore, one has the choice whether to stay in the $\mathsf{SO}(2,1)$ setting and then lift the results to the covering group via formula (25) or work directly in $\mathsf{SU}(1,1)$ using the corresponding adjoint action, split quaternion multiplication and Killing form (see [3] for details). Note that in the above correspondence reflectionless potentials are associated with rotations by an angle $\phi = 2\arg(t)$ about the OZ axis in $\mathbb{R}^{2,1}$, while potentials with real transition coefficient correspond to pure Lorentz boosts with scalar parameter $\tau = |r|$. In particular, their invariant axis is aligned along OX if the reflection coefficient r happens to be real as well, and respectively, along OY, if r is purely imaginary. Choosing these basic directions in a given decomposition is equivalent to modelling the compound scattering potential with a sequence of simpler ones. Next, we are going to illustrate this idea with an example. Consider a potential corresponding to a pure boost ($t \in \mathbb{R}$) and a XZX decomposition setting. Note that in this case the boost invariant vector may be aligned with OX with a single rotation by an angle $\phi_2 = \arg r - \pi$ about the OZ axis (in this case $\phi_1 = 0$). The third (middle) transformation is a boost about OX with a scalar parameter $\tau_3 = \tau = |r|$. Hence, the $\mathsf{SO}(2,1)$ decomposition scheme considered above has a natural lift to the spin cover $\mathsf{SU}(1,1) \cong \mathsf{SL}(2,\mathbb{R})$ with a matrix representation

$$\frac{1}{t}\begin{pmatrix} 1 & -\bar{r} \\ -r & 1 \end{pmatrix} = \frac{1}{t}\begin{pmatrix} e^{i(\pi-\arg r)} & 0 \\ 0 & e^{i(\arg r-\pi)} \end{pmatrix}\begin{pmatrix} 1 & -|r| \\ -|r| & 1 \end{pmatrix}\begin{pmatrix} e^{i(\arg r-\pi)} & 0 \\ 0 & e^{i(\pi-\arg r)} \end{pmatrix}.$$

This is an effective factorization of $\mathscr{M}$ into a pair of mutually inverse phase-shifting terms and a phase preserving scatterer in the middle. Differentiating in this setting is almost trivial—the two end terms are affected only by varying the phase of the compound reflection coefficient r, while the middle one depends on both t and $|r|$. For the angles (rapidities) this yields

$$\dot{\phi}_2 = \frac{\Re(r)\Im(\dot{r}) - \Im(r)\Re(\dot{r})}{\Re^2(r)+\Im^2(r)}, \qquad \dot{\phi}_3 = \frac{\Re(\dot{r})+\Im(\dot{r})}{\sqrt{\Re^2(r)+\Im^2(r)}}. \tag{27}$$

[2]Here the different sings correspond to the two sheets of the spin cover (see [2]).

Although the above example is certainly oversimplified, an identical treatment is applicable to far more complicated configurations as well.

4 Extension to the 3 + 1 Dimensional Minkowski Space

The local isomorphism

$$\mathsf{SO}^+(3,1) \cong \mathsf{SO}(3,\mathbb{C})$$

allows for extending the vector-parameter construction to the six-dimensional proper Lorentz group via complexification, i.e., $\mathbf{c} \in \mathbb{CP}^3$ (see [5] for details). Let us note that the correspondence between vector-parameters and Lorentz transformations in Minkowski space is explicitly given as (see [5, 6] for details)

$$\Lambda(\mathbf{c}) = \frac{1}{|1+\mathbf{c}^2|}\begin{pmatrix} 1 - |\mathbf{c}|^2 + \mathbf{c}\otimes\bar{\mathbf{c}}^t + \bar{\mathbf{c}}\otimes\mathbf{c}^t + (\mathbf{c}+\bar{\mathbf{c}})^\times & \mathrm{i}(\bar{\mathbf{c}} - \mathbf{c} + \bar{\mathbf{c}}\times\mathbf{c}) \\ \mathrm{i}(\bar{\mathbf{c}} - \mathbf{c} - \bar{\mathbf{c}}\times\mathbf{c})^t & 1 + |\mathbf{c}|^2 \end{pmatrix}.$$

In the reverse direction we denote $\tilde{\Lambda} = \Lambda - \tilde{\eta}\,\Lambda^t\,\tilde{\eta}$, where $\tilde{\eta} = \mathrm{diag}(1,1,1,-1)$ stands for the flat Lorentz metric in $\mathbb{R}^{3,1}$, and hence derive the correspondence

$$\mathbf{c} = \frac{1}{\mathrm{tr}\Lambda}\begin{pmatrix} \tilde{\Lambda}_{32} + \mathrm{i}\tilde{\Lambda}_{14} \\ \tilde{\Lambda}_{13} + \mathrm{i}\tilde{\Lambda}_{24} \\ \tilde{\Lambda}_{21} + \mathrm{i}\tilde{\Lambda}_{34} \end{pmatrix}, \qquad \bar{\mathbf{c}} = \frac{1}{\mathrm{tr}\Lambda}\begin{pmatrix} \tilde{\Lambda}_{32} - \mathrm{i}\tilde{\Lambda}_{14} \\ \tilde{\Lambda}_{13} - \mathrm{i}\tilde{\Lambda}_{24} \\ \tilde{\Lambda}_{21} - \mathrm{i}\tilde{\Lambda}_{34} \end{pmatrix}. \tag{28}$$

Since the Minkowski space-time $\mathbb{R}^{3,1}$ is even-dimensional, the existence of invariant axes is ensured only by the Plücker relations

$$(\Re\,\mathbf{c}, \Im\,\mathbf{c}) = (\Re\,\hat{\mathbf{c}}_k, \Im\,\hat{\mathbf{c}}_k) = 0, \qquad k = 1,2,3. \tag{29}$$

On the other hand, decomposability with real scalar parameters demands

$$(\Re\,\mathbf{c}, \Im\,\hat{\mathbf{c}}_k) + (\Re\,\hat{\mathbf{c}}_k, \Im\,\mathbf{c}) = 0 \tag{30}$$

which effectively projects the kinematics to a three-dimensional (space-like, time-like or light-like) hyperplane. In the generic setting, however, the absence of invariant axes is an obstacle to the proper formulation of the problem itself. The classical Wigner decomposition in this case turns out to be much more convenient. It is performed by conjugating a rotation with a pure boost, i.e.,

$$\Lambda(\mathbf{c}) = \Lambda(-\tilde{\mathbf{c}})\Lambda(\mathbf{c}_0)\Lambda(\tilde{\mathbf{c}}), \qquad \Re\,\tilde{\mathbf{c}} = \Im\,\mathbf{c}_0 = 0 \tag{31}$$

which yields a representation of the Wigner little group of a massive particle. More precisely, considering the momentum in the rest frame, we see that the left-hand side needs to preserve the time direction. This is only possible if $(\boldsymbol{\alpha}, \boldsymbol{\beta}) = 0$, where the notation $\boldsymbol{\alpha} = \Re\,\mathbf{c}$ and $\boldsymbol{\beta} = \Im\,\mathbf{c}$ is used, i.e., $\mathbf{c} = \boldsymbol{\alpha} + \mathrm{i}\boldsymbol{\beta}$. In the decomposition (31) one may choose $\mathbf{c}_0$ proportional to the real (rotational) component of $\mathbf{c}$, i.e., $\mathbf{c}_0 = \mu\boldsymbol{\alpha}$. Following the idea of [1], one obtains

$$\tilde{\mathbf{c}} = \frac{1}{\mathbf{c}_0^2 + (\mathbf{c}, \mathbf{c}_0)}\,(\,\mathbf{c}\times\mathbf{c}_0 + \lambda(\mathbf{c}+\mathbf{c}_0)\,)\,, \qquad \lambda \in \mathbb{CP}^1 \tag{32}$$

and taking into account that $\mathbf{c}_0^2 = \mathbf{c}^2 = \boldsymbol{\alpha}^2 - \boldsymbol{\beta}^2$ $(\boldsymbol{\alpha} \perp \boldsymbol{\beta})$, it is easy to derive

$$\tilde{\mathbf{c}} = \frac{1}{\mu(\mu+1)\boldsymbol{\alpha}^2}\,(\lambda(\mathbf{c} + \mu\boldsymbol{\alpha}) + \mathrm{i}\,\mu\boldsymbol{\beta}\times\boldsymbol{\alpha})\,, \qquad \mu = \sqrt{1 - \frac{\boldsymbol{\beta}^2}{\boldsymbol{\alpha}^2}}$$

for which an invariant axis exists as long as $(\Re\,\tilde{\mathbf{c}}, \Im\,\tilde{\mathbf{c}}) = 0$, hence, $\Re(\lambda^2) = 0$. If we choose $\lambda \in \mathbb{R}$, demanding that the conjugation is performed with a pure boost (as in the classical Wigner setting) yields $\lambda \equiv 0$, so we finally obtain

$$\mathbf{c}_0 = \mu\boldsymbol{\alpha}, \qquad \tilde{\mathbf{c}} = \frac{\mathrm{i}}{\mu+1}\,\frac{\boldsymbol{\beta}\times\boldsymbol{\alpha}}{\boldsymbol{\alpha}^2}, \qquad \mu = \sqrt{1 - \frac{\boldsymbol{\beta}^2}{\boldsymbol{\alpha}^2}}\,. \tag{33}$$

The case of purely imaginary λ seemingly leads to an arbitrary choice of the parameter, but unless we set $\lambda \equiv 0$, we end up with a non-trivial rotational contribution due to the real counterpart $\Re\,\tilde{\mathbf{c}} \sim \mathrm{i}\lambda\boldsymbol{\beta}$, which effectively alters $\mathbf{c}_0$. Note that the solution (33) may be derived from the vector-parameter composition law demanding only orthogonality, i.e., $\mathbf{c}_0 = \mu\boldsymbol{\alpha}$ and $\tilde{\mathbf{c}} = \mathrm{i}\nu\boldsymbol{\beta}\times\boldsymbol{\alpha}$ with $\boldsymbol{\alpha} \perp \boldsymbol{\beta}$. Instead of formula (1), however, we use its (complexified) Euclidean analogue (cf. [3]). The decomposition (31) is then written as $\langle\mathbf{c}_0, \tilde{\mathbf{c}}\rangle = \langle\tilde{\mathbf{c}}, \mathbf{c}\rangle$, which yields our solution in the equivalent form

$$\mu + \nu\boldsymbol{\beta}^2 = (1+\mu)\nu\boldsymbol{\alpha}^2 = 1 \qquad \Longrightarrow \qquad \nu = \frac{1-\mu}{\boldsymbol{\beta}^2}, \qquad \mu = \sqrt{1 - \frac{\boldsymbol{\beta}^2}{\boldsymbol{\alpha}^2}}\,.$$

To illustrate the method we decompose the Lorentz transformation given by

$$\Lambda = \frac{1}{5}\begin{pmatrix} -4 & 12 & 3 & 12 \\ 12 & -11 & -4 & -16 \\ -3 & 4 & -4 & 4 \\ -12 & 16 & 4 & 21 \end{pmatrix} \qquad \Longrightarrow \qquad \mathbf{c} = \begin{pmatrix} 4 \\ 3 \\ 4\mathrm{i} \end{pmatrix}$$

where we use Eq. (28) to obtain $\mathbf{c}$. Since $\boldsymbol{\alpha} \perp \boldsymbol{\beta}$, formula (33) yields $\mathbf{c}_0 = \frac{1}{5}(12,\ 9,\ 0)^t$ and $\tilde{\mathbf{c}} = \frac{\mathrm{i}}{10}(-3,\ 4,\ 0)^t$, so we factorize $\Lambda = \tilde{\Lambda}^{-1}\Lambda_0\,\tilde{\Lambda}$ with

$$\Lambda_0 = \frac{1}{125}\begin{pmatrix} 44 & 108 & 45 & 0 \\ 108 & -19 & -60 & 0 \\ -45 & 60 & -100 & 0 \\ 0 & 0 & 0 & 1 \end{pmatrix}, \qquad \tilde{\Lambda} = \frac{1}{75}\begin{pmatrix} 93 & -24 & 0 & -60 \\ -24 & 107 & 0 & 80 \\ 0 & 0 & 1 & 0 \\ -60 & 80 & 0 & 125 \end{pmatrix}.$$

Certainly, there are alternative boost-rotation decompositions, for example

$$\boldsymbol{\alpha} \perp \boldsymbol{\beta} \;\Rightarrow\; \Lambda(\boldsymbol{\alpha}+\mathrm{i}\boldsymbol{\beta}) = \Lambda(\mathrm{i}\tilde{\boldsymbol{\beta}}_+)\Lambda(\boldsymbol{\alpha}) = \Lambda(\boldsymbol{\alpha})\Lambda(\mathrm{i}\tilde{\boldsymbol{\beta}}_-), \qquad \tilde{\boldsymbol{\beta}}_\pm = \frac{\mathscr{I} \pm \boldsymbol{\alpha}^\times}{1+\boldsymbol{\alpha}^2}\,\boldsymbol{\beta}.$$

One may also consider the isotropic case by choosing a light-like direction $\mathbf{c}_0$ and thus study the corresponding Wigner little groups for the case of massless particles. However, we restrain from such a discussion here as it is far from the primary objective of the present section—merely to provide a simple example.

5 Final Remarks

The non-negative discriminant condition (10) that guarantees real solutions (12) may be interpreted as a condition for the existence of a Lorentz transformation[3] $\tilde{\mathscr{P}} = \mathscr{P}_2\mathscr{P}_1 :\ \mathbf{n} \to \hat{\mathbf{c}}_3$, which is decomposable into a pair of pseudo-rotations with invariant axes $\hat{\mathbf{c}}_1$ and $\hat{\mathbf{c}}_2$, respectively. Of course, the compound pseudo-rotation $\mathscr{P}$ needs to be decomposable into five factors in the first place. The condition for this is slightly more complicated compared to the Euclidean case due to the non-compactness of $\mathsf{SO}(2,1)$. This issue, thoroughly examined in [7], may be avoided by choosing the decomposition settings carefully.

There is a peculiar geometric interpretation of our method, which we explain by analogy with the Euclidean case. Namely, it is obvious that a special orthogonal matrix is always conjugated to a rotation about a given axis, say OZ. The conjugation may be decomposed according to the standard Euler ZXZ setting, which effectively reduces the factors to five due to the equality

$$\mathscr{R} = \mathscr{R}_1(-\alpha)\mathscr{R}_2(-\beta)\mathscr{R}_3(-\gamma)\mathscr{R}_3(\phi)\mathscr{R}_3(\gamma)\mathscr{R}_2(\beta)\mathscr{R}_1(\alpha) = \mathscr{R}_1^{-1}\mathscr{R}_2^{-1}\mathscr{R}_3\,\mathscr{R}_2\,\mathscr{R}_1.$$

The latter naturally extends to the hyperbolic case and allows for deriving the condition (10) in an alternative way (see [3] for details). On the other hand, since the solution is actually independent on the γ-parameter, gimbal lock is no longer an issue, although the light-cone singularity is still possible.

---

[3] This is certainly not possible unless $\hat{\mathbf{c}}_3$ has the same geometric type as $\mathbf{n}$, i.e., $\varepsilon_3 = \varepsilon$.

As for the six-dimensional case, similar arguments hold for the Lie groups $\mathsf{SO}(4)$, $\mathsf{SO}(2,2)$ and $\mathsf{SO}^*(4)$. Thus, one may naturally exploit a generalized Wigner construction, as well as alternative factorization techniques (see [6] for examples).

References

1. Piña, E.: A new parametrization of the rotation matrix. Am. J. Phys. **51**, 375–379 (1983)
2. Brezov, D., Mladenova, C., Mladenov, I.: Vector-parameters in classical hyperbolic geometry. J. Geom. symmetry Phys. **30**, 21–50 (2013)
3. Brezov, D., Mladenova, C., Mladenov, I.: A decoupled solution to the generalized Euler decomposition problem in $\mathbb{R}^3$ and $\mathbb{R}^{2,1}$. J. Geom. Symmetry Phys. **33**, 47–78 (2014)
4. Arnold, V.: Geometrical Methods in the Theory of Ordinary Differential Equations. Springer, New York (1983)
5. Fedorov, F.: The Lorentz Group (in Russian). Science, Moscow (1979)
6. Brezov, D., Mladenova, C., Mladenov, I.: Generalized Euler decompositions of some six-dimensional Lie groups. AIP Conf. Proc. **1631**, 282–291 (2014)
7. Koch, R., Lowenthal, F.: Uniform finite generation of Lie groups locally isomorphic to $\mathsf{SL}(2,\mathbb{R})$. Rocky Mountain J. Math. **7**, 707–724 (1977)

Decision Support Tool in the Project LANDSLIDE

Nina Dobrinkova and Pierluigi Maponi

Abstract In the framework of the LANDSLIDE project has been fulfilled field work on the test areas in Bulgaria, Italy, Poland and Greece. The project goal is to create a software tool helping the decision makers in cases of landslides with up to 20 m depth which can estimate the soil movement by calculating soil moisture and meteorological data conditions which in the literature are considered as main triggers in such natural hazards.

Keywords Landslide hazard · Soil moisture · Limit equilibrium analysis

1 Introduction

The project LANDSLIDE risk assessment model for disaster prevention and mitigation (acronym: LANDSLIDE) is a European project, co-financed by the Directorate General Humanitarian Aid and Civil Protection of the European Commission, with the aim to develop an innovative risk assessment tool to predict and evaluate landslide hazards. The project is with duration 24 months starting from 1st January 2015 and ending on 31st December 2016. The project has been proposed, because landslides occur in many different geological and environmental settings across Europe and are a major hazard in most mountainous and hilly regions [1]. Every year landslides cause fatalities and large damage to infrastructure and property. One of the reasons land-slides to activate its potential is intense or long lasting rainfalls. This is the most frequent trigger of landslides occurrence in Europe [2], and is expected to increase in

N. Dobrinkova (✉)
Institute of Information and Communication Technologies—Bulgarian Academy of Sciences, Sofia, Bulgaria
e-mail: nido@math.bas.bg

P. Maponi
University of Camerino, Camerino, Italy
e-mail: pierluigi.maponi@unicam.it

K. Georgiev et al. (eds.), *Advanced Computing in Industrial Mathematics*,
Studies in Computational Intelligence 681, DOI 10.1007/978-3-319-49544-6_2

the future due to climate change. Methods for landslide hazard evaluation are today mainly based on scientific literature of geomorphologic studies and of historical landslide events, see [3, 4] and the references therein for an example; these studies do not consider or underestimate the impact of climate change. Therefore, it is important that interdisciplinary approach for landslide monitoring and prediction between the nowadays ICT (Information and Communications Technology) solutions and landslide experts has to start. This cooperation is useful in order to provide new tools that can adapt to the new conditions by correctly evaluating and predicting landslide hazards which is a fundamental prerequisite for accurate risk mapping and assessment and for the consequent implementation of appropriate prevention measures.

A physical approach to this problem can be based on the so-called Limit Equilibrium Analysis and the Mohr-Coulomb relation for the shear strengths of the materials along the potential failure surface, see [5] for details. However, the practical application of this theory to the landslide hazard evaluation requires the knowledge of soil moisture. The main scientific contribution of the project is given by a new procedure to evaluate landslide hazard level, where the main components are the soil moisture dynamics and the slope stability analysis. This physics-based procedure allows the evaluation of the hazard level directly from weather forecast data. The resulting hazard evaluation system works in continuous-time and provides hazard maps every daily. These maps depend on the relevant geomorphological features of the test area and of the weather data, so the statistical analysis of a long series of such maps may also provide the impact of climate change on the study area.

The LANDSLIDE project, coordinated by the University of Camerino, is made up of 6 partner organizations coming from Italy, Bulgaria, Greece and Poland:

- University of Camerino (project Coordinator), Italy
  - Institute of Information and Communication Technologies—Bulgarian Academy of Sciences, Bulgaria
  - National Observatory of Athens—Institute of Geodynamics, Greece
  - Province of Ancona, Italy
  - Regional Government of Smolyan, Bulgaria
  - Bielsko-Biala District, Poland

This partnership aims jointly to develop a Landslide Hazard Assessment Model that will be tested in four hydrographic basins selected as test sites. This model and software will be able to make completely automatic predictions of landslide hazards which may occur in soil up to 20th meter depth. The predictions can be automated on a day to day basis, as well as to correctly evaluate the impact of climate change, in a medium long term. The system, which is still under development, focus on landslides with dept up to 20 m, because the meteorological conditions usually influence on such landslides. Deeper soil anomalies are thought by literature as more complex than just weather conditions driven natural hazards.

2 Description of the Dynamic Evaluation of Soil-Moisture Content in LANDSLIDE Project Software Tool

The main components in the method for the computation of landslide hazard are: the soil moisture dynamics and the slope stability analysis. The following subsections give a detailed description of these two components.

2.1 *Soil Moisture Dynamics*

The dynamics of soil-moisture content is a complex phenomenon depending on atmospheric conditions, geological features of the region under study, and the corresponding land use. It can be formally described by diffusion equation models arising from Darcys law, and mass continuity law. These models depend on several parameters that should be chosen on the basis of the geological features of the region under study. When these parameters are set, a numerical approximation method must be used for the computation of the soil-moisture dynamics from the weather data inputs. This model is usually called the Richards equation:

$$\left(C(\psi) - S\frac{\theta(\psi)}{n}\right)\frac{\partial h}{\partial t} = \nabla\left(K(\psi)\nabla h\right) + W(x, y, z, t) - ET(x, y, z, t),$$
$$(x, y, z) \in B,\ t > 0 \qquad (1)$$

where $h = \psi + z$ is the hydraulic head and ψ is the pressure head, K is a diagonal matrix describing the hydraulic conductivity, which measures the ability of water to flow in the porous isotropic medium, C is the specific moisture capacity, S is the storage coefficient, n is the porosity of the soil, W is the recharge and it is related to the rate of precipitation, ET is the evapo-transpiration and it represents the loss of water due to the evaporation and transpiration of plants. Note that in (1) appears also function θ, that is the water content of the soil; it can be computed from the pressure head ψ through the Van Genuchten formula [1]. So, the soil moisture content θ can be obtained from the knowledge of the solution h of the Eq. (1). See [2, 6] for a detailed description of the Richards equation. The spatial domain B, where Eq. (1) is defined, gives a three-dimensional description of the basin under study. In particular, B is given by a slice of soil beneath the slope under study; so, the boundary ∂B of B is constituted by a top surface S_T, describing the soil surface, a bottom surface S_B describing the depth where the soil moisture content is analyzed, and a vertical surface S_V joining the boundary of S_T and S_B. The two source terms W and ET appearing in Eq. (1) depend on the weather data and on the land use. These functions have a narrow support that extend beneath top surface S_T; moreover, W is computed from the precipitation data, and ET from the so called Penman-Montieth equation [6], that gives an evaluation of evapo-transpiration by using vegetation data and weather data. Appropriate initial-boundary conditions must be considered with Eq. (1) in order to define a unique solution h, see [2] for details. The resulting initial-boundary value problem constitutes the proposed model for the soil moisture dynamics. The numerical solution of this problem is computed by a finite difference method resembling the well-known Cranck-Nicolson method for diffusion problems, see [3] for details.

2.2 *Slope Stability Analysis*

This analysis must give an evaluation of the resistance of inclined surfaces to failure by sliding or collapsing. The hazard degree can be expressed by the factor of safety F, which is the ratio between the forces that prevent the slope from failing and those that make the slope fail; the Mohr-Coulomb criterion [4] is used for this evaluation. Of course, when the factor of safety is less than one the slope should be considered unstable. Different methods can be used for the evaluation of F. The Infinite Slope Model is probably the simplest possible method: the soil surface is approximated by an infinite inclined plane. In this case the following approximation of F can be obtained:

$$F = \frac{C + (z\gamma - z_w\gamma_w)\cos^2(\beta)\tan(\phi)}{z\gamma\sin(\beta)\cos(\beta)} \tag{2}$$

where C is the effective cohesion, γ is the weight of soil, γ_w is the weight of water soil, z is the depth of failure surface, z_w is the depth of water table, β is the slope surface inclination, ϕ is the angle of internal friction, [4] for details. The factor of safety F can be easily computed from formula (2) in every point of the slope under study. In the LANDSLIDE project this is used as a first estimate of the landslide hazard index, that is automatically produced by the software tool. The corresponding risk map is delivered to the competent territorial authority, that, on the base of this map, can also require a more detailed analysis on small portions of the slope under study. The accurate estimation of the factor of safety F is obtained by a three-dimensional version of the method of slices. In this method the slab of soil beneath the slope is discretized by several vertical columns, where it is considered the forces equilibrium and the corresponding moments equilibrium along the three coordinate directions. For sake of brevity a detailed description of this method must be omitted, however an exhaustive explanation of the method of slices is provided in [7].

3 Test Beds in the Project LANDSLIDE

The areas where the software developed under the project LANDSLIDE is going to be tested and validated are located on the territories of Greece, Italy, Bulgaria and Poland. Short descriptions and fulfilled work in order the software requirements to be implemented will be briefly introduced.

The first test area is located on Peloponnesus peninsula in Greece. The responsible partner for that area is the National Observatory of Athens (NOA) implementing all LANDSLIDE activities. Peloponnesus is a very mountainous area with over 1,000,000 inhabitants, including also remote villages at risk of landslides. The territory is administratively divided between the Region of Peloponnesus with Tripoli as the capital city and the Region of Western Greece with Patras as the capital city. The area of Panagopoula was chosen as the test site, because past landslide events have cut rail and road connection from the Greek capital Athens to Patras for weeks. Patras is the most important port of western Greece. At the moment new rail and road connections from Athens to Patras (co-financed by European Union) are being constructed under Panagopoula as the morphology of Northern Peloponnesus does not allow any alternatives. Any major landside event may have tremendous consequences for the economy of western Greece. The Panagopoula landslide area is situated in the northern part of the Prefecture of Achaia (northwestern Peloponnesus, about 15 km east of Patras). NOA has selected the pilot area very carefully. All measurements on the field have been completed and 10 GIS maps were elaborated. To correctly collect data, a meteorological station has been acquired and installed in a safe position and started transmitting data in real time in the weather meteorological stations network of NOA.

The second test area is located in Province of Ancona Italy. The pilot area selected for the test and implementation of the LANDSLIDE model and software in Italy is a hydrographic basin of 11.69 sq. km located in the mid part of the Esino River basin (i.e. in the central part of the territory of the Province of Ancona). The test area was chosen as it is representative, also of the other hydrographic basins of the Province of Ancona: in effect about 30 % of the whole test area is concerned by landslides (i.e. 3.55 sq. km) and a landslide has recently damaged and interrupted the provincial road connecting 3 municipalities. Furthermore, several meteorological stations are located in the test area and previous geological and geotechnical studies have already been carried out. Having identified the hydrographic basin, drilling and sampling operations are being implemented according to the protocol agreed among project partners, as well as geological and geomorphological data of the area, are being collected to elaborate GIS maps. It is crucial to note that more than 300 sq. km (15.6 % of the whole provincial territory) are afflicted by landslides. According to the Basin Authority of the Marche Region: out of a total number of 18815 landslides that occurred on the territory of the Marche Region, 5676 are landslides occurred only in the Province of Ancona. The worst landslide ever occurred on the territory is the so called Ancona big landslide, a very large and deep soil movement (slipping surface was identified more than 100 m deep in the Pliocene clays) collapsed on 13th December 1982 after a very rainy period. The area concerned by the landslide, affected and damaged two hospitals, one university building, 280 houses, the railway and more than 2.5 km of the main Adriatic road. 3661 people were evacuated.

The third test area used by the project model and software is the Smolyan Region in Bulgaria. The hydrographic basin selected for the test and implementation of the LANDSLIDE project in Bulgaria is a landslide called "Smolyan Lakes". It is located northwest of the town of Smolyan. Its borders match the Kriva River to the west,

Muneva River to the East, Cherna River to the South, and the steep rock cliff, which is located south of the peak Snejanka. The length of the landslide is 5.35 km, average width—1.38 km and the depth of the landslide area—35–80 m. Its total area is 7.4 sq. km. The biggest landslide in Bulgaria was registered in 1923. It is located in the western suburbs of the very town of Smolyan. It has a width of 1 km and length of 5 km. It is active, with speed of movement of 5–25 cm per year. Its depth is 80 m. The landslide is moving slowly, however if no measures will be taken it could affect the main road of the Region, disconnecting the access to Smolyan. Smolyan Region covers an area of 3192 km^2, it has a permanent population of 120,456 people and is located in South Bulgaria in the central part of the Rhodope Mountains. The regional territory is characterized by mountains and several landslides occur every year; in effect, 87 landslides have been registered and one of these is the biggest landslide ever occurred in Bulgaria.

The fourth test area is located in Poland called Bielsko-Biala District. The hydrographic basin has been set within the Small Beskid mountain range, and field studies are taking place in an inactive sandstone quarry in Kozy. Landslides are among the most common geodynamic threats, often having the characteristics of a natural disaster. The floods which occurred in the southern Poland throughout the last 20 years resulted in the fact that many residents of the cataclysm area were affected by the phenomena of landslides combined with various losses (area degradation, destruction of residential buildings together with road network, sewerage system, telecommunication lines, electric lines, gas pipelines, crops and forests). The first deformation processes within the Small Beskid area were already registered in 1968 in the quarry located in Kozy. The surface mass movements have been a continuous phenomenon, and their development have been constantly observed. In 2010 a total number of 437 landslides were registered (active, periodically active and inactive). Bielsko-Biala is located in the southern part of Poland, in the province of Silesia. Geographically, it is located in the eastern part of the Silesian Foothills of the outer part of the Western Carpathians, and in terms of morphology it is situated in the massif of Small Beskid and Silesian Beskid mountain ranges. In the area of Bielsko-Biala District, can be observed numerous mass movements, which is part of the responsibility of the Head of a district.

4 E-platform Architecture and Design in LANDSLIDE Project

During the first year of the project lifetime the team from IICT-BAS has started collection of all data in one GIS database. This was done in order the model developed by the University of Camerino in Italy to be able to start validation and tests of its functionalities. The software development is following the description of the Fig. 1 workflow.

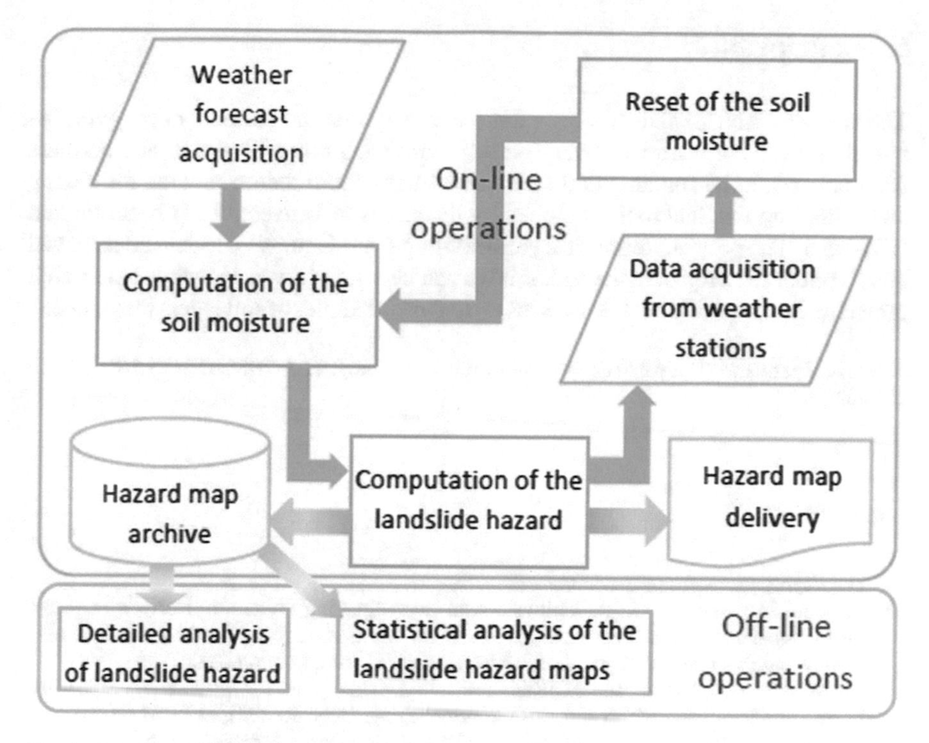

Fig. 1 Data flow in the computational system

The data processing in the system a continuous-time system, where every day gets the weather forecast data (for the next 24 h); from these data, computes the soil moisture content and the corresponding Safety Factor by the using Infinite Slope Model, that is used to create the landslide hazard maps; at the end of the day, the system acquires the weather data from the weather stations on the territory and refines the soil moisture content. Then it restarts the forecast procedure by taking into account the weather forecast data for the next day. An off-line functionally is also provided. This is mainly composed of two facilities: (i) a refined analysis of the hazard level by using a three-dimensional version of the method of slices, (ii) a statistical analysis of the hazard maps.

The main actors are divided in three general groups in the system. The first group is dedicated to users from the project, the second is dedicated to the GIS administrators of the system and the third which is public is oriented to the WEB users of the system. The system architecture is made in a way that the WEB site can dynamically generate map content and present any kind of spatial data or attribute query for it. The architecture is done in a way to be compatible with any kind of platforms (UNIX, Windows etc.), with J2EE compliant server. The tool can be run on any JAVA enabled Internet browser or standalone, on any platform with JRE 1.1.8 or greater for SUN Microsystems.

5 Conclusion

The project LANDSLIDE has been done in a way that the partners developing the model and the software work very closely with each other. The end user partners from Greece, Italy, Bulgaria and Poland will have more than 6 months for testing and validating the final tool developed by the teams of University of Camerino and IICT-BAS. The project lifetime is 2 years starting from January 2015, but all achieved results under the project in the end of 2016 can be a step forward towards better civil protection in cases of landslide hazards up to the 20th meter of soil layers movements.

Acknowledgements This paper has been supported by the project LANDSLIDE DG ECHO/SUB/2014/693902.

References

1. http://geology.com/usgs/landslides/
2. Polemio, M., Petrucci, O.: Rainfall as a landslide triggering factor: an overview of recent international research. In: Bromhead, E., Dixon, N., Ibsen, M.-L. (eds.) Landslides in Research, Theory and Practice, vol. 3, pp. 1219–1226. Thomas Telford, London (2000)
3. van Westen, C.J., Castellanos, E., Kuriakose, S.L.: Spatial data for landslide susceptibility, hazard, and vulnerability assessment: an overview. Eng. Geol. **102**(34), 112–131 (2008)
4. Guzzetti, F., Reichenbach, P., Cardinali, M., Galli, M.: Francesca Ardizzone probabilistic landslide hazard assessment at the basin scale. Geomorphology **72**(14), 272–299 (2005)
5. Duncan, J.M., Wright, S.G., Brandon, T.L.: Soil Strength and Slope Stability, 2nd edn. Wiley (2014)
6. Huggel, C., Khabarov, N., Korup O., Obersteiner, M.: Landslides ypes, Mechanisms and Modeling. In: Clague, J.J., Stead, D. (eds.). Cambridge University Press (2012)
7. Van Genuchten, M.: A closed form equation for predicting the hydraulic conductivity of unsatured soil. Soil Sci. Soc. Am. **44**, 892 (1980)

Generalized Nets as a Tool for Modelling of Railway Networks

Stefka Fidanova, Krassimir Atanassov and Ivan Dimov

Abstract The oldest public transport, which is used now days, is the railway. There exist different kinds of transportation models. The importance and role of each type of models is discussed in relation of its function. Some of the models are concentrated on scheduling. Other models are focused on simulation to analyze the level of utilization of different types of transportation. There are models which goal is optimal transportation network design. In this work we propose a model of the railway transport with Generalized Nets. It is shown that Generalized nets can be used as a tool for modeling of railway networks. An example of a generalized net of a part of the railway network in Southern Bulgaria, is given.

Keywords Generalized nets · Railway network

1 Introduction

Generalized Nets (GNs) [1–3] are a tool for modeling of parallel processes, including as partial cases the standard Petri nets and all their modifications and extensions (as Time Petri nets, E-nets, Color Petri nets, Predicative-Transition Petri nets, Fuzzy Petri nets, etc.). The apparatus of the GNs is used for modeling of different processes in the areas of Artificial Intelligence, medicine and biology, economics, industry,

S. Fidanova (✉) · I. Dimov
Institute of Information and Communication Technology,
Bulgarian Academy of Science, Sofia, Bulgaria
e-mail: stefka@parallel.bas.bg

K. Atanassov
Institute of Biophysics and Biomedical Engineering,
Bulgarian Academy of Science, Sofia, Bulgaria
e-mail: krat@bas.bg

K. Atanassov
Intelligent Systems Laboratory, Prof. Asen Zlatarov University,
8000 Bourgas, Bulgaria

K. Georgiev et al. (eds.), *Advanced Computing in Industrial Mathematics*,
Studies in Computational Intelligence 681, DOI 10.1007/978-3-319-49544-6_3

and many others. By the moment, there are GN-models of processes of bus and air transport [4].

In this paper for the first time, GNs are used as a tool for modeling of processes in the railway transport. There exist different kinds of transportation models [5]. The importance and role of each type of models is discussed in relation of its function. Several models are concentrated on scheduling [6]. Exist models which focused on analysis of the level of utilization of different types of transportation [7]. The goal of some of the models is optimal transportation network design [8].

In this paper we propose a model of the railway transport using Generalized nets. This model shows current situation in the railway network and gives ideas how it can be optimized and how to proceed if some problem arizes.

The rest of the paper is organized as follows. In Sect. 2, we give short remarks from GN-theory, in Sect. 3—description of some components of the railway network of a country. In Sect. 4, we describe a concrete example with a part of the railway network in Southern Bulgaria.

2 Introduction to the Theory of the Generalized Nets

In this work we propose a model of the railway transport with Generalized Nets. Generalized Nets are extension and generalization of Petri nets. For a first time they was proposed in 1991 [2]. Later they was used by a lot of scientists for describing and modeling different processes and algorithms [9–11]. They are powerful tool for universal description of models of complex systems with many different and in most of the cases not homogeneous components, with simultaneous activities. The static structure consists of objects called **transitions**, which have input and output **places**. Two transitions can share a place, but every place can be an input of at most one transition and can be an output of at most one transition.

The dynamic structure consists of **tokens**, which act as information carriers and can occupy a single place at every moment of the GN execution. The tokens pass through the transition from one input to another output place; such an ordered pair of places is called **transition arc**. The tokens' movement is governed by conditions (predicates), contained in the **predicate** matrix of the transition.

The information carried by a token is contained in its **characteristics**, which can be viewed as an associative array of characteristic names and values. The values of the token characteristics change in time according to specific rules, called **characteristic functions**. Every place possesses at most one characteristic function, which assigns new characteristics to the incoming tokens. Apart from movement in the net and change of the characteristics, tokens can also split and merge in the places. A transition can contain m input and n output places where $n, m \geq 1$.

The GN are expandable. Every place can be replaced with new GN. Thus the process described by GN can be developed and complicated on more deeper level. The GN can help us to understand the processes and to see possibilities for their optimization.

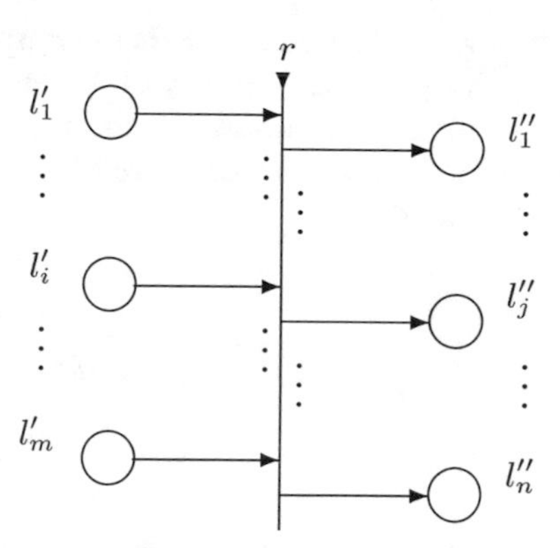

Fig. 1 The form of transition

In our model every station will be represented by transition. Several tokens will entry the transition with different characteristics, representing the possibilities to continue to other station.

The GNs are defined in a way that is different in principle from the ways of defining the other types of PNs [2, 3].

The first basic difference between GNs and the ordinary PNs is the "place—transition" relation. Here, the transitions are objects of more complex nature. A transition may contain m input places and n output places, where the integers m, $n \geq 1$.

Formally, every transition is described by a seven-tuple (Fig. 1):

$$Z = \langle L', L'', t_1, t_2, r, M, \square \rangle,$$

where:

(a) L' and L'' are finite, non-empty sets of places (the transition's input and output places, respectively); for the transition in Fig. 1 these are $L' = \{l'_1, l'_2, \ldots, l'_m\}$ and $L'' = \{l''_1, l''_2, \ldots, l''_n\}$;
(b) t_1 is the current time-moment of the transition's firing;
(c) t_2 is the current value of the duration of its active state;
(d) r is the transition's *condition* determining which tokens will pass (or *transfer*) from the transition's inputs to its outputs; it has the form of an Index Matrix (IM; see [12]):

$$r = \begin{array}{c|ccccc} & l''_1 & \ldots & l''_j & \ldots & l''_n \\ \hline l'_1 & & & & & \\ \vdots & & & r_{i,j} & & \\ l'_m & & & & & \end{array};$$

$r_{i,j}$ is the predicate that corresponds to the i-th input and j-th output place ($1 \leq i \leq m, 1 \leq j \leq n$). When its truth value is "*true*", a token from the i-th input place transfers to the j-th output place; otherwise, this is not possible;

(e) M is an IM of the capacities $m_{i,j}$ of transition's arcs, where $m_{i,j} \geq 0$ is a natural number:

$$M = \begin{array}{c|ccccc} & l''_1 & \dots & l''_j & \dots & l''_n \\ \hline l'_1 & & & & & \\ \vdots & & & m_{i,j} & & \\ l'_m & & & & & \end{array};$$

(f) $\square$ is an object of a form similar to a Boolean expression. It contains as variables the symbols that serve as labels for a transition's input places, and $\square$ is an expression built up from variables and the Boolean connectives $\wedge$ and $\vee$. When the value of a type (calculated as a Boolean expression) is "*true*", the transition can become active, otherwise it cannot.

The ordered four-tuple

$$E = \langle\langle A, \pi_A, \pi_L, c, f, \theta_1, \theta_2\rangle, \langle K, \pi_K, \theta_K\rangle, \langle T, t^o, t^*\rangle, \langle X, \Phi, b\rangle\rangle$$

is called a GN if:

(a) A is a set of transitions;
(b) π_A is a function giving the priorities of the transitions, i.e., $\pi_A : A \to N$, where $N = \{0, 1, 2, \dots\} \cup \{\infty\}$;
(c) π_L is a function giving the priorities of the places, i.e., $\pi_L : L \to N$, where $L = pr_1 A \cup pr_2 A$, and $pr_i X$ is the i-th projection of the n-dimensional set, where $n \in N, n \geq 1$ and $1 \leq k \leq n$ (obviously, L is the set of all GN—places);
(d) c is a function giving the capacities of the places, i.e., $c : L \to N$;
(e) f is a function that calculates the truth values of the predicates of the transition's conditions (for the GN described here, let the function f have the value "*false*" or "*true*", that is, a value from the set $\{0, 1\}$);
(f) θ_1 is a function which indicates the next time-moment when a certain transition Z can be activated, that is, $\theta_1(t) = t'$, where $pr_3 Z = t, t' \in [T, T + t^*]$ and $t \leq t'$. The value of this function is calculated at the moment when the transition ceases to function;
(g) θ_2 is a function which gives the duration of the active state of a certain transition Z, i.e., $\theta_2(t) = t'$, where $pr_4 Z = t \in [T, T + t^*]$ and $t' \geq 0$. The value of this function is calculated at the moment when the transition starts to function;
(h) K is the set of the GN's tokens.
(i) π_K is a function which gives the priorities of the tokens, that is, $\pi_K : K \to N$;
(j) θ_K is a function which gives the time-moment when a given token can enter the net, that is, $\theta_K(\alpha) = t$, where $\alpha \in K$ and $t \in [T, T + t^*]$;

(k) T is the time-moment when the GN starts to function. This moment is determined with respect to a fixed (global) time-scale;
(l) t^o is an elementary time-step, related to the fixed (global) time-scale;
(m) t^* is the duration of the functioning of the GN;
(n) X is the set of all initial characteristics which the tokens can obtain on entering the net;
(o) Φ is the characteristic function that assigns new characteristics to every token when it makes the transfer from an input to an output place of a given transition.
(p) b is a function which gives the maximum number of characteristics a given token can obtain, that is, $b : K \rightarrow N$.

A given GN may not have some of the above components. In these cases, any redundant component will be omitted. The GNs of this kind form a special class of GNs called "*reduced GNs*".

3 Generalized Net Models of Some Components of the Railway Network of a Country

Having in mind that the railway network of any country has the form of a (non-oriented) graph, we can represent it by a GN. Each train station is represented by a transition with the form of Fig. 2.

In it, the train station X has direct connections with n other stations and let us denote them by $Y_1, \ldots, Y_n$. Here, places l_{2i-1} and l_{2i} correspond to railway lines from and to station Y_i for $i = 1, 2, \ldots, n$. The index matrix [12] of the transition conditions has the form

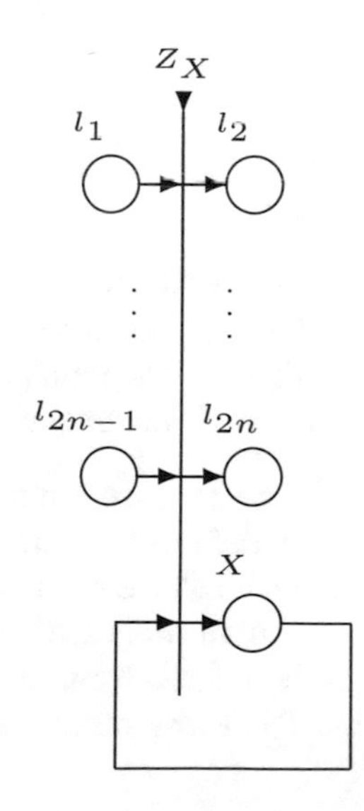

Fig. 2 Representation of train station

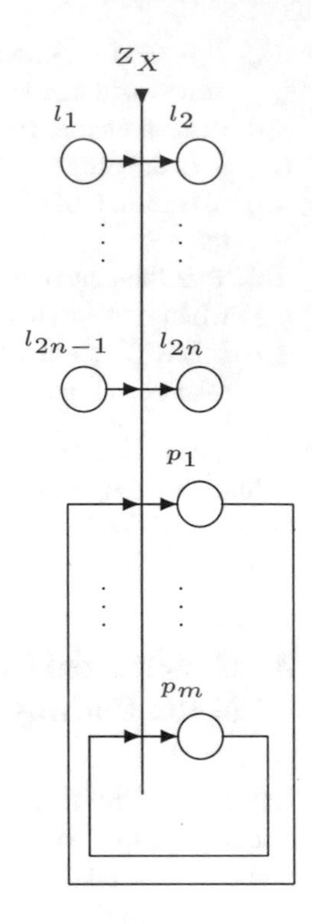

Fig. 3 Detailed representation of station

	l_2	l_4	...	l_{2n}	X
l_1	$false$	$false$	...	$false$	$true$
l_3	$false$	$false$	...	$false$	$true$
$\vdots$	$\vdots$	$\vdots$	$\vdots$	$\vdots$	$\vdots$
l_{2n-1}	$false$	$false$	...	$false$	$true$
X	$W_{X,2}$	$W_{X,4}$	...	$W_{X,2n}$	$W_{X,X}$

,

where

$W_{X,2}$ = "the train is directed to train station Y_1",
$W_{X,4}$ = "the train is directed to train station Y_2", ...,
$W_{X,2n}$ = "the train is directed to train station Y_n",
$W_{X,X}$ = "the train must shunt in train station X".

If we like, we can detailize the GN-model, using some of the hierarchical operators, defined over the GNs (see [2, 4]) and it can obtain the form from Fig. 3, in which all above notations are preserved.

In it, the train station has m parallel lines and let place p_j correspond to the j-th of these lines. Now, we can model the shutting of the trains in the region of station X. The above index matrix of the transition conditions has similar form, but now, it contains the additional information about the connections between the separate (parallel) lines in the train station.

So, we can describe the train moves to and from a given train station, and stay at it.

In the next Section, we illustrate the possibility to represent the railway network of a given country by GNs, using as an example, the railway network in Southern Bulgaria between Sofia and Burgas. In the model that follows, we describe only the most important train connections.

4 Generalized Net Model of a Part of the Railway Network in Southern Bulgaria

The railway possibilities between the station Sofia and Burgas are represented in this section. They are vary important part of the railway network in Southern Bulgaria. The GN-model contains 12 transitions and 56 places. Its tokens represent separate trains that will move from a train station to the next one or with shunt in a train station. The places denoted by alphabetic indices represent the following train stations between Sofia and Burgas: So—Sofia, Ya—Yana, EP—Elin Pelin, Kl—Karlovo, Pl—Plovdiv, Tu—Tulovo, SZ—Stara Zagora, Du—Dubovo, NZ—Nova Zagora, Zi—Zimnica, Ka—Karnobat, Bu—Burgas. These places are simultaneously inputs and outputs of the transitions. The places, denoted by numerical indices represent the activities of the trains related to entering or leaving the respective train station.

Each token, entering a place, obtains as a current characteristic information about the time for realization of the previous activity. In places, denoted by alphabetic indices, the token will obtain as additional information the train station from where it arrives and the number of passengers that leave the respective train or catch it (Fig. 4).

The GN-transitions are the following.

$$Z_{So} = \langle \{l_1, l_2, l_3, l_9, l_{14}, l_{So}\}, \{l_4, l_5, l_6, l_7, l_8, l_{So}\},$$

$$\begin{array}{c|cccccc} & l_4 & l_5 & l_6 & l_7 & l_8 & l_{So} \\ \hline l_1 & false & false & false & false & false & true \\ l_2 & false & false & false & false & false & true \\ l_3 & false & false & false & false & false & true \\ l_9 & false & false & false & false & false & true \\ l_{14} & false & false & false & false & false & true \\ l_{So} & W_{So,4} & W_{So,5} & W_{So,6} & W_{So,7} & W_{So,8} & W_{So,So} \end{array} \rangle,$$

where

$W_{So,4}$ = “the train is directed to train station Yana”,
$W_{So,5}$ = “the train is directed other station in Southern Bulgaria”,

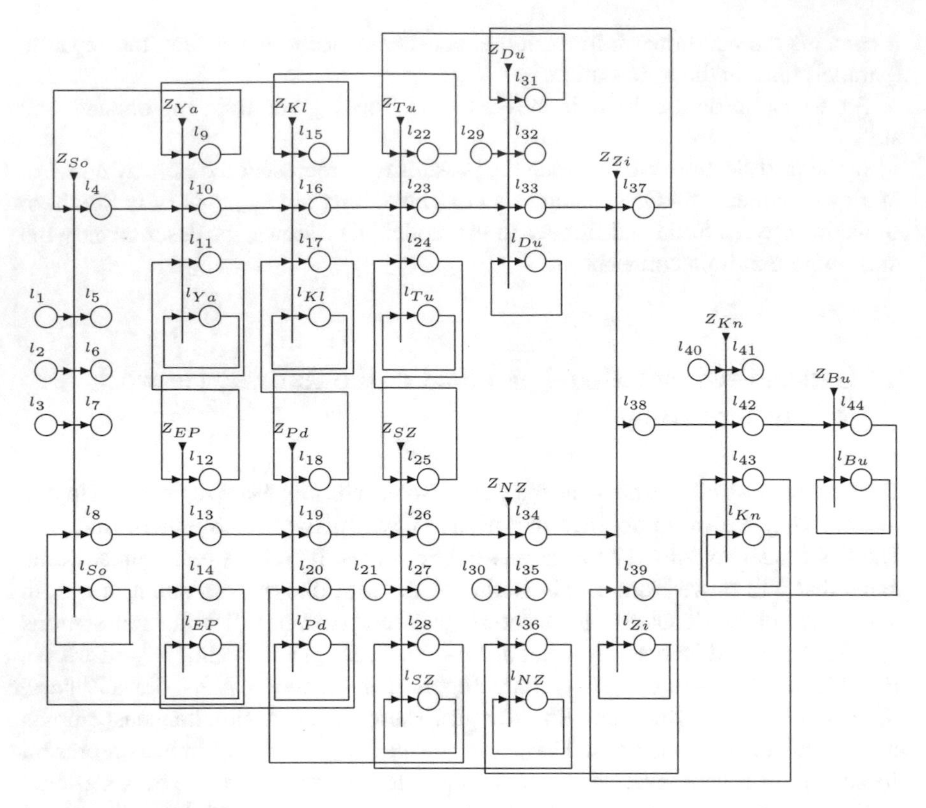

Fig. 4 Railway network

$W_{So,6}$ = "the train is directed to Western Bulgaria",
$W_{So,7}$ = "the train is directed to Northern Bulgaria",
$W_{So,8}$ = "the train is directed to train station Elin Pelin".
$W_{So,So}$ = "the train must shunt in train station of Sofia".

$$Z_{Ya} = \langle \{l_4, l_{12}, l_{15}, l_{Ya}\}, \{l_9, l_{10}, l_{11}, l_{Ya}\},$$

	l_9	l_{10}	l_{11}	l_{Ya}
l_4	*false*	*false*	*false*	*true*
l_{12}	*false*	*false*	*false*	*true*
l_{15}	*false*	*false*	*false*	*true*
l_{Ya}	$W_{Ya,9}$	$W_{Ya,10}$	$W_{Ya,11}$	$W_{Ya,Ya}$

$\rangle$,

where

$W_{Ya,9}$ = "the train is directed to train station Sofia",
$W_{Ya,10}$ = "the train is directed to train station Karlovo",

$W_{Ya,11}$ = "the train is directed to train station Elin Pelin",
$W_{Ya,Ya}$ = "the train must shunt in train station of Yana".

$$Z_{EP} = \langle\{l_8, l_{11}, l_{20}, l_{EP}\}, \{l_{12}, l_{13}, l_{14}, l_{EP}\},$$

$$\begin{array}{c|cccc} & l_{12} & l_{13} & l_{14} & l_{EP} \\ \hline l_8 & false & false & false & true \\ l_{11} & false & false & false & true \\ l_{20} & false & false & false & true \\ l_{EP} & W_{EP,12} & W_{EP,13} & W_{EP,14} & W_{EP,EP} \end{array} \rangle,$$

where

$W_{EP,12}$ = "the train is directed to train station Yana",
$W_{EP,13}$ = "the train is directed to train station Plovdiv",
$W_{EP,14}$ = "the train is directed to train station Sofia",
$W_{EP,EP}$ = "the train must shunt in train station of Elin Pelin".

$$Z_{Kl} = \langle\{l_{10}, l_{18}, l_{22}, l_{Kl}\}, \{l_{15}, l_{16}, l_{17}, l_{Kl}\},$$

$$\begin{array}{c|cccc} & l_{15} & l_{16} & l_{17} & l_{Kl} \\ \hline l_{10} & false & false & false & true \\ l_{18} & false & false & false & true \\ l_{22} & false & false & false & true \\ l_{Kl} & W_{Kl,15} & W_{Kl,16} & W_{Kl,17} & W_{Kl,Kl} \end{array} \rangle,$$

where

$W_{Kl,15}$ = "the train is directed to train station Yana",
$W_{Kl,16}$ = "the train is directed to train station Tulovo",
$W_{Kl,17}$ = "the train is directed to train station Plovdiv",
$W_{Kl,Kl}$ = "the train must shunt in train station of Karlovo".

$$Z_{Pd} = \langle\{l_{13}, l_{17}, l_{28}, l_{Pd}\}, \{l_{18}, l_{19}, l_{20}, l_{Pd}\},$$

$$\begin{array}{c|cccc} & l_{18} & l_{19} & l_{20} & l_{Pd} \\ \hline l_{13} & false & false & false & true \\ l_{17} & false & false & false & true \\ l_{28} & false & false & false & true \\ l_{Pd} & W_{Pd,18} & W_{Pd,19} & W_{Pd,20} & W_{Pd,Pd} \end{array} \rangle,$$

where

$W_{Pd,18}$ = "the train is directed to train station Karlovo",
$W_{Pd,19}$ = "the train is directed to train station Stara Zagora",
$W_{Pd,20}$ = " the train is directed to train station Elin Pelin",
$W_{Pd,Pd}$ = "the train must shunt in train station of Plovdiv".

$$Z_{Tu} = \langle \{l_{16}, l_{25}, l_{31}, l_{Tu}\}, \{l_{22}, l_{23}, l_{24}, l_{Tu}\},$$

$$\begin{array}{c|cccc} & l_{22} & l_{23} & l_{24} & l_{Tu} \\ \hline l_{16} & false & false & false & true \\ l_{25} & false & false & false & true \\ l_{31} & false & false & false & true \\ l_{Tu} & W_{Tu,22} & W_{Tu,23} & W_{Tu,24} & W_{Tu,Tu} \end{array} \rangle,$$

where

$W_{Tu,22}$ = "the train is directed to train station Karlovo",
$W_{Tu,23}$ = "the train is directed to train station Dubovo",
$W_{Tu,24}$ = "the train is directed to train station Stara Zagora",
$W_{Tu,Tu}$ = "the train must shunt in train station of Tulovo".

$$Z_{SZ} = \langle \{l_{19}, l_{21}, l_{24}, l_{36}, l_{SZ}\}, \{l_{25}, l_{26}, l_{27}, l_{28}, l_{SZ}\},$$

$$\begin{array}{c|ccccc} & l_{25} & l_{26} & l_{27} & l_{28} & l_{SZ} \\ \hline l_{19} & false & false & false & false & true \\ l_{21} & false & false & false & false & true \\ l_{24} & false & false & false & false & true \\ l_{36} & false & false & false & false & true \\ l_{SZ} & W_{SZ,25} & W_{SZ,26} & W_{SZ,27} & W_{SZ,28} & W_{SZ,SZ} \end{array} \rangle,$$

where

$W_{SZ,25}$ = "the train is directed to train station Tulovo",
$W_{SZ,26}$ = "the train is directed to train station Nova Zagora",
$W_{SZ,27}$ = "the train is directed to other station in Southern Bulgaria",
$W_{SZ,28}$ = "the train is directed to train station Plovdiv",
$W_{SZ,SZ}$ = "the train must shunt in train station of Stara Zagora".

$$Z_{Du} = \langle \{l_{23}, l_{29}, l_{37}, l_{Du}\}, \{l_{31}, l_{32}, l_{33}, l_{Du}\},$$

$$\begin{array}{c|cccc} & l_{31} & l_{32} & l_{33} & l_{Du} \\ \hline l_{23} & false & false & false & true \\ l_{29} & false & false & false & true \\ l_{37} & false & false & false & true \\ l_{Du} & W_{Du,31} & W_{Du,32} & W_{Du,33} & W_{Du,Du} \end{array} \rangle,$$

where

$W_{Du,31}$ = "the train is directed to train station Tulovo",
$W_{Du,32}$ = "the train is directed to Northern Bulgaria",
$W_{Du,33}$ = "the train is directed to train station Zimnica",
$W_{Du,Du}$ = "the train must shunt in train station of Dubovo"

$$Z_{NZ} = \langle \{l_{26}, l_{30}, l_{39}, l_{NZ}\}, \{l_{34}, l_{35}, l_{36}, l_{NZ}\},$$

$$\begin{array}{c|cccc} & l_{34} & l_{35} & l_{36} & l_{NZ} \\ \hline l_{26} & false & false & false & true \\ l_{30} & false & false & false & true \\ l_{39} & false & false & false & true \\ l_{NZ} & W_{NZ,34} & W_{NZ,35} & W_{NZ,36} & W_{NZ,NZ} \end{array} \rangle,$$

where

$W_{NZ,34}$ = "the train is directed to train station Zimnica",
$W_{NZ,35}$ = "the train is directed to other station in Southern Bulgaria",
$W_{NZ,36}$ = "the train is directed to train station Stara Zagora",
$W_{NZ,NZ}$ = "the train must shunt in train station of Nova Zagora"

$$Z_{Zi} = \langle \{l_{33}, l_{34}, l_{43}, l_{Zi}\}, \{l_{37}, l_{38}, l_{39}, l_{Zi}\},$$

$$\begin{array}{c|cccc} & l_{37} & l_{38} & l_{39} & l_{Zi} \\ \hline l_{33} & false & false & false & true \\ l_{34} & false & false & false & true \\ l_{43} & false & false & false & true \\ l_{Zi} & W_{Zi,37} & W_{Zi,38} & W_{Zi,39} & W_{Zi,Zi} \end{array} \rangle,$$

where

$W_{Zi,37}$ = "the train is directed to train station Dubovo",
$W_{Zi,38}$ = "the train is directed to train station Karnobat",
$W_{Zi,39}$ = "the train is directed to train station Nova Zagora",
$W_{Zi,Zi}$ = "the train must shunt in train station of Zimnica".

$$Z_{Kn} = \langle\{l_{38}, l_{40}, l_{44}, l_{Kn}\}, \{l_{41}, l_{42}, l_{43}, l_{Kn}\},$$

$$\begin{array}{c|cccc} & l_{41} & l_{42} & l_{43} & l_{Kn} \\ \hline l_{38} & false & false & false & true \\ l_{40} & false & false & false & true \\ l_{44} & false & false & false & true \\ l_{Kn} & W_{Kn,41} & W_{Kn,42} & W_{Kn,43} & W_{Kn,Kn} \end{array} \rangle,$$

where

$W_{Kn,41}$ = "the train is directed to Northern Bulgaria",
$W_{Kn,42}$ = "the train is directed to train station Burgas",
$W_{Kn,43}$ = "the train is directed to train station Zimnica",
$W_{Kn,Kn}$ = "the train must shunt in train station of Karnobat".

$$Z_{Bu} = \langle\{l_{42}, l_{Bu}\}, \{l_{44}, l_{Bu}\},$$

$$\begin{array}{c|cc} & l_{44} & l_{Bu} \\ \hline l_{42} & false & true \\ l_{Bu} & W_{Bu,44} & W_{Bu,Bu} \end{array} \rangle,$$

where

$W_{Bu,44}$ = "the train is directed to train station of Karnobat",
$W_{Bu,Bu}$ = "the train must shunt in train station of Burgas".

The represented GN can be extended by replacing some of the places with transitions or with GN. For example the places l_5, l_6, l_7 and l_8 can be replaced with GN and thus can be represented connection of Sofia with the railway network of the south-west Bulgaria. The places l_{27}, l_{32} and l_{35} can be replaced with GN and thus to be represented connection of the line Sofia-Plovdiv-Burgas with the railway network of the south-east Bulgaria and with north-Bulgaria. Other details can be included by replacing some of the places or some of the transitions with GN. The index matrices can be made more detailed and thus the schedule of the trains can be included in the model. If some problem occurs, the GN model can help us for some solution. For example, let here is problem to pass from Elin Pelin to Plovdiv. Then the problem can be bypassed, redirected the trains through Yana, Karlovo and then Plovdiv, or usinb busses between Elin Pelin and Plovdiv.

5 Conclusion

The so constructed GN-model can be used for simulation of different situations that can occur between the trains. In a result, we can obtain ideas e.g., for keeping or changing of the schedule of concrete train(s) in the current moment, or to redirect some of the trains to other line to bypass the problematic section. The constructed GN can be extended including other types of transportation, for example buses, or including passenger flow.

Acknowledgements Work presented here is partially supported by the Bulgarian National Scientific Fund under the grants DFNI-I02/20 Efficient Parallel Algorithms for Large Scale Computational Problems.

References

1. Alexieva, J., Choy, E., Koycheva, E.: Review and bibliography on generalized nets theory and applications. In: Choy, E., Krawczak, M., Shannon, A., Szmidt, E. (eds.) A Survey of Generalized Nets, Raffles KvB Monograph No. 10, pp. 207–301 (2007)
2. Atanassov, K.: Generalized Nets. World Scientific, London (1991)
3. Atanassov, K.: On Generalized Nets Theory. Prof. M. Drinov Academic Publishing House, Sofia (2007)
4. Atanassov, K.: Applications of Generalized Nets. World Scientific, London (1993)
5. Assad, A.A.: Models for rail transportation. Transp. Res. Part A Gen. **143**, 205–220 (1980)
6. El Amaraoui, A., Mesghouni, K.: Train scheduling networks under time duration uncertainty. In: Proceedings of the 19th World Congress of the International Federation of Automatic Control, pp. 8762–8767 (2014)
7. Woroniuk, C., Marinov, M.: Simulation modelling to analyze the current level of utilization of sections along rail rout. J. Transp. Lit. **7**(2), 235–252 (2013)
8. Jin, J.G., Zhao, J., Lee, D.H.: A column generation based approach for the train network design optimization problem. J. Transp. Res. **50**(1), 1–17 (2013)
9. Krawczak, M.: A novel modeling methodology: generalized nets. Artificial Intelligence and Soft Computing, LNCS 4029, pp. 1160–1168. Springer (2006)
10. Peneva, D., Tasseva, V., Kodogiannis, V., Sotirova, E.: Generalized nets as an instrument for description of the process of expert system construction. IEEE Intell. Syst. 755–759 (2006)
11. Shannon, A., Sorsich, J., Atanassov, K.: Generalized Nets in Medicine. Prof. M. Drinov Academic Publishing House, Sofia (1996)
12. Atanassov, K.: Index Matrices: Towards an Augmented Matrix Calculus. Springer, Cham (2014)

On Nonlocal Models of Kulish-Sklyanin Type and Generalized Fourier Transforms

V.S. Gerdjikov

Abstract A special class of multicomponent NLS equations, generalizing the vector NLS and related to the **BD.I**-type symmetric are shown to be integrable through the inverse scattering method (ISM). The corresponding fundamental analytic solutions are constructing thus reducing the inverse scattering problem to a Riemann-Hilbert problem. We introduce the minimal sets of scattering data $\mathfrak{T}$ which determines uniquely the scattering matrix and the potential Q of the Lax operator. The elements of $\mathfrak{T}$ can be viewed as the expansion coefficients of Q over the 'squared solutions' that are natural generalizations of the standard exponentials. Thus we demonstrate that the mapping $\mathfrak{T} \to Q$ is a generalized Fourier transform. Special attention is paid to two special representatives of this MNLS with three-component and five components which describe spinor ($F = 1$ and $F = 2$, respectively) Bose-Einstein condensates.

1 Introduction

The integrable multicomponent NLS (MNLS) equations are naturally related to the symmetric spaces [1, 2]. Formally the corresponding MNLS can be written as:

$$i\frac{\partial \mathbf{u}}{\partial t} + \frac{1}{2}\frac{\partial^2 \mathbf{u}}{\partial x^2} + \mathbf{u}\mathbf{u}^\dagger \mathbf{u} = 0, \tag{1}$$

see [3, 4]. Here $\mathbf{u}$ may be generic $k \times n$ rectangular matrix. It is well known that these MNLS are related to the A.III class of symmetric spaces in Cartan classification.

V.S. Gerdjikov (✉)
Institute of Nuclear Research and Nuclear Energy,
Bulgarian Academy of Sciences, 72 Tsarigradsko Chausee, 1784 Sofia, Bulgaria
e-mail: gerjikov@inrne.bas.bg

V.S. Gerdjikov
Institute of Mathematics and Informatics, Bulgarian Academy of Sciences,
Acad. Georgi Bonchev Str., Block 8, 1113 Sofia, Bulgaria

K. Georgiev et al. (eds.), *Advanced Computing in Industrial Mathematics*,
Studies in Computational Intelligence 681, DOI 10.1007/978-3-319-49544-6_4

Of course, one should consider also the numerous MNLS that can be obtained from (1) by applying Mikhailov reductions [5], see [4, 6–10].

Some of these MNLS have applications to physics. Most of them are related to the vector NLS, i.e. $k = 1$ and $\mathbf{u}$ is an n-component vector; for $n = 2$ this is the famous Manakov model [2], see also [11, 12].

Another very interesting class of MNLS has been discovered by Kulish and Sklyanin [13]. The simplest nontrivial Kulish-Sklyanin (KS) model is a 3-component one

$$\begin{aligned} &i\partial_t\Phi_1 + \partial_x^2\Phi_1 + 2(|\Phi_1|^2 + 2|\Phi_0|^2)\Phi_1 + 2\Phi_{-1}^*\Phi_0^2 = 0, \\ &i\partial_t\Phi_0 + \partial_x^2\Phi_0 + 2(|\Phi_{-1}|^2 + |\Phi_0|^2 + |\Phi_1|^2)\Phi_0 + 2\Phi_0^*\Phi_1\Phi_{-1} = 0, \\ &i\partial_t\Phi_{-1} + \partial_x^2\Phi_{-1} + 2(|\Phi_{-1}|^2 + 2|\Phi_0|^2)\Phi_{-1} + 2\Phi_1^*\Phi_0^2 = 0. \end{aligned} \tag{2}$$

Its integrability, both in classical and quantum sense, was demonstrated in [13], see also [9, 14, 15].

The next member in this class is a 5-component one:

$$\begin{aligned} i\partial_t\Phi_{\pm 2} + \partial_{xx}\Phi_{\pm 2} &= -2\varepsilon(\Phi, \Phi^*)(x,t)\Phi_{\pm 2} + \varepsilon\Theta(x,t)\Phi_{\mp 2}^*, \\ i\partial_t\Phi_{\pm 1} + \partial_{xx}\Phi_{\pm 1} &= -2\varepsilon(\Phi, \Phi^*)(x,t)\Phi_{\pm 1} - \varepsilon\Theta(x,t)\Phi_{\mp 1}^*, \\ i\partial_t\Phi_0 + \partial_{xx}\Phi_0 &= -2\varepsilon(\Phi, \Phi^*)(x,t)\Phi_0 + \varepsilon\Theta(x,t)\Phi_0^*, \end{aligned} \tag{3}$$

where $\varepsilon = \pm 1$ and

$$\begin{aligned} (\Phi, \Phi^*)(x,t) &= \sum_{\alpha=-2}^{2} \Phi_\alpha\Phi_\alpha^*, \\ \Theta(x,t) &= (\Phi, s_0\Phi) = 2\Phi_2\Phi_{-2} - 2\Phi_1\Phi_{-1} + \Phi_0^2. \end{aligned} \tag{4}$$

Both KS models find important physical applications in describing spin-1 and spin-2 Bose-Einstein condensates (BEC). Indeed, BEC of alkali atoms in the $F = 1$ hyperfine state, elongated in x direction and confined in the transverse directions y, z by purely optical means are described by a 3-component normalized spinor wave vector $\Phi(x,t) = (\Phi_1, \Phi_0, \Phi_{-1})^T(x,t)$ satisfying the Eq. (2), see [12, 16–19]:

The assembly of atoms in the hyperfine state of spin F can be described by a normalized spinor wave vector with $2F + 1$ components

$$\Phi(x,t) = (\Phi_F(x,t), \Phi_{F-1}(x,t), \dots, \Phi_{-F}(x,t))^T,$$

whose components are labeled by the values of $m_F = F, \dots, 1, 0, -1, \dots, -F$. So the spinor BEC with $F = 2$ (taken for rather specific choices of the scattering lengths) in dimensionless coordinates takes the form (3) [9, 19]. For those who are interested in the physics of spinor BEC we provide some more relevant references [17, 18, 20–23].

In the last decade a new trend was started in nonlinear optics in attempt to explain artificial heterogenic media. Such media exhibit new properties, due to the resonance type of interaction of the media and light are observed in photonic crystals, random lasers, etc. (for a review, see [24]). Some of them can be modeled by the so-called $\mathcal{P}$- and $\mathcal{PT}$-symmetric (parity-time) symmetric systems [11, 25–33].

The initial interest in such systems was motivated by quantum mechanics [28, 31]. In [28] it was shown that quantum systems with a non-hermitian Hamiltonian admit states with real eigenvalues, i.e. the hermiticity of the Hamiltonian is not a necessary condition to have real spectrum. Using such Hamiltonians one can build up new quantum mechanics [28, 29, 31, 32]. Starting point is the fact that in the case of a non-Hermitian Hamiltonian with real spectrum, the modulus of the wave function for the eigenstates is time-independent even in the case of complex potentials. All this naturally lead to the development of a special class of non-local versions of the NLS equation and its multicomponent versions [26, 30, 34]. The nonlocality introduced is due to the reductions.

The aim of the present paper is to analyze a special type of MNLS equations of KS type and to show that they preserve integrability also when nonlocal reductions are applied. For $r = 3$ and $r = 5$ and with the standard (local) reductions they are characterized by the Gross-Pitaevsky energy functionals (see Eqs. (6) and (7)) and correspond to integrable MNLS models related to symmetric spaces [1] of **BD.I**-type $\simeq$ SO(2r + 1)/SO(2) $\times$ SO(2r − 1). Our expose will treat in parallel both reductions.

In Sect. 2 we give preliminaries about the BEC in one dimension. We also formulate the Lax representations for the KS-type equations for any r Sect. 3 deals with the direct and inverse scattering problem for the Lax operators. More specifically, we outline the construction of the fundamental analytic solutions (FAS) of L which allows us to reduce the inverse scattering problem (ISP) for L to a Riemann-Hilbert problem (RHP). Such approach allows one to use the Zakharov-Shabat dressing method for calculating the soliton solutions of the KS equations. All these considerations are valid for Lax operators of generic form, i.e. without any reductions imposed. In Sect. 4 we formulate the expansions of $q(x, t)$ and its variation $\delta q(x, t)$ over the squared solutions of L for the simplest nontrivial case when L has no discrete eigenvalues. We will see below, that these expansions are compatible with both the local and non-local $\mathbb{Z}_2$-reductions. Their expansions coefficients are provided by the minimal sets of scattering data and their variations. They allow one to generalize the idea of [35] also to the multicomponent KS-type equations with both local and nonlocal reductions. Thus we demonstrate that in all these case the ISM is a generalized Fourier transform. In Sect. 5 we outline the fundamental properties of these NLEE of KS type. In Sect. 6 we recall Mikhailov's reduction group which can naturally be applied also to nonlocal reductions. We derive the constraints on the scattering data imposed by each of these reductions.

2 Preliminaries

2.1 The BEC in One Dimension

The main tool for investigating BEC is the Gross-Pitaevski (GP) equation and the GP functional. In the one-dimensional approximation the GP equation in 1D x-space becomes:

$$i\frac{\partial \Phi}{\partial t} = \frac{\delta E_{\rm GP}[\Phi]}{\delta \Phi^*}. \tag{5}$$

where for $F = 1$ the GP energy functional is given by:

$$\begin{aligned} E_{\rm GP} = \int dx \Big\{ \frac{\hbar^2}{2m}|\partial_x \Phi|^2 + \bar{c}\Big[|\Phi_1|^4 + |\Phi_{-1}|^4 + 2|\Phi_0|^2(|\Phi_1|^2 + |\Phi_{-1}|^2)\Big] \\ + (\bar{c}_0 - \bar{c}_2)|\Phi_1|^2|\Phi_{-1}|^2 + \frac{\bar{c}_0}{2}|\Phi_0|^4 + \bar{c}_2(\Phi_1^*\Phi_{-1}^*\Phi_0^2 + \Phi_0^{*2}\Phi_1\Phi_{-1})\Big\}. \end{aligned} \tag{6}$$

For $F = 2$ the energy functional is defined by [19, 21, 23]

$$E_{\rm GP}[\Phi] = \int_{-\infty}^{\infty} dx \left(\frac{\hbar^2}{2m}|\partial_x \Phi|^2 + \frac{\varepsilon c_0}{2}n^2 + \frac{c_2}{2}\mathbf{f}^2 + \frac{\varepsilon c_4}{2}|\Theta(x,t)|^2 \right), \tag{7}$$

where $\varepsilon = \pm 1$. The number density n and the singlet-pair amplitude Θ are defined in [19, 23]

These two sets of vector NLS equations can be viewed as members of another class of MNLS equations related to the BD.I type of symmetric spaces. They can be written as [7, 14, 15]:

$$i\mathbf{q}_t + \mathbf{q}_{xx} + 2(\mathbf{q},\mathbf{q}^*)\mathbf{q} - (\mathbf{q}, s_0\mathbf{q})s_0\mathbf{q}^* = 0, \tag{8}$$

where $\mathbf{q}$ is $2r-1$-component vector and the constant matrix s_0 has nonvanishing elements ± 1 only on the second diagonal, see Eq. (10).

2.2 Lax Representation for BD.I-Type MNLS Equations

The symmetric spaces of the series **BD.I** are isomorphic to $SO(2r+1)/(SO(2)\otimes SO(2r-1)$, see [36]. The local coordinates on them are provided by the co-adjoint orbits of the algebras $so(2r+1)$ passing through $J = \mathrm{diag}\,(1, 0, \dots, 0, -1)$. These local coordinates are provided by the matrices $q(x,t) = \sum_{\alpha\in\Delta_1^+}(q_\alpha E_\alpha + p_\alpha E_{-\alpha})$ where the set of roots $\Delta_1^+ = \{e_1 - e_2, \dots, e_1 - e_r, e_r, e_1 + e_r, \dots, e_1 + e_2\}$. For the typical representation we have the matrix form:

$$q(x,t) = \begin{pmatrix} 0 & \mathbf{q}^T & 0 \\ \mathbf{p} & 0 & s_0\mathbf{q} \\ 0 & \mathbf{p}^T s_0 & 0 \end{pmatrix}, \qquad J = \mathrm{diag}(1, 0, \dots 0, -1). \tag{9}$$

The $2r-1$-component vectors $\mathbf{q} = (q_2, \dots, q_{2r})^T$ are formed by the coefficients q_α as follows: $q_k \equiv q_{e_1-e_k}$, $q_{r+1} \equiv q_{e_1}$ and $q_{2r+1-k} \equiv q_{e_1+e_k}$, $k = 2, \dots, r$; the vector $\mathbf{p} = (p_2, \dots, p_n)^T$ is formed analogously. The matrix $s_0 = S_0^{(n)}$ enters in the definition of $so(n)$, i.e. $X \in so(n)$, if $X + S_0^{(n)} X^T S_0^{(n)} = 0$, and for $n = 2r+1$:

$$S_0^{(n)} = \sum_{s=1}^{n} (-1)^{s+1} E_{s,n+1-s}^{(n)}, \tag{10}$$

With this definition of orthogonality the Cartan subalgebra generators are represented by diagonal matrices. By $E_{sp}^{(n)}$ above we mean $n \times n$ matrix whose matrix elements are $(E_{sp}^{(n)})_{ij} = \delta_{si}\delta_{pj}$.

The MNLS equations allow Lax representation $[L, M] = 0$ as follows

$$L\psi(x,t,\lambda) \equiv i\partial_x\psi + (q(x,t) - \lambda J)\psi(x,t,\lambda) = 0. \tag{11}$$

$$M\psi(x,t,\lambda) \equiv i\partial_t\psi + (V_0(x,t) + \lambda V_1(x,t) - \lambda^2 J)\psi(x,t,\lambda) = 0, \tag{12}$$

$$V_1(x,t) = q(x,t), \quad V_0(x,t) = i\,\mathrm{ad}_J^{-1}\frac{dq}{dx} + \frac{1}{2}\left[\mathrm{ad}_J^{-1} q, q(x,t)\right]. \tag{13}$$

In terms of these notations the generic MNLS type equations connected to **BD.I.** acquire the form

$$\begin{aligned} &i\mathbf{q}_t + \mathbf{q}_{xx} + 2(\mathbf{q}, \mathbf{p})\mathbf{q} - (\mathbf{q}, s_0\mathbf{q})s_0\mathbf{p} = 0, \\ &i\mathbf{p}_t - \mathbf{p}_{xx} - 2(\mathbf{q}, \mathbf{p})\mathbf{p} + (\mathbf{p}, s_0\mathbf{p})s_0\mathbf{q} = 0, \end{aligned} \tag{14}$$

This equation allows two types of reductions. The first one—the typical reduction $\mathbf{p}(x,t) = \mathbf{q}^*(x,t)$ is well studied by now, see [1, 13, 37]. The corresponding Hamiltonian for the Eq. (14) is given by

$$H_{\mathrm{MNLS}} = \int_{-\infty}^{\infty} dx \left((\partial_x\mathbf{q}, \partial_x\mathbf{q}^*) - (\mathbf{q}, \mathbf{q}^*)^2 + (\mathbf{q}, s_0\mathbf{q})(\mathbf{q}^*, s_0\mathbf{q}^*) \right), \tag{15}$$

For $r = 2$ we introduce the variables $\Phi_1 = q_2$, $\Phi_0 = q_3/\sqrt{2}$, $\Phi_{-1} = q_4$; for $r = 3$ we set $\Phi_2 = q_2$, $\Phi_1 = q_3$, $\Phi_0 = q_4$, $\Phi_{-1} = q_5$ and $\Phi_{-2} = q_6$. This reproduces the action functionals E_{GP} for $F = 1$ and $F = 2$.

The second reduction is a non-local one $\mathbf{p}(x,t) = -\mathbf{q}^*(-x,t)$ and is the main topic of the present paper. As a result we obtain the nonlocal NLS model of **BD.I**-type:

$$i\mathbf{q}_t + \mathbf{q}_{xx} - 2(\mathbf{q}(x,t), \mathbf{q}^*(-x,t))\mathbf{q}(x,t) + (\mathbf{q}(x,t), s_0\mathbf{q}(x,t))s_0\mathbf{q}^*(-x,t) = 0. \tag{16}$$

3 The Direct and the Inverse Scattering Problem

Here we will outline the solution of the direct scattering problem and the construction of the fundamental analytic solutions (FAS) [38]. The construction goes true for both choices of involutions: local and nonlocal. Following [39] we reduce it to a RHP.

3.1 The Direct Scattering Problem

Solving the direct scattering problem for L uses the Jost solutions which are defined by, see [6] and the references therein

$$\lim_{x\to-\infty} \phi(x,t,\lambda)e^{i\lambda Jx} = \mathbb{1}, \qquad \lim_{x\to\infty} \psi(x,t,\lambda)e^{i\lambda Jx} = \mathbb{1} \tag{17}$$

and the scattering matrix $T(\lambda,t) \equiv \psi^{-1}\phi(x,t,\lambda)$. The choice of J and the fact that the Jost solutions and $T(\lambda,t)$ take values in the group $SO(2r+1)$ means that we can use the following block-matrix structure of $T(\lambda,t)$

$$T(\lambda,t) = \begin{pmatrix} m_1^+ & -\mathbf{B}^{-T} & c_1^- \\ \mathbf{b}^+ & \mathbf{T}_{22} & -s_0\mathbf{b}^- \\ c_1^+ & \mathbf{B}^{+T}s_0 & m_1^- \end{pmatrix}, \qquad \hat{T}(\lambda,t) = \begin{pmatrix} m_1^- & \mathbf{b}^{-T} & c_1^- \\ -\mathbf{B}^+ & s_0\mathbf{T}_{22}s_0 & s_0\mathbf{B}^- \\ c_1^+ & -\mathbf{b}^{+T}s_0 & m_1^+ \end{pmatrix}, \tag{18}$$

where $\mathbf{b}^{\pm}(\lambda,t)$ and $\mathbf{B}^{\pm}(\lambda,t)$ are $2r-1$-component vectors, $\mathbf{T}_{22}(\lambda)$ and $\boldsymbol{m}^{\pm}(\lambda)$ are $2r-1\times 2r-1$ block matrices, and $m_1^{\pm}(\lambda)$, $c_1^{\pm}(\lambda)$ are scalars. The matrix elements of $T(\lambda,t)$ satisfy a number of relations which ensure that $T(\lambda)$ belongs to $SO(2r+1)$ and that $T(\lambda)\hat{T}(\lambda) = \mathbb{1}$. Some of them take the form:

$$\begin{aligned} &m_1^+m_1^- + \mathbf{B}^{-T}\mathbf{B}^+ + c_1^+c_1^- = 1, \qquad \mathbf{b}^+\mathbf{b}^{-T} + T_{22}s_0T_{22}^Ts_0 + s_0\mathbf{b}^-\mathbf{b}^{+T}s_0 = \mathbb{1}, \\ &m_1^+m_1^- + \mathbf{b}^{+T}\mathbf{b}^- + c_1^+c_1^- = 1, \qquad \mathbf{B}^+\mathbf{B}^{-T} + s_0T_{22}^Ts_0T_{22} + s_0\mathbf{B}^-\mathbf{B}^{+T}s_0 = \mathbb{1}. \end{aligned} \tag{19}$$

3.2 The Fundamental Analytic Solutions

It is well known that the Jost solutions satisfy a system of Volterra-type integral equations. Indeed, if we introduce

$$Y_+(x,t,\lambda) = \psi(x,t,\lambda)e^{iJ\lambda x}, \qquad Y_-(x,t,\lambda) = \phi(x,t,\lambda)e^{iJ\lambda x}, \tag{20}$$

then $Y_{\pm}(x,t,\lambda)$ must satisfy:

$$Y_{\pm;jk}(x,t,\lambda) = \delta_{jk} + i\int_{\pm\infty}^{x} dy\; e^{-i\lambda(a_j-a_k)(x-y)}\left([J,Q(y,t)]Y_{\pm}(x,t,\lambda)\right)_{jk}. \tag{21}$$

Here we have used the notation $J = \mathrm{diag}\,(a_1, a_2, \dots, a_{2r}, a_{2r+1})$; i.e. $a_1 = 1$, $a_2 = a_3 = \cdots = a_{2r} = 0$, $a_{2r+1} = -1$, (see Eq. (9)).

The Volterra equations (20) always have solution for real λ. Analytic extension for $\lambda \in \mathbb{C}_+$ (resp. for $\lambda \in \mathbb{C}_-$) is possible only for the first column of $Y_-(x, t, \lambda)$ and for the last column of $Y_+(x, t, \lambda)$ (resp. for the last column of $Y_-(x, t, \lambda)$ and for the first column of $Y_+(x, t, \lambda)$. Following Shabat's method [38] we consider two sets of integral equations:

$$\xi^+_{jk}(x, t, \lambda) = \delta_{jk} + i \int_{\varepsilon_{jk}\infty}^{x} dy\, e^{-i\lambda(a_j - a_k)(x-y)} \left([J, Q(y, t)]\xi^+(y, t, \lambda) \right)_{jk}. \quad (22)$$

$$\xi^-_{jk}(x, t, \lambda) = \delta_{jk} + i \int_{-\eta_{jk}\infty}^{x} dy\, e^{-i\lambda(a_j - a_k)(x-y)} \left([J, Q(y, t)]\xi^-(y, t, \lambda) \right)_{jk}, \quad (23)$$

where

$$\varepsilon_{jk} = \begin{cases} 1 & \text{for } j \prec k, \\ -1 & \text{for } j \succeq k, \end{cases}, \qquad \eta_{jk} = \begin{cases} -1 & \text{for } j \preceq k, \\ 1 & \text{for } j \succ k, \end{cases}. \quad (24)$$

Here we used the notation

$$\begin{array}{llllll} j \prec k & \text{iff} & a_j > a_k; & j \preceq k & \text{iff} & a_j \geq a_k; \\ j \succ k & \text{iff} & a_j < a_k; & j \succeq k & \text{iff} & a_j \leq a_k. \end{array} \quad (25)$$

Then one can prove that the Eq. (22) (resp. (23)) possess solutions $\xi^+(x, t, \lambda)$ (resp. $\xi^-(x, t, \lambda)$) which allow analytic extension for $\lambda \in \mathbb{C}_+$ (resp. for $\lambda \in \mathbb{C}_-$). The solutions $\xi^\pm(x, t, \lambda)$ can be viewed also as solutions to a RHP

$$\xi^+(x, t, \lambda) = \xi^-(x, t, \lambda) G(x, t, \lambda), \qquad G(x, t, \lambda) = e^{i\lambda J x} G_0(t, \lambda) e^{-i\lambda J x} \quad (26)$$

with canonical normalization, i,e, $\lim_{\lambda \to \infty} \xi^\pm(x, t, \lambda) = \mathbb{1}$.

If we denote by $\chi^\pm(x, t, \lambda) = \xi^\pm(x, t, \lambda) e^{-i\lambda J x}$ then $\chi^+(x, t, \lambda)$ will be the FAS of L [38, 40, 41]. Below we will use two equivalents sets of FAS:

$$\begin{aligned} \chi^\pm(x, t, \lambda) &= \psi(x, t, \lambda) T^\mp_J(t, \lambda) D^\pm_J(\lambda), & \chi^\pm(x, t, \lambda) &= \phi(x, t, \lambda) S^\pm_J(t, \lambda), \\ \tilde{\chi}^\pm(x, t, \lambda) &= \phi(x, t, \lambda) S^\pm_J(t, \lambda) \hat{D}^\pm_J(\lambda), & \tilde{\chi}^\pm(x, t, \lambda) &= \psi(x, t, \lambda) T^\mp_J(t, \lambda), \end{aligned} \quad (27)$$

where $S^\pm_J$, $T^\pm_J$ and $D^\pm_J$ are generalized Gauss factors of the scattering matrix, see [3, 10, 40–42]:

$$T(\lambda, t) = T^-_J D^+_J \hat{S}^+_J = T^+_J D^-_J \hat{S}^-_J, \quad (28)$$

where

$$
\begin{aligned}
T_J^-(\lambda,t) &= \begin{pmatrix} 1 & 0 & 0 \\ \boldsymbol{\rho}^+ & \mathbb{1} & 0 \\ \tilde{c}_1^- & \boldsymbol{\rho}^{+T} s_0 & 1 \end{pmatrix}, & T_J^+(\lambda,t) &= \begin{pmatrix} 1 & -\boldsymbol{\rho}^{-,T} & \tilde{\tilde{c}}_1^+ \\ 0 & \mathbb{1} & -s_0\boldsymbol{\rho}^- \\ 0 & 0 & 1 \end{pmatrix}, \\
S_J^+(\lambda,t) &= \begin{pmatrix} 1 & \boldsymbol{\tau}^{+T} & \tilde{c}_1^+ \\ 0 & \mathbb{1} & s_0\boldsymbol{\tau}^+ \\ 0 & 0 & 1 \end{pmatrix}, & S_J^-(\lambda,t) &= \begin{pmatrix} 1 & 0 & 0 \\ -\boldsymbol{\tau}^- & \mathbb{1} & 0 \\ \tilde{\tilde{c}}_1^- & -\boldsymbol{\tau}^{-T} s_0 & 1 \end{pmatrix}, \\
D_J^+(\lambda) &= \begin{pmatrix} m_1^+ & 0 & 0 \\ 0 & \mathbf{m}_2^+ & 0 \\ 0 & 0 & 1/m_1^+ \end{pmatrix}, & D_J^-(\lambda) &= \begin{pmatrix} 1/m_1^- & 0 & 0 \\ 0 & \mathbf{m}_2^- & 0 \\ 0 & 0 & m_1^- \end{pmatrix},
\end{aligned}
\tag{29}
$$

We have made use of the following notations above:

$$
\begin{aligned}
\boldsymbol{\rho}^\pm &= \frac{\mathbf{b}^\pm}{m_1^\pm}, & \boldsymbol{\tau}^\pm &= \frac{\mathbf{B}^\mp}{m_1^\pm}, & \tilde{c}_1^+ &= \frac{1}{2}(\boldsymbol{\tau}^{+T} s_0 \boldsymbol{\tau}^+), \\
\tilde{c}_1^- &= \frac{1}{2}(\boldsymbol{\rho}^{+T} s_0 \boldsymbol{\rho}^+), & \tilde{\tilde{c}}_1^+ &= \frac{1}{2}(\boldsymbol{\rho}^{-T} s_0 \boldsymbol{\rho}^-), & \tilde{\tilde{c}}_1^- &= \frac{1}{2}(\boldsymbol{\tau}^{-T} s_0 \boldsymbol{\tau}^-), \\
c_1^+ &= \frac{(\mathbf{b}^{+T} s_0 \mathbf{b}^+)}{2m_1^+}, & c_1^- &= \frac{(\mathbf{B}^{-T} s_0 \mathbf{B}^-)}{2m_1^+}.
\end{aligned}
\tag{30}
$$

3.3 The Inverse Scattering Problem (ISP)

An important tool for reducing the ISP to a Riemann-Hilbert problem (RHP) are the fundamental analytic solution (FAS) $\chi^\pm(x,t,\lambda)$ and $\tilde{\chi}^\pm(x,t,\lambda)$.

The Lax representation (11) and (12) ensures that if $q(x,t)$ evolves according to (14) then the scattering matrix and its elements satisfy the following linear evolution equations

$$
i\frac{d\vec{\rho}^\pm}{dt} \pm \lambda^2 \vec{\rho}^\pm(t,\lambda) = 0, \qquad i\frac{d\vec{\tau}^\pm}{dt} \mp \lambda^2 \vec{\tau}^\pm(t,\lambda) = 0, \qquad i\frac{dD^\pm}{dt} = 0, \tag{31}
$$

so the block-diagonal matrices $D^\pm(\lambda)$ can be considered as generating functionals of the integrals of motion. The fact that all $(2r-1)^2$ matrix elements of $\boldsymbol{m}_2^\pm(\lambda)$ for $\lambda \in \mathbb{C}_\pm$ generate integrals of motion reflect the superintegrability of the model and are due to the degeneracy of the dispersion law of (14). We remind that $D_J^\pm(\lambda)$ allow analytic extension for $\lambda \in \mathbb{C}_\pm$ and that their zeroes and poles determine the discrete eigenvalues of L.

Given the solutions $\chi^\pm(x,t,\lambda)$ one recovers $q(x,t)$ via the formula

$$
q(x,t) = \lim_{\lambda\to\infty} \lambda \left(J - \chi^\pm J \widehat{\chi}^\pm(x,t,\lambda) \right). \tag{32}
$$

The main goal of the dressing method [10, 37, 41–43] is, starting from a known solutions $\chi_0^{\pm}(x,t,\lambda)$ of $L_0(\lambda)$ with potential $q_{(0)}(x,t)$ to construct new singular solutions $\chi_1^{\pm}(x,t,\lambda)$ of L with a potential $q_{(1)}(x,t)$ with two (or more) additional singularities located at prescribed positions $\lambda_1^{\pm}$. It is related to the regular one by a dressing factor $u(x,t,\lambda)$, for details see [6, 43, 44].

4 The Generalized Fourier Transforms for Non-regular J

The generalized Fourier transforms (GFT) for the NLEE are based on the completeness relation for the 'squared solutions' of L. These completeness relations for the case of generic J have been proved in [42], see also [3, 30]. In our case J is highly degenerate: $2r-1$ of its eigenvalues are vanishing. This fact substantially changes the two important steps in the construction:

(i) split the algebra $\mathfrak{g} \simeq so(2r+1)$ into two subspaces: $\mathfrak{g} = \mathcal{O}_J \oplus \mathcal{O}_J^{\perp}$. Here $\mathcal{O}_J$ is the image of the operator ad_J and provides the co-adjoint orbit in $\mathfrak{g}$ passing through J. In our case $\mathcal{O}_J \equiv \mathrm{span}\,\{E_\alpha, E_{-\alpha}, \alpha \in \delta_1^+\}$. $\mathcal{O}_J^{\perp}$ is the complementary space orthogonal to $\mathcal{O}_J$ with respect to the Killing form. In what follows we will introduce the operator $\pi_J = \mathrm{ad}_J^{-1}\mathrm{ad}_J$ which projects any element of $\mathfrak{g}$ onto $\mathcal{O}_J$;

(ii) split each of the 'squared solutions' $e_\alpha^{\pm}(x,\lambda) = \chi^{\pm}(x,\lambda)E_\alpha\hat{\chi}^{\pm}(x,\lambda)$ and $\tilde{e}_\alpha^{\pm}(x,\lambda) = \tilde{\chi}^{\pm}(x,\lambda)E_\alpha\hat{\tilde{\chi}}^{\pm}(x,\lambda)$ into two parts:

$$e_\alpha^{\pm}(x,\lambda) = \mathbf{e}_\alpha^{\pm}(x,\lambda) + e_\alpha^{\pm,\perp}(x,\lambda), \qquad \tilde{e}_\alpha^{\pm}(x,\lambda) = \tilde{\mathbf{e}}_\alpha^{\pm}(x,\lambda) + \tilde{e}_\alpha^{\pm,\perp}(x,\lambda), \tag{33}$$

where $\mathbf{e}_\alpha^{\pm}(x,\lambda)$, $\tilde{\mathbf{e}}_\alpha^{\pm}(x,\lambda)$ belong to $\mathcal{O}_J$, $\mathbf{e}_\alpha^{\pm,\perp}(x,\lambda)$ and $\tilde{\mathbf{e}}_\alpha^{\pm,\perp}(x,\lambda)$ belong to $\mathcal{O}_J^{\perp}$.

We can view $q(x,t) \in \mathcal{O}_J$ as a generic element of the co-adjoint orbit. The rest of the idea for the GFT is based on the analyticity properties of the 'squared solutions' and on the completeness relation of $\mathbf{e}_\alpha^{\pm}(x,\lambda)$ and $\tilde{\mathbf{e}}_\alpha^{\pm}(x,\lambda)$, $\alpha \in \delta_1^+ \cup (-\delta_1^+)$ on $\mathcal{O}_J$. and is a natural generalization of the proof for generic J [3, 6, 42]. Skipping the details we formulate the expansions for $q(x)$ and $\mathrm{ad}_J^{-1}\delta q(x)$. Of course, for the sake of brevity we treat the case when the Lax operator L has no discrete eigenvalues.

$$\begin{aligned} q(x) &= -\frac{i}{\pi}\int_{-\infty}^{\infty} d\lambda \sum_{\alpha\in\delta_1^+} \left(\tau_\alpha^+(\lambda)\mathbf{e}_\alpha^+(x,\lambda) - \tau_\alpha^-(\lambda)\mathbf{e}_{-\alpha}^-(x,\lambda)\right) \\ &= \frac{i}{\pi}\int_{-\infty}^{\infty} d\lambda \sum_{\alpha\in\delta_1^+} \left(\rho_\alpha^+(\lambda)\tilde{\mathbf{e}}_{-\alpha}^+(x,\lambda) - \rho_\alpha^-(\lambda)\tilde{\mathbf{e}}_\alpha^-(x,\lambda)\right). \end{aligned} \tag{34}$$

Lemma 1 *Let the potential $q(x,t)$ be such that the Lax operator L has no discrete eigenvalues. Then as minimal set of scattering data which determines uniquely the scattering matrix $T(\lambda,t)$ and the corresponding potential $q(x,t)$ one can consider either one of the sets $\mathfrak{T}_i$, $i=1,2$*

$$\mathfrak{T}_1 \equiv \{\rho_\alpha^+(\lambda,t), \rho_\alpha^-(\lambda,t), \quad \alpha \in \delta_1^+\}, \qquad \mathfrak{T}_2 \equiv \{\tau_\alpha^+(\lambda,t), \tau_\alpha^-(\lambda,t), \quad \alpha \in \delta_1^+\},$$

for $\lambda \in \mathbb{R}$. In other words, the minimal sets of scattering data consist of the expansion coefficients of $q(x)$ over the 'squared solutions'.

Similar expansions hold true also for the variation of $q(x)$ [3, 40, 42]:

$$\begin{aligned} \mathrm{ad}_J^{-1}\delta q(x) &= \frac{i}{\pi}\int_{-\infty}^{\infty} d\lambda \sum_{\alpha\in\Delta_1^+} \left(\delta\tau_\alpha^+(\lambda)\mathbf{e}_\alpha^+(x,\lambda) + \delta\tau_\alpha^-(\lambda)\mathbf{e}_{-\alpha}^-(x,\lambda)\right) \\ &= \frac{i}{\pi}\int_{-\infty}^{\infty} d\lambda \sum_{\alpha\in\Delta_1^+} \left(\delta\rho_\alpha^+(\lambda)\tilde{\mathbf{e}}_{-\alpha}^+(x,\lambda) + \delta\rho_\alpha^-(\lambda)\tilde{\mathbf{e}}_{-\alpha}^-(x,\lambda)\right). \end{aligned} \tag{35}$$

If we consider the special type of variations: $\delta q(x) \simeq \frac{\partial q}{\partial t}\delta t + \mathcal{O}((\delta t)^2)$, then the expansions (35) go into

$$\begin{aligned} \mathrm{ad}_J^{-1}\frac{\partial q}{\partial t} &= \frac{i}{\pi}\int_{-\infty}^{\infty} d\lambda \sum_{\alpha\in\Delta_1^+} \left(\frac{\partial\tau_\alpha^+}{\partial t}\mathbf{e}_\alpha^+(x,\lambda) + \frac{\partial\tau_\alpha^-}{\partial t}\mathbf{e}_{-\alpha}^-(x,\lambda)\right) \\ &= \frac{i}{\pi}\int_{-\infty}^{\infty} d\lambda \sum_{\alpha\in\Delta_1^+} \left(\frac{\partial\rho_\alpha^+}{\partial t}\tilde{\mathbf{e}}_{-\alpha}^+(x,\lambda) + \frac{\partial\rho_\alpha^-}{\partial t}\tilde{\mathbf{e}}_{-\alpha}^-(x,\lambda)\right). \end{aligned} \tag{36}$$

To complete the analogy between the standard Fourier transform and the expansions over the 'squared solutions' we need the generating operators $\Lambda_\pm$:

$$\Lambda_\pm X(x) \equiv \mathrm{ad}_J^{-1}\left(i\frac{dX}{dx} + i\left[q(x), \int_{\pm\infty}^{x} dy\,[q(y), X(y)]\right]\right). \tag{37}$$

for which the 'squared solutions' are eigenfunctions:

$$(\Lambda_+ - \lambda)\tilde{\mathbf{e}}_{\mp\alpha}^\pm(x,\lambda) = 0, \qquad (\Lambda_- - \lambda)\mathbf{e}_{\pm\alpha}^\pm(x,\lambda) = 0, \qquad \alpha \in \delta_1^+. \tag{38}$$

5 Fundamental Properties of the MNLS Equations

The expansions (34) and (35) and the explicit form of $\Lambda_\pm$ and Eq. (38) are basic for deriving the fundamental properties of all MNLS type equations related to the Lax operator L. Each of these NLEE is determined by its dispersion law which we choose to be of the form $F(\lambda) = f(\lambda)J$, where $f(\lambda)$ is polynomial in λ. The corresponding NLEE becomes:

$$i\,\mathrm{ad}_J^{-1}q_t + f(\Lambda_\pm)q(x,t) = 0. \tag{39}$$

Theorem 1 *The NLEE (39) are equivalent to: (i) the Eq. (31) and (ii) the following evolution equations for the generalized Gauss factors of $T(\lambda)$:*

$$i\frac{dS_J^+}{dt} + [F(\lambda), S_J^+] = 0, \qquad i\frac{dT_J^-}{dt} + [F(\lambda), T_J^-] = 0, \qquad \frac{dD_J^+}{dt} = 0. \tag{40}$$

or, equivalently. to:

$$i\frac{d\vec{\tau}^\pm}{dt} \mp f(\lambda)\vec{\tau}^\pm(t,\lambda) = 0, \qquad i\frac{d\vec{\rho}^\pm}{dt} \pm f(\lambda)\vec{\rho}^\pm(t,\lambda) = 0. \tag{41}$$

The principal series of integrals is generated by the asymptotic expansion of $\ln m_1^+(\lambda) = \sum_{k=1}^\infty I_k \lambda^{-k}$. The first integrals of motion are of the form:

$$I_1 = -\frac{i}{2}\int_{-\infty}^\infty dx\, \langle q(x), q(x)\rangle, \qquad I_2 = \frac{1}{2}\int_{-\infty}^\infty dx\, \langle q_x(x), \mathrm{ad}_J^{-1} q(x)\rangle, \tag{42}$$

Now iI_1 can be interpreted as the density of the particles, I_2 is the momentum. The third one $I_3 = iH_{\mathrm{MNLS}}$ provides the Hamiltonian. Indeed, the Hamiltonian equations of motion given by $H_{(0)} = -iI_3$ with the Poisson brackets

$$\{q_k(y,t), p_j(x,t)\} = i\delta_{kj}\delta(x-y), \tag{43}$$

coincide with the MNLS equations (14). The above Poisson brackets are dual to the canonical symplectic form:

$$\Omega_0 = i\int_{-\infty}^\infty dx\, \mathrm{tr}\left(\delta \mathbf{p}(x) \underset{'}{\wedge} \delta \mathbf{q}(x)\right) = \frac{1}{2i}[\![\mathrm{ad}_J^{-1}\delta q(x) \underset{'}{\wedge} \mathrm{ad}_J^{-1}\delta q(x)]\!],$$

where $\underset{'}{\wedge}$ means that taking the scalar or matrix product we exchange the usual product of the matrix elements by wedge-product.

The Hamiltonian formulation of Eq. (14) with Ω_0 and H_0 is just one member of the hierarchy of Hamiltonian formulations provided by:

$$\Omega_k = \frac{1}{i}[\![\mathrm{ad}_J^{-1}\delta Q \underset{'}{\wedge} \Lambda^k \mathrm{ad}_J^{-1}\delta Q]\!], \qquad H_k = i^{k+3} I_{k+3}. \tag{44}$$

where $\Lambda = \frac{1}{2}(\Lambda_+ + \Lambda_-)$. We can also calculate Ω_k in terms of the scattering data variations. Imposing the reduction $q(x) = q^\dagger(x)$ we get:

$$\begin{aligned}\Omega_k &= \frac{1}{2\pi i}\int_{-\infty}^\infty d\lambda\, \lambda^k \left(\Omega_0^+(\lambda) - \Omega_0^-(\lambda)\right) \\ &= \frac{1}{2\pi}\int_{-\infty}^\infty d\lambda\, \lambda^k \mathrm{Im}\left(m_1^+(\lambda)\left(\hat{\boldsymbol{m}}_2{}^+ \delta\boldsymbol{\rho}^+(\lambda) \underset{'}{\wedge} \delta\boldsymbol{\tau}^+(\lambda)\right)\right).\end{aligned}$$

This allows one to prove that if we are able to cast Ω_0 in canonical form, then all Ω_k will also be cast in canonical form and will be pair-wise equivalent.

6 The Consequences of the Involutions

6.1 Mikhailov's Group of Reductions

The notion of the reduction group for the integrable NLEE was introduced by Mikhailov in the beginning of the 1980s [5].

The reduction group G_R is a finite group which preserves the Lax representation (11) and (12). This means that the reduction constraints are automatically compatible with the evolution. Mikhailov proposed that G_R must act on the Lax pair with its two realizations simultaneously: (i) $G_R \subset \mathrm{Aut}\mathfrak{g}$ and (ii) $G_R \subset \mathrm{Conf}\,\mathbb{C}$, i.e. as conformal mappings of the complex λ-plane. To each $g_k \in G_R$ we relate a reduction condition for the Lax pair as follows [5]:

$$C_k(L(\Gamma_k(\lambda))) = \eta_k L(\lambda), \quad C_k(M(\Gamma_k(\lambda))) = \eta_k M(\lambda), \tag{45}$$

where $C_k \in \mathrm{Aut}\,\mathfrak{g}$ and $\Gamma_k(\lambda) \in \mathrm{Conf}\,\mathbb{C}$ are the images of g_k and $\eta_k = 1$ or -1 depending on the choice of C_k. Since G_R is a finite group then for each g_k there exist an integer N_k such that $g_k^{N_k} = \mathbb{1}$. In all the cases below $N_k = 2$ and the reduction group is isomorphic to $\mathbb{Z}_2$.

More specifically the automorphisms C_k, $k = 1, \dots, 4$ listed above lead to the following reductions for the matrix-valued functions

$$U(x,t,\lambda) = [J, Q(x,t)] - \lambda J, \qquad V(x,t,\lambda) = V_0(x,t) + \lambda V_1(x,t) - \lambda^2 J, \tag{46}$$

of the Lax representation:

$$\begin{array}{lll} (1) & C_1(U^\dagger(\kappa_1(\lambda))) = U(\lambda), & C_1(V^\dagger(\kappa_1(\lambda))) = V(\lambda), \\ (2) & C_2(U^T(\kappa_2(\lambda))) = -U(\lambda), & C_2(V^T(\kappa_2(\lambda))) = -V(\lambda), \\ (3) & C_3(U^*(\kappa_1(\lambda))) = -U(\lambda), & C_3(V^*(\kappa_1(\lambda))) = -V(\lambda), \\ (4) & C_4(U(\kappa_2(\lambda))) = U(\lambda), & C_4(V(\kappa_2(\lambda))) = V(\lambda), \end{array} \tag{47}$$

For the nonlocal involutions we change also $x \to -x$ and find:

$$\begin{array}{lll} (1) & C_1(U^\dagger(\kappa_1(\lambda))) = -U(\lambda), & C_1(V^\dagger(\kappa_1(\lambda))) = V(\lambda), \\ (2) & C_2(U^T(\kappa_2(\lambda))) = U(\lambda), & C_2(V^T(\kappa_2(\lambda))) = -V(\lambda), \\ (3) & C_3(U^*(\kappa_1(\lambda))) = U(\lambda), & C_3(V^*(\kappa_1(\lambda))) = -V(\lambda), \\ (4) & C_4(U(\kappa_2(\lambda))) = -U(\lambda), & C_4(V(\kappa_2(\lambda))) = V(\lambda), \end{array} \tag{48}$$

Both types of involutions impose constraints on the scattering matrix and on its Gauss factors that are listed below.

6.2 The Local Involution Case

The involution:

$$U^{\dagger}(x,t,\kappa_1\lambda^*) = U(x,t,\lambda), \qquad \Leftrightarrow \qquad q(x,t) = q^{\dagger}(x,t), \qquad \kappa_1 = 1, \tag{49}$$

On the Jost solutions we have

$$\phi^{\dagger}(x,t,\lambda^*) = \phi^{-1}(x,t,\lambda), \qquad \psi^{\dagger}(x,t,\lambda^*) = \psi^{-1}(x,t,\lambda), \tag{50}$$

so for the scattering matrix we have

$$T^{\dagger}(t,\lambda^*) = T^{-1}(t,\lambda), \tag{51}$$

and for the Gauss factors:

$$S^{-\dagger}(\lambda^*) = \hat{S}^{+}(\lambda), \qquad T^{-\dagger}(\lambda^*) = \hat{T}^{-}(\lambda), \qquad D^{-\dagger}(\lambda^*) = \hat{D}^{+}(\lambda), \tag{52}$$

Note that the FAS can be used to define the kernel of the resolvent of L by $R^{\pm}(x.y.\lambda) = -i\chi^{\pm}(x,\lambda)\Theta^{\pm}(x-y)\hat{\chi}^{\pm}(y,\lambda)$, where the functions $\Theta^{\pm}(x-y)$ satisfy the equation $\frac{\partial}{\partial x}\Theta^{\pm}(x-y) = \delta(x-y)\mathbb{1}$ [6, 40]. Next, one can fix up $\Theta^{\pm}(x-y)$ in such a way that $R^{\pm}(x,y,\lambda)$ fall off exponentially for $x, y \to +\infty$. So, if $D^{+}(\lambda)$ (or $D^{-}(\lambda)$) have a zero or a pole at $\lambda = \lambda_1^{+}$ (or at $\lambda = \lambda_1^{-}$) then $\lambda_1^{\pm}$ will be poles of $R^{\pm}(x,y,\lambda)$ and consequently, discrete eigenvalues of L.

If we have local reduction, then

$$\tau^{+}(\lambda) = -\tau^{-,*}(\lambda), \qquad \rho^{+}(\lambda) = -\rho^{-,*}(\lambda), \tag{53}$$

6.3 The Nonlocal Involution Case

Now the involution is:

$$U^{\dagger}(x,t,\lambda^*) = -U(-x,t,-\lambda), \qquad \Leftrightarrow \qquad q(x,t) = q^{\dagger}(-x,t). \tag{54}$$

On the Jost solutions we have

$$\phi^{\dagger}(x,t,\lambda^*) = \psi^{-1}(-x,t,-\lambda), \qquad \psi^{\dagger}(x,t,\lambda^*) = \phi^{-1}(x,t,-\lambda), \tag{55}$$

so for the scattering matrix we have

$$T^{\dagger}(t,-\lambda^*) = T(t,\lambda), \tag{56}$$

As a consequence for the Gauss factors we get:

$$T^{-\dagger}(-\lambda^*) = \hat{S}^+(\lambda), \qquad T^{+\dagger}(-\lambda^*) = \hat{S}^-(\lambda), \qquad D^{\pm\dagger}(\lambda^*) \quad = \hat{D}^{\pm}(-\lambda). \quad (57)$$

In analogy with the local reductions, the kernel of the resolvent has poles at the at the points $\lambda_2^{\pm}$ at which $D^{\pm}(\lambda)$ have poles or zeroes. In particular, if λ_2^+ is an eigenvalue, then $-\lambda_2^+$ is also an eigenvalue. For the reflection coefficients we obtain the constraints:

$$\tau^+(-\lambda) = -\rho^{+,*}(\lambda), \qquad \tau^-(-\lambda) = -\rho^{-,*}(\lambda), \qquad (58)$$

7 Conclusion

We demonstrated that the results concerning the GFT for nonlocal reductions hold true also for the MNLS cases, in particular for the Kulish-Sklyanin type models. The results are natural extensions of the ones in [30] to the multicomponent cases.

References

1. Fordy, A.P., Kulish, P.P.: Nonlinear Schrodinger equations and simple lie algebras. Commun. Math. Phys. **89**, 427–443 (1983)
2. Manakov, S.V.: On the theory of two-dimensional stationary self-focusing of electromagnetic waves. Zh. Eksp. Teor. Fiz. 65, 1392 (1973). (English translation). Sov. Phys. JETP **38**, 248 (1974) (in Russian)
3. Gerdjikov, V.S.: Complete integrability, gauge equivalence and lax representations of the inhomogeneous nonlinear evolution equations. Theor. Math. Phys. **92**, 374–386 (1992)
4. Gerdjikov, V.S., Grahovski, G.G., Kostov, N.A.: On the multi-component NLS type equations on symmetric spaces and their reductions. Theor. Math. Phys. **144**(2), 1147–1156 (2005)
5. Mikhailov, A.V.: The reduction problem and the inverse scattering problem. Phys. D **3**, 73–117 (1981)
6. Gerdjikov, V.S.: Basic aspects of soliton theory. In: Mladenov, I.M., Hirshfeld, A.C. (eds.) Geometry, Integrability and Quantization, pp. 78–125. Softex, Sofia (2005). nlin.SI/0604004
7. Gerdjikov, V.S., Grahovski, G.G.: Multi-component NLS Models on symmetric spaces: spectral properties versus representations theory. SIGMA **6**, 044, 29 pp (2010). arXiv:1006.0301 [nlin.SI]
8. Gerdjikov, V.S., Kostov, N.A., Valchev, T.I.: N-wave equations with orthogonal algebras: $\mathbb{Z}_2$ and $\mathbb{Z}_2 \times \mathbb{Z}_2$ reductions and soliton solutions. SIGMA **3**, paper 039, 19 pp (2007). arXiv:nlin.SI/0703002
9. Gerdjikov, V.S., Kostov, N.A., Valchev, T.I.: Solutions of multi-component NLS models and Spinor Bose-Einstein condensates. Phys. D **238** 1306-1310 (2009). arXiv:0802.4398 [nlin.SI]
10. Gerdjikov, V.S., Grahovski, G.G., Ivanov, R.I., Kostov, N.A.: N-wave interactions related to simple Lie algebras. $\mathbb{Z}_2$-reductions and Soliton Solutions. Inverse Prob. **17**, 999–1015 (2001)
11. Barashenkov, I.V., Zezyulin, D.A., Konotop, V.V.: Exactly solvable Wadati potentials in the PT-symmetric Gross-Pitaevskii equation. arXiv:1511.06633
12. Doktorov, E.V., Wang, J., Yang, J.: Perturbation theory for bright spinor Bose-Einstein condensate solitons. Phys. Rev. A **77**, 043617 (2008)

13. Kulsh, P.P., Sklyanin, E.K.: $0(n)$-invariant nonlinear Schrödinger equation a new completely integrable system. Phys. Lett. A **84A**, 349–352 (1981)
14. Gerdjikov, V.S., Kostov, N.A., Valchev, T.I.: Bose-Einstein condensates with $F = 1$ and $F = 2$. Reductions and soliton interactions of multi-component NLS models. In: Saltiel, S.M., Dreischuh, A.A., Christov, I.P. (eds.) Proc. SPIE **7501**, 7501W (2009). arXiv:1001.0168 [nlin.SI]
15. Kostov, N.A., Atanasov, V.A., Gerdjikov, V.S., Grahovski, G.G.: On the soliton solutions of the spinor Bose-Einstein condensate. Proc. SPIE **6604**, 66041T (2007). Atanasov, P.A., Dreischuh, T.N., Gateva, S.V., Kovachev, L.M. (eds.)
16. Nistazakis, H.E., Frantzeskakis, D.J., Kevrekidis, P.G., Malomed, B.A., Carretero-Gonzalez, R.: Bright-dark soliton complexes in spinor Bose-Einstein condensates. Phys. Rev. A **77**, 033612 (2008)
17. Ieda, J., Miyakawa, T., Wadati, M.: Matter-wave solitons in an $F = 1$ spinor Bose-Einstein condensate. J. Phys. Soc. Jpn. **73**, 2996 (2004)
18. Uchiyama, M., Ieda, J., Wadati, M.: Dark solitons in $F = 1$ spinor Bose-Einstein condensate. J. Phys. Soc. Jpn. **75**, 064002 (2006)
19. Uchiyama, M., Ieda, J., Wadati, M.: Multicomponent bright solitons in $F = 2$ spinor Bose-Einstein condensates. J. Phys. Soc. Jpn. **76**(7), 74005 (2007)
20. Klausen, N. N., Bohn, J. L., Greene, C. H.: Nature of spinor Bose-Einstein condensates in rubidium. Phys. Rev. A **64**, 053602 (2001)
21. Ohmi, T., Machida, K.: Bose-Einstein condensation with internal degrees of freedom in alkali atom gases. J. Phys. Soc. Jpn. **67**, 1822 (1998)
22. Uchino, S., Otsuka, T., Ueda, M.: Dynamical symmetry in spinor Bose-Einstein condensates. arXiv:0710.5210
23. Ueda, M., Koashi, M.: Theory of spin-2 Bose-Einstein condensates: spin correlations, magnetic response, and excitation spectra. Phys. Rev. A **65**, 063602 (2002)
24. Zyablovsky, A.A., Vinogradov, A.P., Pukhov, A.A., Dorofeenko, A.V., Lisyansky, A.A.: PT-symmetry in optics. Phys.-Uspekhi **57**(11), 1063 (2014)
25. Ablowitz, M., Bakirtas, I., Ilan, B.: Wave collapse in a class of nonlocal nonlinear Schrddinger equation. Phys. D **207**, 230–253 (2005)
26. Ablowitz, M., Musslimani, Z.: Integrable nonlocal nonlinear Schrödinger equation. Phys. Rev. Lett. **110**, 064105(5) (2013)
27. Ablowitz, M.J., Musslimani, Z.H.: Inverse scattering transform for the integrable nonlocal nonlinear Schrödinger equation. Under review (2015)
28. Bender, C.M., Boettcher, S.: Real spectra in non-hermitian Hamiltonians having PT symmetry. Phys. Rev. Lett. **80**, 5243–5246 (1998); Bender, C.M., Boettcher, S., Meisinger, P.N.: PT-symmetric quantum mechanics. J. Math. Phys. **40**, 2201–2229 (1999)
29. Bender, C.M.: Making sense of non-hermitian Hamiltonians. Rep. Prog. Phys. **70** 947–1018 (2007). (E-print: hep-th/0703096)
30. Gerdjikov, V.S., Saxena, A.: Complete integrability of nonlocal nonlinear Schrödinger equation. arXiv:1510.00480v1 [nlin.SI]
31. Mostafazadeh, A.: Pseudo-hermiticity versus PT-symmetry I, II, III. J. Math. Phys. **43**, 205–214 (2002). (E-print: math-ph/0107001); 2814–2816 (E-print: math-ph/0110016); 3944–3951 (E-print: math-ph/0203005)
32. Mostafazadeh, A.: Pseudo-hermiticity and Generalized PT- and PT-symmetries. J. Math. Phys. **44**, 974–989 (2003). (E-print: math-ph/0209018); Mostafazadeh, A.: Exact PT-symmetry is equivalent to hermiticity. J. Phys. A: Math. Gen. **36**, 7081–7091 (2003). (E-print: quant-ph/0304080)
33. Valchev, T.: On a nonlocal nonlinear Schrödinger equation. In: Slavova, A. (ed.) Mathematics in Industry, pp. 36–52. Cambridge Scholars Publications (2014). ISBN 978-1-4438-6401-5 (Proceedings of 8-th Annual Meeting of the Bulgarian Section of SIAM, 18–19 December, 2013, Sofia, Bulgaria)
34. Valchev, T.I.: On Mikhailov's reduction group. Phys. Lett. A **379**, 1877–1880 (2015)
35. Ablowitz, M.J., Kaup, D.J., Newell, A.C., Segur, H.: The inverse scattering transform-fourier analysis for nonlinear problems. Stud. Appl. Math. **53**, 249–315 (1974)

36. Helgasson, S.: Differential geometry, lie groups and symmetric spaces. Graduate Studies in Mathematics, vol. 34. AMS, Providence, Rhode Island (2001)
37. Grahovski, G.G., Gerdjikov, V.S., Kostov, N.A., Atanasov, V.A.: New integrable multi-component NLS type equations on symmetric spaces: Z_4 and Z_6 reductions. In: Mladenov, I., De Leon, M. (eds.) Geometry, Integrability and Quantization VII, pp. 154–175. Softex, Sofia (2006)
38. Shabat, A.B.: The inverse scattering problem for a system of differential equations. Funct. Ann. Appl. **9**(3), 75 (1975) (In Russian); Shabat, A.B.: The inverse scattering problem. Differ. Equ. **15**, 1824 (1979) (In Russian)
39. Zakharov, V.E., Shabat, A.B.: A scheme for integrating nonlinear evolution equations of mathematical physics by the inverse scattering method. I & II. Funkts. Anal. Prilozhen. **8**, 43–53 (1974), **13**(3), 13–22 (1979)
40. Gerdjikov, V.S.: Algebraic and analytic aspects of N-wave type equations. Contemp. Math. **301**, 35–68 (2002). nlin.SI/0206014
41. Zakharov, V.E., Manakov, S.V., Novikov, S.P., Pitaevskii, L.I.: Theory of Solitons. The Inverse Scattering Method. Plenum Press (Consultant Bureau), N.Y (1984)
42. Gerdjikov, V.S.: Generalized Fourier transforms for the soliton equations. Gauge covariant formulation. Inverse Prob. **2**(1), 51–74 (1986)
43. Ivanov, R.I.: On the dressing method for the generalized Zakharov-Shabat system. Nucl. Phys. B **694**, 509–524 (2004)
44. Gerdjikov, V.S., Kostov, N.A., Valchev, T.I.: Solutions of multi-component NLS models and Spinor Bose-Einstein condensates. Phys. D. **238**, 1306–1310 (2009). arXiv:0802.4398 [nlin.SI]

Standing Waves in Systems of Perturbed Sine-Gordon Equations

Radoslava Hristova and Ivan Hristov

Abstract Systems of 2D perturbed Sine-Gordon equations are solved numerically by a leapfrog difference scheme. OpenMP parallel programm is realized and tested on the computational cluster of IICT-BAS. As a result a clear three-dimensional standing wave pattern is seen for certain parameters. This result agrees with previous studies of other authors. The observed numerical solutions well explain the experiments on powerful THz radiations from mesas of BSCCO crystals.

1 Introduction

The Josephson effect in Bismuth Strontium Calcium Copper Oxide (BSCCO) crystals was discovered in 1992 [1], i.e. it was discovered that BSCCO crystals are natural stacks of thousands Josephson junctions. A picture of the layered structure of a BSCCO crystal is given in Fig. 1. Powerful THz radiation from BSCCO crystals was immediately predicted as a result of possible synchronization of the phases in different layers (junctions) in the crystal. Soon, in 1993 the mathematical model was deduced and verified [2]. The mathematical model is a system of perturbed Sine-Gordon equations and will be given in the next section. The first powerful THz radiation was achieved after 14 years—in 2007 [3]. It was understand that standing waves of plasma oscillation have been built in the cavity formed by the side surfaces of the crystal, exactly as in a laser. In 2008 the standing waves were seen for the first time in a computer simulation [4]. Although the leading role of the excitation of cavity oscillations was well understand [5], the mechanism of the powerful THz radiation still is not fully explained. For further investigation of the

R. Hristova · I. Hristov (✉)
Faculty of Mathematics and Informatics, St. Kliment Ohridski University of Sofia,
5 James Bourchier Blvd, 1164 Sofia, Bulgaria
e-mail: christov_ivan@abv.bg

R. Hristova
e-mail: radoslava@fmi.uni-sofia.bg

K. Georgiev et al. (eds.), *Advanced Computing in Industrial Mathematics*,
Studies in Computational Intelligence 681, DOI 10.1007/978-3-319-49544-6_5

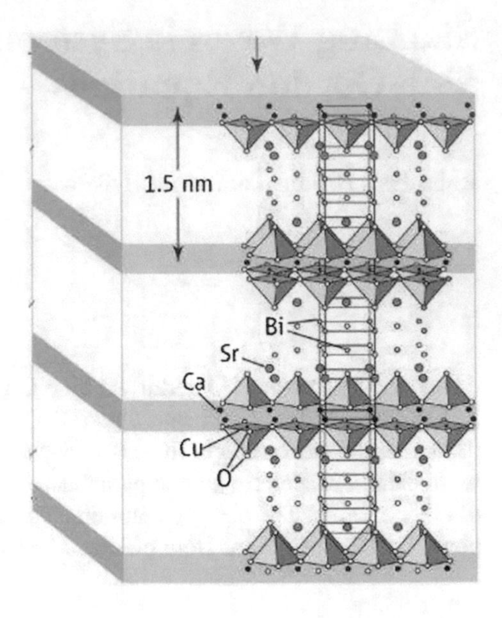

Fig. 1 The layered structure of a BSCCO crystal. Different layers correspond to different Josephson junctions and respectively different differential equation in the system of perturbed Sine-Gordon equations

mechanism, simulations with very long time integration of a quasi 3D model (system of 2D equations) are needed. The investigation with respect to the parameters is also important and computationally intensive. The main aim of this paper is to test a parallel programm realization of a leapfrog algorithm for solving systems of 2D perturbed Sine-Gordon equations.

We want to mention that solving the problem is interesting not only for applications, but from pure mathematical point of view because it is a generalization of the classical perturbed 1D Sine-Gordon equation. As a result of considering a system of equations, new type of solutions without analogues in one equation case, are possible. New 2D effects are also expected.

2 Mathematical Model

We consider systems of 2D perturbed Sine-Gordon equations:

$$S(\varphi_{tt} + \alpha\varphi_t + \sin\varphi - \gamma) = \Delta\varphi, \quad (x, y) \in \Omega \subset R^2 \tag{1}$$

Here Δ is the Laplace operator. Ω is a given domain in R^2. S is the $N_{eq} \times N_{eq}$ cyclic tridiagonal matrix:

$$S = \begin{pmatrix} 1 & s & 0 & . & 0 & s \\ s & 1 & s & 0 & . & 0 \\ . & . & . & . & . & . \\ . & . & . & . & . & . \\ 0 & . & 0 & s & 1 & s \\ s & . & 0 & 0 & s & 1 \end{pmatrix}$$

where $-0.5 < s \leq 0$ and N_{eq} is the number of equations. The unknown is the column vector $\varphi(x, y, t) = (\varphi_1, ..., \varphi_{N_{eq}})^T$. Neumann boundary conditions are considered:

$$\frac{\partial \varphi}{\partial \overrightarrow{n}}|\partial \Omega = 0 \tag{2}$$

In (2) $\overrightarrow{n}$ denotes the exterior normal to the boundary $\partial \Omega$. To close the problem (1), (2) appropriate initial conditions are posed. The choice of the initial conditions is not a trivial task and will be discussed in a future work. The model (1), (2) describes very well the dynamics in N_{eq} periodically stacked Josephson junctions in BSCCO crystals. The parameter s represents the inductive coupling between adjacent Josephson junctions, α is the dissipation parameter, γ is the external current. All the units are normalized as in [2].

3 Numerical Scheme

We solve the problem numerically in rectangular domains by using second order central finite differences with respect to all derivatives in Eq. (1). Let τ be the step in time and h be the mesh size in both x and y space directions, $n + 1$—the number of points in x direction and $m + 1$—the number of points in y direction. Let the mesh points be:

$$x_k = kh, \quad k = 0, \ldots, n, \quad y_i = ih, \quad i = 0, \ldots, m, \quad t_j = j\tau, \quad j = 0, 1, \ldots.$$

and the approximate solution be $\widetilde{\varphi}$.
By using the notations:

$$z^l_{k,i} = \widetilde{\varphi}_l(x_k, y_i, t_j), \quad \hat{z}^l_{k,i} = \widetilde{\varphi}_l(x_k, y_i, t_{j+1}), \quad \check{z}^l_{k,i} = \widetilde{\varphi}_l(x_k, y_i, t_{j-1}), \quad l = 1, \ldots, N_{eq},$$

$$\overline{z}_{k,i} = (z^1_{k,i}, z^2_{k,i}, ..., z^{N_{eq}}_{k,i})^T, \quad \hat{\overline{z}}_{k,i} = (\hat{z}^1_{k,i}, \hat{z}^2_{k,i}, ..., \hat{z}^{N_{eq}}_{k,i})^T, \quad \check{\overline{z}}_{k,i} = (\check{z}^1_{k,i}, \check{z}^2_{k,i}, ..., \check{z}^{N_{eq}}_{k,i})^T,$$

$$z^l_{\bar{x}x,k,i} + z^l_{\bar{y}y,k,i} = \frac{z^l_{k+1,i} + z^l_{k-1,i} + z^l_{k,i+1} + z^l_{k,i-1} - 4z^l_{k,i}}{h^2}, \quad l = 1, \ldots, N_{eq}$$

$$\overline{z}_{\bar{x}x,k,i} + \overline{z}_{\bar{y}y,k,i} = (z^1_{\bar{x}x,k,i} + z^1_{\bar{y}y,k,i}, ..., z^{N_{eq}}_{\bar{x}x,k,i} + z^{N_{eq}}_{\bar{y}y,k,i})^T$$

the finite difference equations that approximate (1) reads as follows:

$$S\hat{\overline{z}}_{k,i} = (S(2\overline{z}_{k,i} + (0.5\alpha\tau - 1)\check{\overline{z}}_{k,i} - \tau^2(\sin\overline{z}_{k,i} - \gamma)) + \tau^2(\overline{z}_{\bar{x}x,k,i} + \overline{z}_{\bar{y}y,k,i}))/(1 + 0.5\alpha\tau). \tag{3}$$

For every $k = 1, \ldots, n-1, i = 1, \ldots, m-1$ a system with the cyclic tridiagonal matrix S is solved with respect to $\hat{z}^l_{k,i}$, $l = 1, \ldots, N_{eq}$, using Gaussian elimination. The diagonally dominance of the matrix S $(-0.5 < s \leq 0)$ ensures the stability of the algorithm. Because of the specific tridiagonal structure of S we need only $9N_{eq} - 12$ floating point operations per solving one system.

To approximate the Neumann boundary conditions (2) third order one-sided finite differences are used. At the left and the right boundaries of the rectangular domain the difference equations are:

$$\hat{\overline{z}}_{0,i} = (18\hat{\overline{z}}_{1,i} - 9\hat{\overline{z}}_{2,i} + 2\hat{\overline{z}}_{3,i})/11, \qquad \hat{\overline{z}}_{n,i} = (18\hat{\overline{z}}_{n-1,i} - 9\hat{\overline{z}}_{n-2,i} + 2\hat{\overline{z}}_{n-3,i})/11.$$

At the top and the bottom boundaries the difference equations are:

$$\hat{\overline{z}}_{k,0} = (18\hat{\overline{z}}_{k,1} - 9\hat{\overline{z}}_{k,2} + 2\hat{\overline{z}}_{k,3})/11, \qquad \hat{\overline{z}}_{k,m} = (18\hat{\overline{z}}_{k,m-1} - 9\hat{\overline{z}}_{k,m-2} + 2\hat{\overline{z}}_{k,m-3})/11.$$

The resulting accuracy of the scheme is $O(h^2 + \tau^2)$. The number of the floating point operations in one time-step for the constructed difference scheme is $O(nmN_{eq})$. This difference scheme is actually a leapfrog difference scheme. The stability condition is:

$$\tau < \frac{\sqrt{1+2s}\; h}{\sqrt{2}} \tag{4}$$

In (4) $\sqrt{1+2s} = \sqrt{\lambda_{min}}$, where λ_{min} is the minimal eigenvalue of S, $\sqrt{2}$ in the denominator corresponds to dimension = 2. The conservativity of the scheme is checked numerically.

4 Numerical Results

The powerful THz radiation from BSCCO crystals corresponds to a new type of solutions which are result of considering systems of equations with strong coupling ($s = -0.4999$ for BSCCO crystals). As numerical results reveal, for certain parameters α and γ the phase $\varphi(x, y, t)$ in a particular equation (junction) is a sum of three terms: a linear with respect to time term vt (resistive term), a static π kink term $pi\_kink(x, y)$ with amplitude $A(\gamma)$ and an oscillating term (with average zero):

$$\varphi(x, y, t) = vt + A(\gamma).pi\_kink(x, y) + oscillating\_term(x, y, t)$$

A clear three-dimensional standing wave pattern is seen as in numerical works [4, 6], i.e. the oscillating term is one and the same standing wave for every equation. The oscillating term is approximately a solution of the linear equation:

$$\varphi_{tt} = \frac{1}{1+2s}\Delta\varphi,\ (x, y) \in \Omega \subset R^2$$

with Neumann boundary conditions. These types of oscillations (standing waves) are called cavity oscillations [5]. For rectangular mesas (rectangular domain with sizes L_x and L_y) oscillation frequencies for cavity mode excitation (M, N) are given explicitly [5]:

$$\omega_{cav}(M, N) = \frac{1}{2\sqrt{1+2s}}\sqrt{\left(\frac{M}{L_x}\right)^2 + \left(\frac{N}{L_y}\right)^2} \tag{5}$$

$$M, N = 0, 1, 2, ..., \quad M + N \geq 1$$

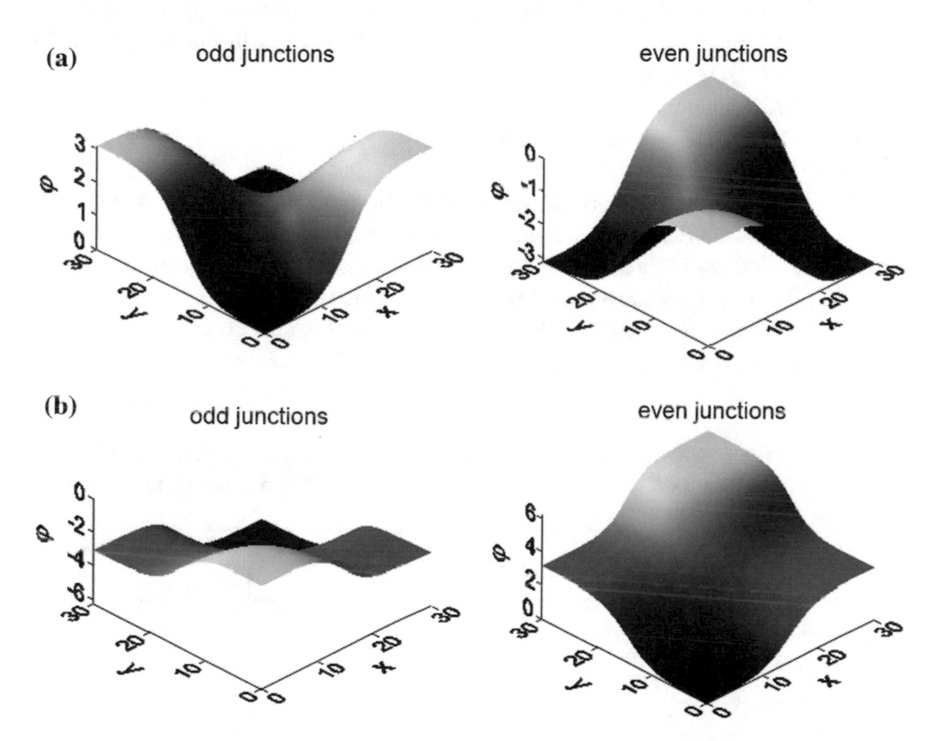

Fig. 2 Two 2D static kink configurations (**a** and **b**) corresponding to excited (1, 1) cavity mode in 30 × 30 square. The cavity mode doesn't determine uniquely the static kink configuration. Number of equations is $N_{eq} = 10$. The parameters are $s = -0.4999$, $\alpha = 0.13$, $\gamma = 1.14$

The oscillating frequency of the solution of problem (1), (2) calculating by FFT (Fast Fourier Transform) and the frequency calculated by formula (5) excellently agree (differ by less then 10^{-5}). This means that the computed solution is a cavity standing wave, exactly as in the experiment [3].

As opposite to the oscillating part, which is the same for each equation (junction), the static part (in our case static kinks) have alternative character, i.e. opposite static kinks alternate in odd and even junctions (equations). For example, two different 2D static kink configurations corresponding to excited (1, 1) cavity mode in 30×30 square are shown in Fig. 2.

5 Computational Aspects

5.1 Parallelization of the Difference Scheme

For solving a problem with large computational domain a parallelization of the difference scheme is needed. OpenMP Fortran realization of the above leapfrog algorithm is made by a straightforward parallelization of all DO loops in the programm by using the OMP DO directive:

```
!$OMP DO [ Clauses ]
Fortran DO loop to be executed in parallel
!$OMP END DO
```

Tests in double precision arithmetic are done on one computational node of the CPU platform of the IICT-BAS cluster, which consists of 2 x $Intel^{®}$ $Xeon^{®}$ Processors X5560 (8 cores, 16 threads with hyper-threading). The result is given in Table 1. The parallel efficiency when 8 OpenMP threads are distributed per each of 8 physical cores, is very good, about 94 %. When the number of threads exceed 8 and we start to run 2 threads per core, parallel efficiency decreases faster. However we have addi-

Table 1 OpenMP realization of the leapfrog algorithm on 2x processors Xeon X5560. Computational domain size is $N_{eq} \times N_x \times N_y = 10 \times 3600 \times 3600$, number of time steps are $N_{time\_steps} = 150$

Number of threads	Time (s)	Speedup	Parallel efficiency (%)
1	411.49	1.00	100.
2	208.92	1.97	98.4
4	105.08	3.92	97.9
6	72.76	5.66	94.3
8	54.58	7.54	94.2
10	49.31	8.34	83.4
12	44.91	9.16	76.3
14	41.07	10.02	71.6
16	37.88	**10.86**	67.9

tional speedup gain of using hyper-threading up to ∼11 when we use all 16 physical threads.

5.2 *Investigation with Respect to the Parameters*

In the case of investigation with respect to the parameters we have the so-called natural parallelism. In our case we have to solve the solutions for different set of parameters (α, γ). In that case we solve many independent problems at the same time on different computational nodes of the CPU platform of the IICT BAS cluster. Each problem is parallelized inside one computational node with OpenMP. The speed up is proportional to the number of used computational nodes.

6 Conclusions

OpenMP realization of a leapfrog algorithm for solving systems of perturbed Sine-Gordon equation is tested on the cluster of IICT-BAS. The numerical results agree with previous studies of other authors. A good base for further computationally intensive investigation of the mechanism of powerful THz radiation in BSCCO crystals is made.

Acknowledgements This work is supported by the National Science Fund of Bulgaria under Grant DFNI-I02/8 and by the National Science Fund of BMSE under grant I02/9/2014.

References

1. Kleiner, R., et al.: Intrinsic Josephson effects in Bi 2 Sr 2 CaCu 2 O 8 single crystals. Phys. Rev. Lett. **68**(15), 2394 (1992)
2. Sakai, S., Bodin, P., Pedersen, N.F.: Fluxons in thin-film superconductor-insulator superlattices. J. Appl. Phys. **73**(5), 2411–2418 (1993)
3. Ozyuzer, L., et al.: Emission of coherent THz radiation from superconductors. Science **318**(5854), 1291–1293 (2007)
4. Lin, S., Hu, X.: Possible dynamic states in inductively coupled intrinsic Josephson junctions of layered high-T c superconductors. Phys. Rev. Lett. **100**(24), 247006 (2008)
5. Tsujimoto, M., et al.: Geometrical resonance conditions for THz radiation from the intrinsic Josephson junctions in Bi 2 Sr 2 CaCu 2 O 8. Phys. Rev. Lett. **105**(3), 037005 (2010)
6. Hu, X., Lin, S.: Three-dimensional phase-kink state in a thick stack of Josephson junctions and terahertz radiation. Phys. Rev. B. **78**(13), 134510 (2008)

Stability Analysis of an Inflation of Internally-Pressurized Hyperelastic Spherical Membranes Connected to Aneurysm Progression

Tihomir B. Ivanov and Elena V. Nikolova

Abstract In this study, we consider a nonlinear second-order ordinary differential equation to describe the inflation of a thin-walled hyperelastic spherical membrane, subjected to an internal distention pressure. We examine the stability of the equilibria of the basic model using three different forms of strain energy functions (SEFs), representing the mechanical properties of elastomers and soft tissues. It is shown that the mechanical stability or instability of the membrane material is associated with the monotonicity or non-monotonicity of the pressure-stretch relation. We define two types of instabilities according to the specific constitutive relation. We prove analytically that a stable inflation of the membrane can retain or can change to an unstable one depending on the specific analytical form of SEF and its material parameters. We derive conditions for the stability/instability of the equilibria of the model with the different SEFs and relate the identified unstable equilibrium states to development and rupture of aneurysms.

1 Introduction

An aneurysm is a localized, blood-filled balloon-like bulge in the wall of a blood vessel [1]. In many cases, its rupture causes massive bleeding with associated high mortality. Recently, one of the main hypotheses, related to aneurysm formation is connected with the appearance of mathematical bifurcations in the quasi-static response

---

T.B. Ivanov (✉)
Institute of Mathematics and Informatics, Bulgarian Academy of Sciences,
Acad. Georgi Bonchev Str., Block 8, 1113 Sofia, Bulgaria
e-mail: tbivanov@fmi.uni-sofia.bg

T.B. Ivanov
Faculty of Mathematics and Informatics, Sofia University,
5 James Bourchier Blvd., 1164 Sofia, Bulgaria

E.V. Nikolova
Institute of Mechanics, Bulgarian Academy of Sciences,
Acad. Georgi Bonchev Str., Block 4, 1113 Sofia, Bulgaria
e-mail: elena@imbm.bas.bg

K. Georgiev et al. (eds.), *Advanced Computing in Industrial Mathematics*,
Studies in Computational Intelligence 681, DOI 10.1007/978-3-319-49544-6_6

of the arterial wall, associated to its material properties, i.e. the corresponding constitutive model [2–8]. In this aspect, the human arteries are modelled as inflated thin-walled, fluid-filled hyperelastic tubes and the hyperelasticity of the arterial wall is presented by different forms of SEFs, which describe the mechanical properties of soft tissues as well as elastomers. In [9] it is pointed out, that the elastic stability or instability of the material is associated with the monotonicity or non-monotonicity of the pressure-stretch (pressure-volume) relation. The loss of stability of rubber-like membranes is explained by a non-monotone inflation, i.e. existence of a maximum in the function relating the induced internal pressure to the corresponding equilibrium stretch of the membrane. The maximum pressure may be either a local or a global maximum, depending on the specific constitutive model of the material (in particular, the concrete form of SEF). The global maximum is connected with a "limit point instability", where a "party balloon effect", which is inherent for rubber-like materials, is observed [10, 11]. For many rubber-like materials, however, the pressure-stretch curve may present a local maximum, followed by a local minimum, which can be followed again by an increasing section of the curve [9, 11]. In this case, when the pressure is increased above the maximum, the stretch "jumps" to a significantly higher value. This phenomenon is defined as an "inflation jump" in [11].

In the present work, we focus on analytical investigation of instabilities, which can appear in the inflation of a thin-walled incompressible hyperelastic spherical membrane (in particular, a perfectly spherical aneurysm) subjected to an internal distention pressure. The main motivation for our study came from the paper of Goriely et al. [11], where the existence of limit point instabilities and inflation jumps in inflated spherical shells was discussed. The authors considered five different forms of SEF and found values of the material parameters, corresponding to a limit point instability, for each of the five SEFs. Unlike their study, we come from another mathematical formulation of the basic model and we consider the instability problem, using three types of SEF, which are different from those, examined in [11].

The remainder of this paper is organized as follows. In Sect. 2 we formulate a mathematical model, describing the problem, mentioned above. Three different types of SEFs are used to represent the hyperelacticy of the membrane. Identification of the equilibrium states of the model and analysis of its stability for each of the three SEFs are discussed in Sect. 3. The main analytical findings are supported by numerical simulations in Sect. 4. Several concluding comments, connected to the interpretation of the basic results in terms of aneurysm progression, are summarized in Sect. 5.

2 Mathematical Formulation of the Basic Model

Let us consider an inflation of a thin-walled spherical membrane (a closed-end spherical aneurysm) under the action of an internal (arterial blood) pressure. The membrane (the aneurysm wall) is modelled as a hyperelastic, isotropic and incompressible material. According to the assumptions for material isotropy and incompressibility, let $\lambda_1 = \lambda_2 = \lambda = r/R$ be the circumferential stretch ratios (r and R are deformed and undeformed membrane radii, respectively), and $\lambda_3 = 1/\lambda^2 = h/H$—the stretch ratio in the thickness direction (h and H are the deformed and the undeformed wall thicknesses, respectivelly). The basic equation of the inflation of the membrane (the aneurysm) can be presented in the following form [10]:

$$\rho \frac{HR}{\lambda^2} \frac{d^2\lambda}{dt^2} = P(t) - \frac{2T(\lambda)}{\lambda R}, \tag{1}$$

where ρ is the mass density of the membrane, $P(t)$ is the arterial blood pressure, and $T_1 = T_2 = T(\lambda)$ is the principal stress resultant in the circumferential direction. The hyperelasticity of the membrane is presented by the following constitutive relation [12]:

$$S = \frac{\partial W(E)}{\partial E} = \frac{1}{\lambda} \frac{\partial W(\lambda)}{\partial \lambda}, \; T = HS, \tag{2}$$

where S are the Piola–Kirchhoff stresses, $E = (\lambda^2 - 1)/2$ are the Green–Lagrange strains, and W is a SEF, respectively.

In our study, we use three well-known SEFs—the SEF of Fung [13], the SEF of Ogden with two parameters [14], and the SEF of Raghavan and Vorp [15]. When $\lambda_1 = \lambda_2 = \lambda$, $\lambda_3 = 1/\lambda^2$, they can reduce to the following forms:

- The SEF of Fung:

$$W^F = a(exp(bE^2) - 1) = a(exp(b((\lambda^2 - 1)/2)^2) - 1), \tag{3}$$

where a and b are material constants.

- The SEF of Ogden with two parameters:

$$W^O = \chi(\lambda_1^\mu + \lambda_2^\mu + \lambda_3^\mu - 3) = \chi(2\lambda^\mu + \lambda^{-2\mu} - 3), \tag{4}$$

where χ and μ are material constants.

- The SEF of Raghavan and Vorp:

$$W^{RV} = \alpha(I_1 - 3) + \beta(I_1 - 3)^2, \tag{5}$$

where $I_1 = 2\lambda^2 + 1/\lambda^4$ and α and β are material constants.

We rewrite the second order ODE (1) as its equivalent first order system

$$\begin{aligned} \frac{d\lambda}{dt} &= \xi, \\ \frac{d\xi}{dt} &= \frac{\lambda^2}{\rho HR} P(t) - \frac{2\lambda S(\lambda)}{\rho R^2}. \end{aligned} \tag{6}$$

3 Stability Analysis

Let us assume that $P(t) = P \equiv const$. Thus, the system (6) is autonomous.

The equilibrium points of the model (6) (if they exist) are in the form $(\tilde{\lambda}, 0)$. Let us denote by $P_{eq}(\tilde{\lambda})$ the constant pressure that corresponds to the equilibrium value $\tilde{\lambda}$. From the second equation in (6) we obtain

$$P_{eq}(\tilde{\lambda}) = \frac{2HS(\tilde{\lambda})}{\tilde{\lambda}R}. \tag{7}$$

Proposition 1 *Let $\tilde{E} = (\tilde{\lambda}, 0)$ be an equilibrium point for the model (6). Then $\tilde{E}$ is a saddle point if $\Delta < 0$ (or $P'_{eq}(\tilde{\lambda}) < 0$) and a center if $\Delta > 0$ (or $P'_{eq}(\tilde{\lambda}) > 0$), where*

$$\Delta = \frac{2S(\tilde{\lambda}) + 2\tilde{\lambda}S'(\tilde{\lambda})}{\rho R^2} - \frac{2\tilde{\lambda}P_{eq}(\tilde{\lambda})}{\rho HR} = \frac{2}{\rho R^2}(\tilde{\lambda}S'(\tilde{\lambda}) - S(\tilde{\lambda}))$$

and $' \equiv \frac{d}{d\lambda}$.

Proof The Jacobian matrix of (6) evaluated at $\tilde{E}$ has the form

$$J(\tilde{\lambda}) = \begin{pmatrix} 0 & 1 \\ -\Delta & 0 \end{pmatrix}$$

and, thus, its characteristic equation is

$$\Lambda^2 + \Delta = 0.$$

If $\Delta < 0$ holds true, then $J(\tilde{\lambda})$ has two real eigenvalues of opposite signs and $\tilde{E}$ is a saddle point. If $\Delta > 0$ is valid, then the eigenvalues of $J(\tilde{\lambda})$ are purely imaginary and $\tilde{E}$ is a center. On the other hand

$$P'_{eq}(\tilde{\lambda}) = \frac{2H}{\tilde{\lambda}^2 R}\left(\tilde{\lambda}S'(\tilde{\lambda}) - S(\tilde{\lambda})\right).$$

Now, it is obvious that $P'_{eq}(\tilde{\lambda})$ and Δ have the same sign and, thus, the proposition follows.

Therefore, in order to study the local stability of the model (6) for the different SEFs, we have to examine the behavior of $P_{eq}(\tilde{\lambda})$ in those cases.

3.1 Equilibrium Pressure Functions

3.1.1 Fung SEF

Now, we shall study the model (6) using the Fung SEF (Eq. 3). In this case, we obtain

$$S^F = \nu(\lambda^2 - 1)exp(\Gamma(\lambda^2 - 1)^2), \tag{8}$$

where $\nu = ab/2$ and $\Gamma = b/4$.

For the equilibrium pressure (7) we have

$$P_{eq}^F(\tilde{\lambda}) = \frac{2H\nu(\tilde{\lambda}^2 - 1)\exp[\Gamma(\tilde{\lambda}^2 - 1)^2]}{\lambda R}. \tag{9}$$

Lemma 1 *$P_{eq}^F(\tilde{\lambda})$ is strictly monotone increasing for every $\tilde{\lambda} > 0$ and has an inflection point at its only zero $\tilde{\lambda} = 1$. Further,*

$$P_{eq}^F(\tilde{\lambda}) \to -\infty, \text{ as } \tilde{\lambda} \to 0^+ \text{ and } P_{eq}^F(\tilde{\lambda}) \to \infty, \text{ as } \tilde{\lambda} \to \infty.$$

Proof For the derivative of $P_{eq}^F(\tilde{\lambda})$ we have

$$\frac{dP_{eq}^F}{d\tilde{\lambda}} = \frac{2H\exp[\Gamma(\tilde{\lambda}^2 - 1)^2]\nu[1 + (1 + 4\Gamma)\tilde{\lambda}^2 - 8\Gamma\tilde{\lambda}^4 + 4\Gamma\tilde{\lambda}^6]}{R\tilde{\lambda}^2}.$$

The sign of the latter is determined by the sign of the polynomial

$$\begin{aligned} Q(\tilde{\lambda}) &= [1 + (1 + 4\Gamma)\tilde{\lambda}^2 - 8\Gamma\tilde{\lambda}^4 + 4\Gamma\tilde{\lambda}^6] \\ &> 4\Gamma\tilde{\lambda}^2(\tilde{\lambda}^4 - 2\tilde{\lambda}^2 + 1) \\ &= 4\Gamma\tilde{\lambda}^2(\tilde{\lambda}^2 - 1)^2 \geq 0. \end{aligned}$$

Now the lemma follows straightforwardly.

We illustrate the behaviour of $P_{eq}^F(\tilde{\lambda})$ in Fig. 1.

3.1.2 Ogden SEF

In the case of the Ogden SEF (Eq. 4) we obtain

$$S = 2\chi\mu(\lambda^{\mu-2} - \lambda^{-2\mu-2}). \tag{10}$$

and for the equilibrium pressure (7):

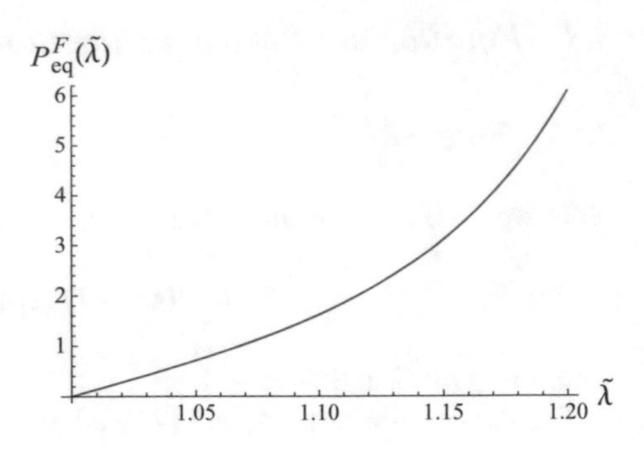

Fig. 1 Equilibrium pressure-stretch ratio relation for the Fung SEF (Eq. 9), where P is in [N/cm^2]

$$P^O_{eq}(\tilde{\lambda}) = \frac{4H\chi\mu\left(\lambda^{\mu-2} - \lambda^{-2\mu-2}\right)}{\lambda R}. \tag{11}$$

Lemma 2 *For the behavior of $P^O_{eq}(\tilde{\lambda})$ the following cases exist.*

- *If $\mu > 3$ holds true, then $P^O_{eq}(\tilde{\lambda})$ is strictcly monotone increasing for every $\tilde{\lambda} > 0$ and its only zero is $\tilde{\lambda} = 1$. Further,*

$$P^O_{eq}(\tilde{\lambda}) \to -\infty, \text{ as } \tilde{\lambda} \to 0^+ \text{ and } P^O_{eq}(\tilde{\lambda}) \to \infty, \text{ as } \tilde{\lambda} \to \infty.$$

- *If $\mu < 3$ is valid, then $P^O_{eq}(\tilde{\lambda})$ is strictly monotone increasing for every $0 < \tilde{\lambda} < \left(\frac{3+2\mu}{3-\mu}\right)^{1/(3\mu)}$ and strictly monotone decreasing for $\tilde{\lambda} > \left(\frac{3+2\mu}{3-\mu}\right)^{1/(3\mu)}$. Further,*

$$P^O_{eq}(\tilde{\lambda}) \to -\infty, \text{ as } \tilde{\lambda} \to 0^+ \text{ and } P^O_{eq}(\tilde{\lambda}) \to 0, \text{ as } \tilde{\lambda} \to \infty$$

and $\tilde{\lambda} = 1$ is its only zero. $P^O_{eq}(\tilde{\lambda})$ has one maximum and it is at $\tilde{\lambda} = \left(\frac{3+2\mu}{3-\mu}\right)^{1/(3\mu)}$.

Proof For the derivative of $P^O_{eq}(\tilde{\lambda})$ we have

$$\frac{dP^F_{eq}}{d\tilde{\lambda}} = \frac{4H\chi\mu\tilde{\lambda}^{-2(\mu+2)}\left(3 + 2\mu + \tilde{\lambda}^{3\mu}(\mu - 3)\right)}{R}.$$

The sign of $\frac{dP^F_{eq}}{d\tilde{\lambda}}$ depends on the sign of the expression $\left(3 + 2\mu + \tilde{\lambda}^{3\mu}(\mu - 3)\right)$. If $\mu > 3$, it is always positive. Otherwise, if $0 < \mu < 3$, it is positive for $0 < \tilde{\lambda} < \left(\frac{3+2\mu}{3-\mu}\right)^{1/(3\mu)}$ and negative for $\tilde{\lambda} > \left(\frac{3+2\mu}{3-\mu}\right)^{1/(3\mu)}$.

We illustrate the behaviour of $P^O_{eq}(\tilde{\lambda})$ for $\mu < 3$ and $\mu > 3$ in Fig. 2.

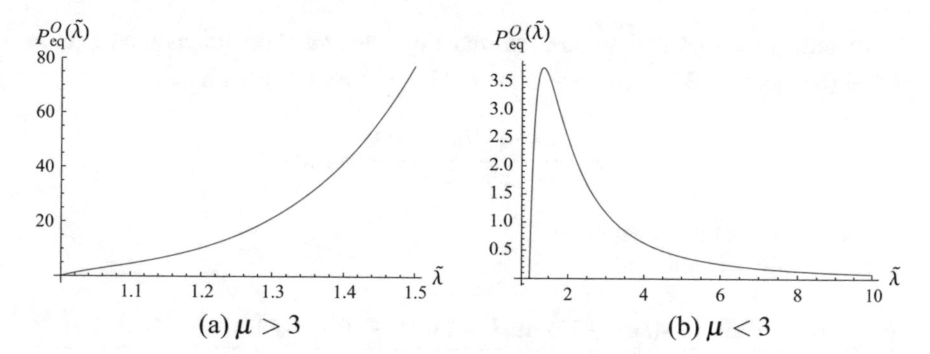

Fig. 2 Equilibrium pressure-stretch ratio relation for the Ogden SEF (Eq. 11), where P is in [N/cm^2]

3.1.3 Raghavan–Vorp SEF

For the Raghavan–Vorp SEF (Eq. 5) we obtain

$$S = 4(\lambda - \lambda^{-5})[\alpha + 2\beta(2\lambda^2 + \lambda^{-4} - 3)]. \tag{12}$$

and from (7) we have

$$P^{RV}_{eq}(\tilde{\lambda}) = \frac{8H\left(1 - \frac{1}{\tilde{\lambda}^6}\right)\left(\alpha + 2\beta\left(\frac{1}{\tilde{\lambda}^4} + 2\tilde{\lambda}^2 - 3\right)\right)}{\tilde{\lambda}R}. \tag{13}$$

Lemma 3 *For the behavior of $P^{RV}_{eq}(\tilde{\lambda})$ two possibilities exist. Let us define $\varphi(\kappa) = \frac{4\kappa^6+10\kappa^3+22}{\kappa^5-7\kappa^2}$.*

- *If $\frac{\alpha}{\beta} < 6 + \varphi\left(\sqrt[3]{3\sqrt{37}+16}\right) \approx 23.5954$, then $P^{RV}_{eq}(\tilde{\lambda})$ is strictly monotone increasing for every $\tilde{\lambda} > 0$.*
- *If $\frac{\alpha}{\beta} > 6 + \varphi\left(\sqrt[3]{3\sqrt{37}+16}\right) \approx 23.5954$, then there exist $0 < \tilde{\lambda}_{\max} < \tilde{\lambda}_{\min}$ such that $P^{RV}_{eq}(\tilde{\lambda})$ is strictly monotone increasing for $0 < \tilde{\lambda} < \tilde{\lambda}_{\max}$ and $\tilde{\lambda} > \tilde{\lambda}_{\min}$ and strictly monotone decreasing for $\tilde{\lambda}_{\max} < \tilde{\lambda} < \tilde{\lambda}_{\min}$.*

Proof For the derivative of $P^{RV}_{eq}(\tilde{\lambda})$ we obtain consecutively

$$\begin{aligned}\frac{dP^{RV}_{eq}}{d\tilde{\lambda}} &= \frac{56\alpha H}{\tilde{\lambda}^8 R} - \frac{8\alpha H}{\tilde{\lambda}^2 R} + \frac{176\beta H}{\tilde{\lambda}^{12} R} - \frac{336\beta H}{\tilde{\lambda}^8 R} + \frac{80\beta H}{\tilde{\lambda}^6 R} + \frac{48\beta H}{\tilde{\lambda}^2 R} + \frac{32\beta H}{R}\\ &= \frac{8H\beta}{R\kappa^6}\left(4\kappa^6 + (6-\gamma)\kappa^5 + 10\kappa^3 + 7(\gamma-6)\kappa^2 + 22\right),\end{aligned}$$

where $\gamma = \alpha/\beta$ and $\kappa = \tilde{\lambda}^2 > 0$.

Therefore the sign of $\frac{dP_{eq}^{RV}}{d\tilde{\lambda}}$ is determined by the sign if the polynomial $g(\kappa) = 4\kappa^6 + (6-\gamma)\kappa^5 + 10\kappa^3 + 7(\gamma - 6)\kappa^2 + 22$. κ is a root of $g(\kappa)$ if

$$\gamma - 6 = \frac{4\kappa^6 + 10\kappa^3 + 22}{\kappa^5 - 7\kappa^2} = \varphi(\kappa).$$

The graph of $\varphi(\kappa)$ is shown in Fig. 3

Taking into account that

$$\frac{d\varphi}{d\kappa} = \frac{4(\kappa^9 - 33\kappa^6 - 45\kappa^3 + 77)}{\kappa^3(\kappa^3-7)^2} \frac{4(\kappa^3-1)(\kappa^3 - 16 - 3\sqrt{37})(\kappa^3 - 16 + 3\sqrt{37})}{\kappa^3(\kappa^3-7)^2},$$

it is easy to see that the minimum of $\varphi(\kappa)$, $\kappa > \sqrt[3]{7}$, is obtained at $\kappa = \sqrt[3]{3\sqrt{37} + 16}$. Now the proposition follows easily.

We illustrate the behavior of $P_{eq}^{RV}(\tilde{\lambda})$ in Fig. 4.

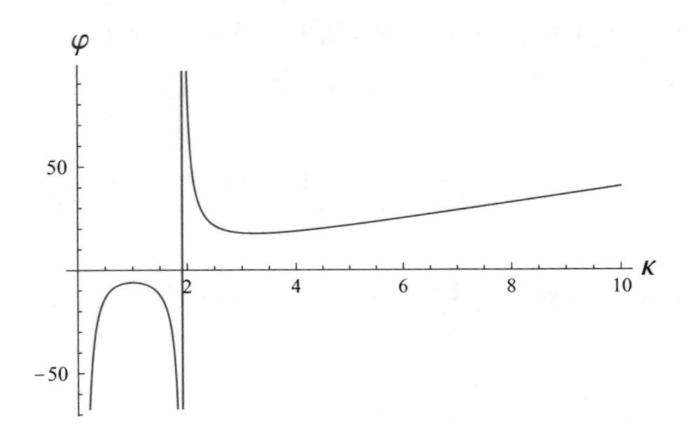

Fig. 3 Graph of $\varphi(\kappa)$

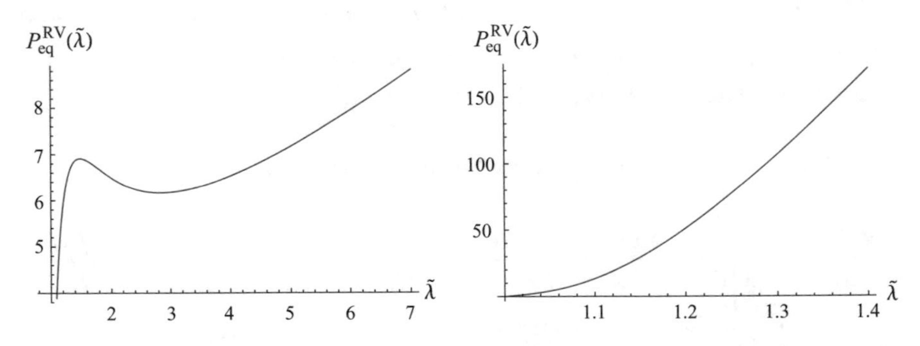

Fig. 4 Equilibrium pressure-stretch ratio relation for the Raghavan-Vorp SEF (Eq. 13), where P is in [N/cm^2]. On the *left*, $\alpha/\beta > 23.5954$ and, on the *right*, $\alpha/\beta < 23.5954$

3.2 Existence and Local Stability of Equilibria

From the above Lemmas and Proposition 1, the next Propositions follow directly.

Proposition 2 *For all positive values of the model parameters, the model (6) with the Fung-type SEF has exactly one equilibrium point in the form* $\tilde{E}^F = (\tilde{\lambda}, 0)$ *and it is a center.*

Proposition 3 *For the equilibria of the model (6) with the Ogden SEF the following holds true.*

- *If* $\mu > 3$ *holds true, then for all positive values of the model parameters the model has exactly one equilibrium point of the form* $\tilde{E}^O = (\tilde{\lambda}, 0)$ *and it is a center.*
- *If* $\mu < 3$ *and* $P < P^O_{eq}\left(\left(\frac{3+3\mu}{3-\mu}\right)^{1/(3\mu)}\right)$, *then the model has two equilibria—*$\tilde{E}^O_1 = (\tilde{\lambda}_1, 0)$ *and* $\tilde{E}^O_2 = (\tilde{\lambda}_2, 0)$ *where* $0 < \tilde{\lambda}_1 < \left(\frac{3+2\mu}{3-\mu}\right)^{1/(3\mu)}$ *and* $\tilde{\lambda}_2 > \left(\frac{3+2\mu}{3-\mu}\right)^{1/(3\mu)}$. *The equilibrium* $\tilde{E}^O_1$ *is a center and* $\tilde{E}^O_2$ *is a saddle point*
- *If* $\mu < 3$ *and* $P > P^O_{eq}\left(\left(\frac{3+3\mu}{3-\mu}\right)^{1/(3\mu)}\right)$, *then the model has no equilibria.*

Proposition 4 *For the model (6) with the Raghavan–Vorp SEF the following possibilities exist. Let* $\varphi(\kappa)$, $\tilde{\lambda}_{\min}$, *and* $\tilde{\lambda}_{\max}$ *be defined as in Lemma 3 and* $P^{RV}_{eq}(\tilde{\lambda})$ *is defined in (4).*

- *If* $\frac{\alpha}{\beta} < 6 + \varphi\left(\sqrt[3]{3\sqrt{37} + 16}\right) \approx 23.5954$, *then there exist one equilibrium point of the model and it is a center.*
- *If* $\frac{\alpha}{\beta} > 6 + \varphi\left(\sqrt[3]{3\sqrt{37} + 16}\right) \approx 23.5954$ *and* $P < P^{RV}_{eq}(\tilde{\lambda}_{\min})$ *or* $P > P^{RV}_{eq}(\tilde{\lambda}_{\max})$, *then the model (6) has exactly one equilibrium point and it is a center.*
- *If* $\frac{\alpha}{\beta} > 6 + \varphi\left(\sqrt[3]{3\sqrt{37} + 16}\right) \approx 23.5954$ *and* $P^{RV}_{eq}(\tilde{\lambda}_{\max}) < P < P^{RV}_{eq}(\tilde{\lambda}_{\min})$, *then the model (6) has three equilibria—*$\tilde{E}^{RV}_i = (\tilde{\lambda}_i, 0)$, $i = 1, 2, 3$. $\tilde{E}_1$ *and* $\tilde{E}_3$ *are centers and* $\tilde{E}_2$ *is a saddle point;* $0 < \tilde{\lambda}_1 < \tilde{\lambda}_{\max} < \tilde{\lambda}_2 < \tilde{\lambda}_{\min} < \tilde{\lambda}_3$.

4 Numerical Simulations

We illustrate the above results by numerical computing the orbits for the system (6) with the different SEFs. In all examples, $\rho = 1050 \times 10^{-6}$ kg/cm^3 (the mass density of the arterial wall), and $H = 0.15$ cm, $R = 3$ cm [15].

Example 1 The results for the system (6) with the Fung-type SEF (Eq. (3)) at the following values of the other model parameters— $\nu = 20.5$N/cm^2, $\Gamma = 1.734$ [12], $P = 2$ N/cm^2—and three different initial conditions, are presented in Fig. 5. All the trajectories are periodic orbits.

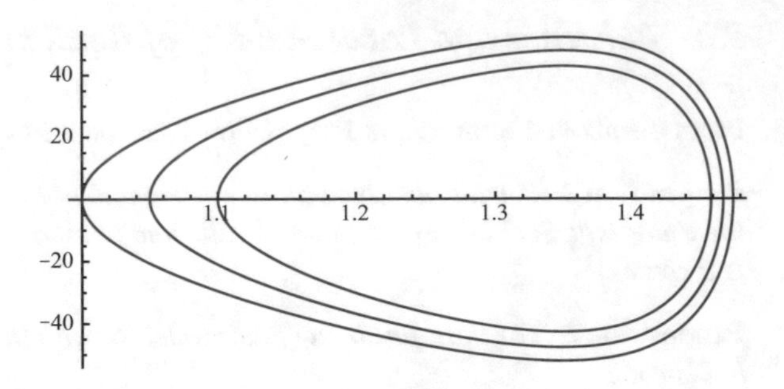

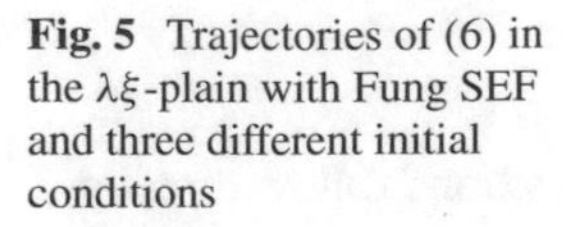
Fig. 5 Trajectories of (6) in the $\lambda\xi$-plain with Fung SEF and three different initial conditions

Example 2 The results for the system (6) with the Ogden SEF (Eq. (4)) at the following values of the other model parameters—$\chi = 0.621\,\mathrm{N/cm^2}$, $\mu = 13.21$ [12] (for the case $\mu > 3$), $P = 2\,\mathrm{N/cm^2}$—and three different initial conditions, are presented in Fig. 6a. For the case $\mu < 3$, we use $\mu = 1.3$, $\chi = 6.3\,\mathrm{N/cm^2}$ [4], $P = 0.2\,\mathrm{N/cm^2}$ (see Fig. 6b), and $\mu = 1.3$, $\chi = 6.3\,\mathrm{N/cm^2}$ [4], $P = 2\,\mathrm{N/cm^2}$ (see Fig. 6c). Those cases correspond to the three cases in Proposition 3. Note that in the case when no equilibria exist the solutions are unbounded.

Example 3 We consider the model (6) with the Raghavan–Vorp SEF (Eq. (5)) at the following values of the material coefficients—$\alpha = 19.29\,\mathrm{N/cm^2}$, $\beta = 55.27\,\mathrm{N/cm^2}$, which are calculated by a best-fit procedure of Eq. (12) to experimental data, obtained by Stoychev et al. (personal communication), $P = 2\,\mathrm{N/cm^2}$—and three different initial conditions (see Fig. 7a). Results for the same model with parameters $\alpha = 17.4\,\mathrm{N/cm^2}$, $\beta = 0.45\,\mathrm{N/cm^2}$, $P = 2\,\mathrm{N/cm^2}$, and $\alpha = 17.4\,\mathrm{N/cm^2}$, $\beta = 0.45\,\mathrm{N/cm^2}$, $P = 6.5\,\mathrm{mN/cm^2}$ are presented in Fig. 7b and c. Those three cases correspond to the three possibilities in Proposition 4.

5 Discussions

The analytical investigation, presented in Sect. 3 of the current article, aims to highlight the key role of the constitutive equation (in particular, the concrete form of SEF) on the stability of an inflated aneurysm wall, subjected to an intravascular blood pressure. Here, we shall summarize the main results, obtained in Sect. 3, and shall translate them to aneurysm progression. According to the results, presented in Sects. 3 and 4, several distinct scenarios are possible:

- The curve $P_{eq}(\tilde{\lambda})$ is monotonically increasing function for every $\tilde{\lambda} > 1$. Then the basic model has exactly one equilibrium point, which is of a center type, i.e. neutrally stable. This behavior is sustained for all material parameters of SEF of Fung, for $\mu > 3$ of SEF of Ogden, and for $\alpha/\beta < 23.5954$ of SEF of Raghavan and Vorp. In these cases, the inflation pressure does not affect the increase of the

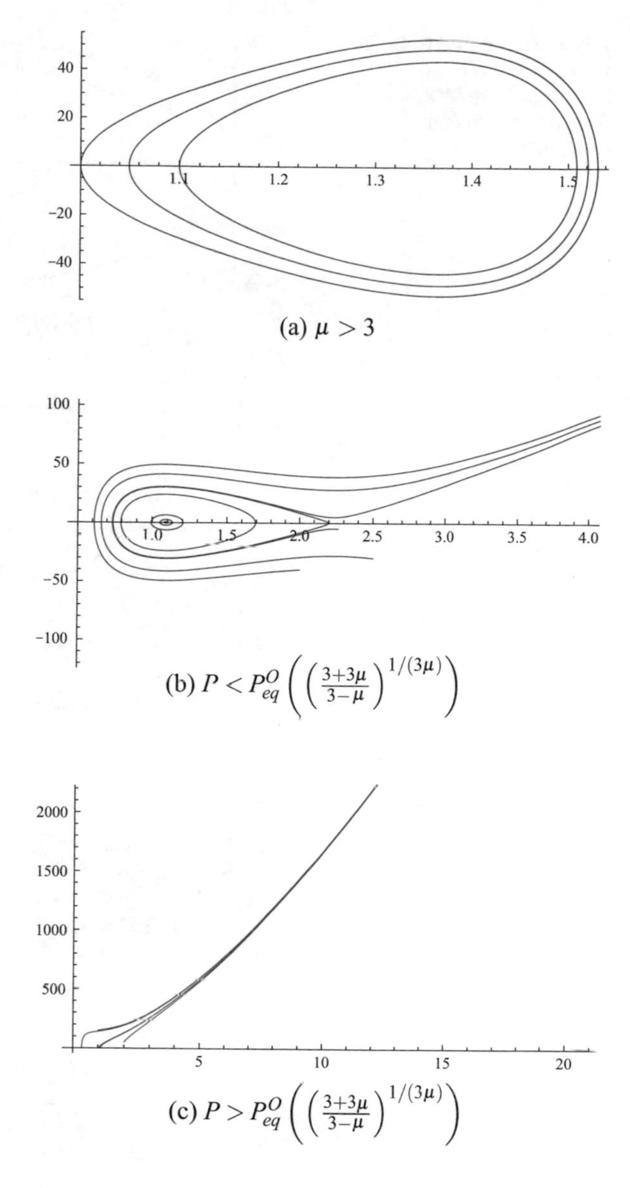

Fig. 6 Trajectories of (6) in the $\lambda\xi$-plain with Ogden SEF and three different initial conditions

stretch, i.e. the aneurysm wall sustains periodic oscillations (vibrations) between its initial states $(1, 0)$ and final ones $(\tilde{\lambda}_{max}, 0)$. From a biomechanical point of view, this means, that the material is less distensible or stiff. This conclusion rejects the appearance of local arterial dilatations (in particular, aneurysms). But, if these abnormal states, such as aneurysms, already exist, on the basis of another physical or medical factors, they do not enlarge and rupture at the conditions, given above.

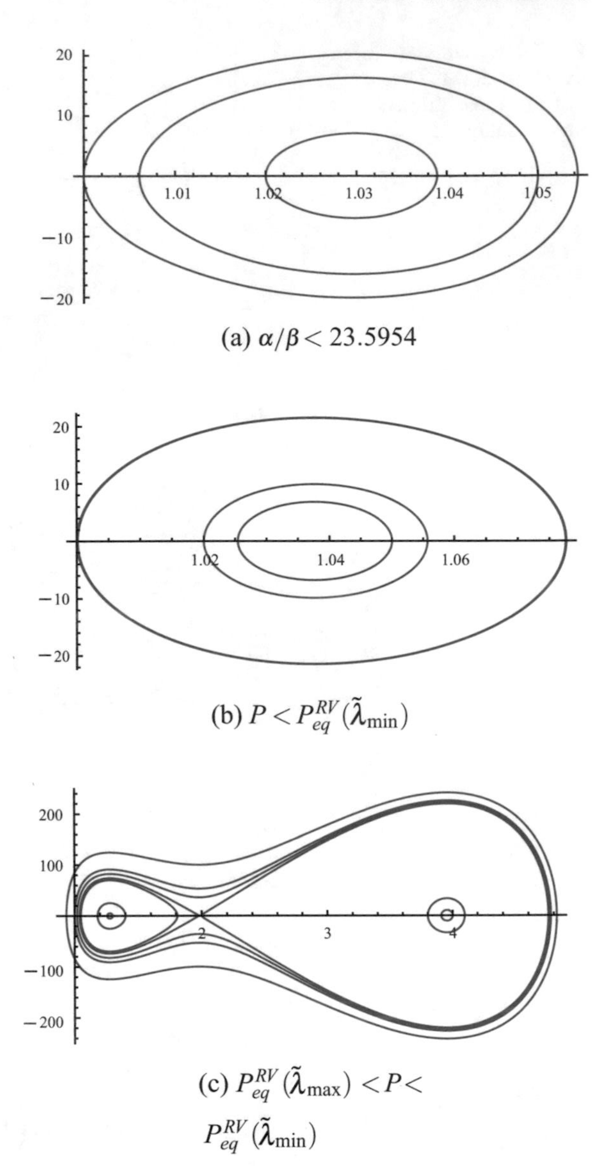

Fig. 7 Trajectories of (6) in the $\lambda\xi$-plain with Raghavan–Vorp SEF and three different initial conditions

- The curve $P_{eq}(\tilde{\lambda})$ demonstrates a local maximum, which determines the critical pressure over which the aneurysm wall cannot hold out in static inflation. In this case the basic model has two equilibrium points, one on the increasing part of the curve, which is stable, and another on the decreasing part, which is unstable. This behavior is established in the model at some restrictions of SEF of Ogden, i.e. at $\mu < 3$ and $P < P_{eq}^{O}\left(\left(\frac{3+3\mu}{3-\mu}\right)^{1/(3\mu)}\right)$, and is associated with the notion of "limit point instability" in [10, 11]. From a biomechanical point of view, this phenomenon can be explained by local stiffening of the material at low stretches, and local softening of the material at higher stretch ratios. In [10], it is pointed that such a behavior is typical for rubber-like materials. Like the "party balloon effect" the aneurysm continues to inflate at decreasing pressure and there is a high risk of its rupture at some critical pressure.

- The curve $P_{eq}(\tilde{\lambda})$ shows a local maximum, followed by a local minimum, and followed again by an increasing curve. In our finding, the basic model demonstrates such a behavior at some restrictions of SEF of Raghavan and Vorp, namely at $\alpha/\beta > 23.5954$ and $P_{eq}^{RV}(\tilde{\lambda}_{\max}) < P < P_{eq}^{RV}(\tilde{\lambda}_{\min})$. Then there are three equilibria, two neutrally stable centers and one unstable saddle. Although this scenario is associated with "jump inflation" in [9, 11], for the behavior of the arterial (aneurysm) wall, it is not physiologically acceptable (mainly in terms of the last increasing branch of the pressure-stretch curve), because the stretch ratios for arterial walls do not go over 1.4–1.6 [10]. In this aspect, like the previous scenario, we focus only on identification of the critical pressure and critical stretch ratio, over which the aneurysm wall cannot withstand in static inflation, i.e. it can ultimately enlarge and rupture.

6 Conclusions

In the present paper, we have presented a theoretical investigation of the elastic response of a perfectly spherical aneurysm subjected to an intravascular blood pressure. We showed that the stability of the equilibria of the aneurysm wall is strongly dependent on the concrete constitutive relation, describing its specific material properties. We have proved analytically that the stable or unstable quasi-static behavior of the aneurysm is determined by the concrete numerical intervals (boundary values) of the material parameters of each of the thee forms of SEF, used in the current study. The instabilities, identified by our analytical analysis, have been interpreted as possible factors for development and rupture of aneurysms.

Acknowledgements The authors thank to the National Science Fund of Bulgarian Ministry of Education and Research: Grant DFNI–I02/3 for the financial support.

References

1. Aneurysm, From Wikipedia, the free encyclopedia. https://en.wikipedia.org/wiki/Aneurysm
2. Alhayani, A.A., Giraldo, J.A., Rodrguez, J., Merodio, J.: Computational modelling of bulging of inflated cylindrical shells applicable to aneurysm formation and propagation in arterial wall tissue. Finite Elem. Anal. Des. **73**, 20–29 (2013)
3. Alhayani, A.A., Rodrguez, J., Merodio, J.: Competition between radial expansion and axial propagation in bulging of inflated cylinders with application to aneurysms propagation in arterial wall tissue. Int. J. Eng. Sci. **85**, 74–89 (2014)

4. Bucchi, A., Hearn, G.E.: Predictions of aneurysm formation in distensible tubes: part B application and comparison of alternative approaches. Int. J. Mech. Sci. **70**, 155–170 (2013)
5. Fu, Y., Xie, Y.: Effects of imperfections on localized bulging in inflated membrane tubes. Phil. Trans. R. Soc. A **370**, 1896–1911 (2012). doi:10.1098/rsta.2011.0297
6. Fu, Y.B., Pearce, S.P., Liu, K.K.: Post-bifurcation analysis of a thin-walled hyperelastic tube under inflation. Int. J. Non-Linear Mech. (2008). doi:10.1016/j.ijnonlinmec.2008.03.003
7. Ilichev, A.T., Fu, Y.B.: Stability of aneurysm solutions in a fluid-filled elastic membrane tube. Acta Mechanica Sinica **28**(4), 1209–1218 (2012)
8. Fu, Y.B., Rogerson, G.A., Zhang, I.Y.T.: Initiation of aneurysms as a mechanical bifurcation. Int. J. Non-linear Mech. **47**, 179–184 (2012)
9. Horny, L., Netuil, M., Hork, Z.: Limit point instability in pressurization of anisotropic finitely extensible hyperelastic thin-walled tube. Int. J. Nonlinear Mech. **77**, 107–114 (2015). doi:10.1016/j.ijnonlinmec.2015.08.003
10. Humphrey, J.D.: In Cardiovascular Solid Mechanics, Cells, Tissues and Organs. Springer, New York (2002)
11. Goriely, A., Destrade, M., Ben Amar, M.: Instabilities in elastomers and in soft tissues. J. Mech. Appl. Math. **59**(4), 615–630 (2006)
12. Stoytchev, S., Antonova, S., Antonova, M.: Dynamical behavior and strain energy function (SEF) of abdominal aorta aneurysms (AAAs). In: 10th Anniversary Conference of the Hellenic Society for Biomaterials, At Athens, Greece (2015). https://www.researchgate.net/publication/289536521_Dynamical_behavior_and_strain_energy_function_SEF_of_abdominal_aorta_aneurysms_AAAs
13. Fung, Y.C.: In Biomechanics: Mechanical Properties of Living Tissue, 2nd edn. Springer, New York (1993)
14. Ogden, R.W.: Large deformation isotropic elasticity on the correlation of theory and experiment for incompressible rubber-like solids. Proc. Roy. Soc. A **326**, 565–584 (1972)
15. Raghavan, M.L., Vorp, D.A.: Toward a biomechanical tool to evaluate rupture potential of abdominal aortic aneurysm: identification of a finite strain constitutive model and evaluation of its applicability. J. Biomech. **33**, 475–482 (2000)

New Bounds for Probability of Error of Coded and Uncoded TQAM in AWGN Channel

Hristo Kostadinov, Liliya Kraleva and Nikolai L. Manev

Abstract We investigate the performance of coded modulation scheme based on the application of integer codes to triangular quadrature amplitude modulation (TQAM). An upper and a lower bounds for symbol error probability (SER) in the case of AWGN channel are derived. These bounds are so closed that it makes the calculation of the exact value of SER unnecessary in practice.

1 Introduction

The term coded modulation means a combination of a scheme of coding and modulation techniques. Nowadays, in modern digital communication systems, high-order modulation is preferred for high-speed data transmission. One of the most popular modulation in commercial communication systems is square quadrature amplitude modulation (SQAM). SQAM scheme with its simple detection procedure is easy for implementation and demonstrates a good performance.

Recently, the triangular quadrature amplitude modulation (TQAM) was proposed. In TQAM constellation the signal points are vertexes of a lattice of equilateral triangles and the constellation is symmetric with respect to the origin. The comparison of TQAM with SQAM given in [7] shows that the former is more power efficient while preserves the low detection complexity of the latter. In [8] a general formula for calculating the average energy per symbol of the TQAM is derived and symbol error rate (SER) and bit error rate (BER) of the TQAM in the presence of additive white Gaussian noise (AWGN) is analyzed.

H. Kostadinov (✉) · L. Kraleva
IMI-BAS, Acad. G. Bonchev St., Bl.8, 1113 Sofia, Bulgaria
e-mail: hristo@math.bas.bg

L. Kraleva
e-mail: liliya@math.bas.bg

N.L. Manev
USEA "Lyuben Karavelov" and Institute of Mathematics and Informatics,
IMI-BAS, Acad. G. Bonchev St., Bl.8, 1113 Sofia, Bulgaria
e-mail: nlmanev@math.bas.bg

K. Georgiev et al. (eds.), *Advanced Computing in Industrial Mathematics*,
Studies in Computational Intelligence 681, DOI 10.1007/978-3-319-49544-6_7

Codes over finite rings and their applications in coding theory have been investigated by many researchers during the past several decades. It is well known that algebraic theory of block codes over finite rings have severe problems with coding for multidimensional constellations. The reason is, that in two or higher dimensions the Hamming distance is inappropriate. One possible solution of that problem is given in [1], where Huber introduced Mannheim distance and has applied codes over Gaussian integers for that distance. The weakness of that construction comes from the fact that based on a given code over Gaussian integers we arrange the points in the constellation.

So called integer codes, codes defined over finite rings of integers, are more useful from practical point of view. In contrast to the traditional block codes they are designed to correct errors of a given type. It means that for a given channel and modulator we can construct integer code capable of correcting the type of errors, which are the most common for this channel.

In this work we shall investigate the performance of coded modulation scheme based on the application of integer codes to TQAM constellation. Necessary definitions and results are given in Sect. 2. In Sect. 3 an upper and a lower bound for symbol error probability (SER) in the case of AWGN channel are derived. These bounds are so closed that it makes the calculation of the exact value of SER unnecessary in practice. The error performance of integer codes in case of TQAM constellation over AWGN channel is discussed in Sect. 4. Section 5 describes the simulations that have been carried out.

2 Preliminaries

2.1 TQAM Constellation

In this paper we will consider TQAM constellation of $M = 2^{2m}$ signal points placed in $L = 2^m$ rows parallel to real axis with L signal point in each row. The points form a lattice of equilateral triangles and the constellation is symmetric with respect to the origin. The power gain of M-ary TQAM over M-ary SQAM in decibels [8] is

$$10\log_{10}\left(\frac{8M-8}{7M-4}\right) \xrightarrow[M\to\infty]{} 0.5799\text{ dB}$$

For $M = 16,\ 64,\ 256$ the power gain is 0.458, 0.5505 and 0.5726, respectively.

An example of 16-ary TQAM is given in Fig. 1. The correspondence between signal points and all 16 sets of four bits presented in the figure is referred to as *Grey mapping*. In the case of SQAM this mapping guarantees that the 4-bit sets corresponding to the closest points differ only in one bit. In the case of TQAM it is impossible to realize such a mapping since the number of closest points can be up to six. But the same property holds for two (of three possible) direction in the

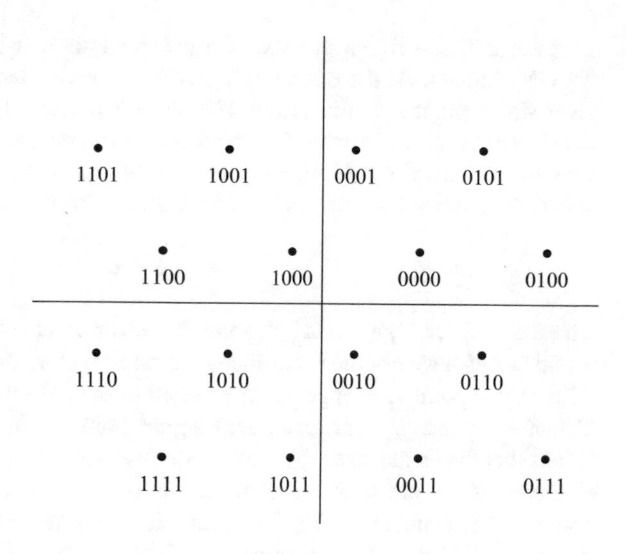

Fig. 1 16-TQAM constellation

constellation. Examples of 16-ary and 64-ary TQAMs with Grey mapping are given in [8], too.

Remark Some authors use (in the case of SQAM) the term Grey codding, but we avoid intentionally this notation. A coded modulation with Grey mapping decreases BER but it does not include decoding procedure and does not affect SER. We refer this case to as *uncoded case with Grey mapping* in contrast to the case of using integer codes.

2.2 Integer Codes

Herein we give some necessary definitions and notations which we shall use in the next sections. More details the reader can find in the mentioned papers.

Integer codes were proposed by Varshamov and Tenengolz [9] in 1965 for correcting single insertion/deletion per codeword, but in [3, 4] it was demonstrated that such classes of codes are very suitable for realization of coded modulation procedures. The applications of integer codes with different modulation schemes, in partial with SQAM, are discussed also in [5, 6].

Definition 2.1 [4] Let $\mathbf{Z}_A$ be the ring of integers modulo A. An *integer code* of length n with parity-check matrix $H \in \mathbf{Z}_A^{m\times n}$, is referred to as a subset of $\mathbf{Z}_A^n$, defined by

$$\mathbf{C}(H, \mathbf{d}) = \{\mathbf{c} \in \mathbf{Z}_A^n \mid \mathbf{c}H^T = \mathbf{d} \bmod A\}$$

where $\mathbf{d} \in \mathbf{Z}_A^m$.

Assume that a signal point s_i of a given signal constellation is sent through an AWGN-channel. At the other end the detector estimates the received signal r_i and gives signal point s_j at the output. If $j \neq i$ the detector has taken a wrong decision. In terms of block codes over $\mathbf{Z}_A$ the aforesaid can be described in the following way. When a codeword $\mathbf{c} \in \mathbf{C}(\mathbf{w}, d)$ is sent through a communication channel (usually noisy) the received vector can be written in the form

$$\mathbf{r} = \mathbf{c} + \mathbf{e},$$

where $\mathbf{e} = (e_1, \ldots, e_n) \in \mathbf{Z}_A^n$ denotes the error vector. It is clear that the different signal points have not the same chance to be a result of decision process. The probability signal point s_j to appear at the output of the detector depends on the Euclidean distance between s_j and really-sent signal point s_i. In terms of codes over $\mathbf{Z}_A$ it means that the elements of $\mathbf{Z}_A$ are not equally probable as a value taken by e_i. Which elements of $\mathbf{Z}_A$ are more probable depends on the chosen indexing of the signal points by the elements of $\mathbf{Z}_A$. Therefore, it makes sense to consider (there is a point in considering) the next definition.

Definition 2.2 [4] The code $\mathbf{C}(H, \mathbf{d})$ is a t-multiple $(\pm e_1, \pm e_2, \ldots, \pm e_s)$-error correctable if it can correct up to t errors with values from the set $\{\pm e_1, \pm e_2, \ldots, \pm e_s\}$ which are occurred in a codeword.

Remark Without loss of generality in the definitions above we can assume that $\mathbf{d} = \mathbf{0}$. For convenience of a notation we shall use $\mathbf{C}$ instead of $\mathbf{C}(H, \mathbf{0})$.

To decode integer codes one can use a hard decoding algorithm [6]. This algorithm uses a look-up table which maps each syndrome value to the corresponding error vector. So, the complexity of the algorithm is linear with respect to the alphabet size A.

3 Bounds for SER and BER

Suppose that a signal point $\mathbf{s}$ is sent through a communication channel. At the other end the detector estimates the received signal $\mathbf{r}$ and has to take a decision: which signal point has been sent. Let the channel be an AWGN channel with power spectral density N_0 watts/hertz. Then

$$\mathbf{r} = \mathbf{s} + \mathbf{n},$$

where $\mathbf{n}$ is two dimensional zero-mean Gaussian random process with variance $\sigma^2 = N_0/2$.

Hence, at any arbitrary time the value of $\mathbf{n}(x, y)$ is statistically characterized by the Gaussian probability density function,

$$p(x, y) = \frac{1}{2\pi\sigma^2} e^{-\frac{x^2+y^2}{2\sigma^2}}.$$

Let q_u denote the probability of right decision of the detector, that is the probability of correct detection per symbol. In the case of uncoded SQAM constellation the detector takes the right decision if the received signal $\mathbf{r}$ belongs to a square with center the sent signal point $\mathbf{s}$ and side equals $2d$, which is the minimal possible distance between two signal points. In this case is more conveniently to consider the noise as $\mathbf{n} = (n_x, n_y)$ where n_x and n_y are independent (orthogonal) zero-mean Gaussian random processes with power spectral density $N_0/2$ (variance $\sigma^2 = N_0/2$). The probability of correct demodulation in the SQAM case is widely treated in the literature and well known (e.g., [2]). In [4] the reader can find also the probability of correct decoding in the case of L^2-SQAM coded with integer codes.

Unfortunately, in the case of TQAM the detection region are hexagon with side equals $a = 2d/\sqrt{3}$ (the regions of contour points are more complex) and the consideration of the noise as a set of two orthogonal and independent noises does not help so much as in the square case.

In order to avoid the complex calculations we suggest a different approach. Our idea is to approximate the detection regions by circles with center the sent signal point. At least for hexagonal region the inscribed and circumscribed circles give very tight lower and upper bounds for the probability of correct detection (upper and lower bounds for the probability of error, respectively). For the contour points circles are also a good approximation because of the part of the region outside the circle has relatively small contribution to the probability.

If $\mathbf{D}$ is the detection region then

$$q_u = \Pr(\mathbf{r} \in \mathbf{D}) = \iint_D \frac{1}{2\pi\sigma^2} e^{-\frac{x^2+y^2}{N_0}} \, dxdy.$$

In the case when $\mathbf{D}$ is a circle with radius λd we get

$$q_u = 1 - e^{-\frac{\lambda^2 d^2}{N_0}}$$

As we mentioned above q_u for L^2-SQAM is well known and it is given by the formula

$$q_u = \frac{1}{L^2}[1 + (L-1)\operatorname{erf}(\gamma)]^2,$$

where $\gamma = \frac{d}{\sqrt{N_0}} = \sqrt{\frac{3}{2(L^2-1)} \cdot \frac{E_s}{N_0}}$, $\operatorname{erf}(x) = \frac{2}{\sqrt{\pi}} \int_0^x e^{-u^2} du$, and E_s is the average energy per signal point, that is, $\mathrm{SNR}_s = E_s/N_0$ is the signal-to-noise ratio per symbol.

This enable us to test our idea. The inscribed and circumscribed circles of the square with side $2d$ have radius d and $d\sqrt{2}$, respectively. Therefore we have to expect that

$$1 - e^{-\frac{d^2}{N_0}} < q_u < 1 - e^{-\frac{2d^2}{N_0}}$$

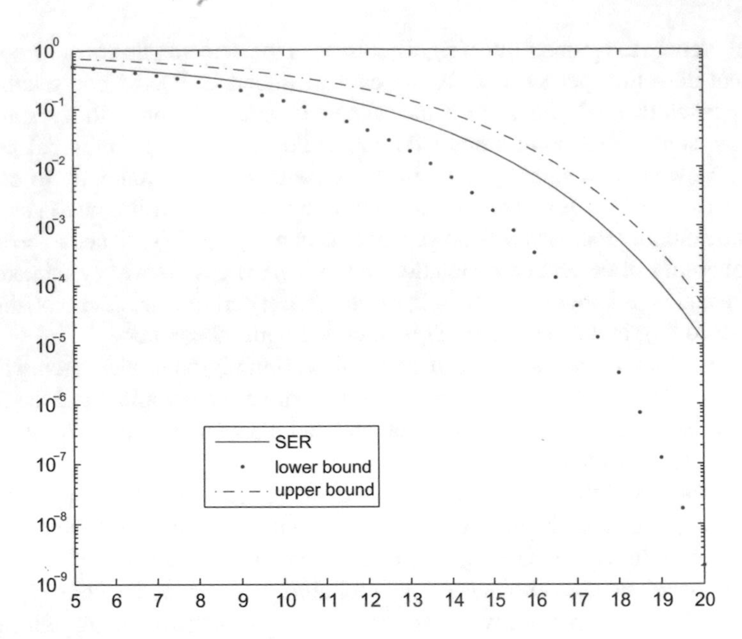

Fig. 2 Probability of symbol error for uncoded 16-SQAM

or for the symbol error (SER) p_u:

$$e^{-\frac{2d^2}{N_0}} < p_u < e^{-\frac{d^2}{N_0}}.$$

In the case of 16-SQAM (i.e., $L = 4$) the inequality for p_u as a function of SNR per symbol has the form

$$e^{-\frac{\mathrm{SNR}_s}{5}} < p_u < e^{-\frac{\mathrm{SNR}_s}{10}}$$

ant it is graphically represented in Fig. 2.

In the case of L^2-TQAM the distance between adjacent symbols is also $2d$, but the distance between horizontal rows is $h = d\sqrt{3}$ (which is the altitude of the of equilateral triangles). Then [8]

$$\frac{d^2}{N_0} = \frac{12}{7L^2 - 4} \cdot \frac{E_s}{N_0} = \frac{12}{7L^2 - 4} \cdot \mathrm{SNR}_s. \tag{1}$$

The radius of the inscribed circle for the hexagon is d, and the one of the circumscribed circle is $a = 2d/\sqrt{3}$ (see Fig. 3). Therefore we have to expect that

$$1 - e^{-\frac{d^2}{N_0}} < q_u < 1 - e^{-\frac{4d^2}{3N_0}}$$

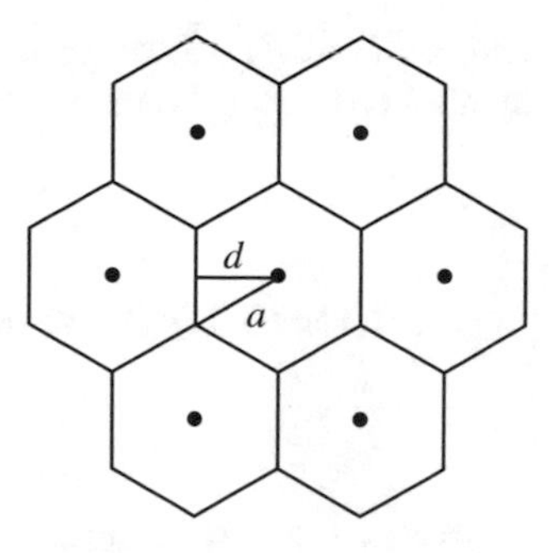

Fig. 3 Inscribed and circumscribed circles' radius

or for the symbol error (SER) p_u:

$$e^{-\frac{4d^2}{3N_0}} < p_u < e^{-\frac{d^2}{N_0}}. \tag{2}$$

Replacing d^2/N_0 by (1) we get

$$e^{-\frac{16}{7L^2-4}\mathrm{SNR}_s} < p_u < e^{-\frac{12}{7L^2-4}\mathrm{SNR}_s}. \tag{3}$$

4 Integer Codes and TQAM

Herein we describe applications of integer codes to 16-TQAM. One possibility is to number the signal points by $\mathbb{Z}_{17}\setminus\{0\} = \{1, 2, \ldots, 16\}$ and to use the integer code with $\mathbf{H} = (1, 2)$ over the ring $\mathbb{Z}_{17}$ of integers modulo 17. This code is a single $(\pm1, \pm3, \pm4, \pm5)$-error correcting code and can correct a wrong detection of the sent point as one of the six neighbor (adjacent) points (see Fig. 4a).

Another possibility is the use of two codes over $\mathbb{Z}_5$ with check matrix $\mathbf{H} = (1, 2)$ and numbering the points by pair of integers as it is shown in Fig. 4b.

Following the idea given in the previous section one can decide to estimate probability of symbol error by a larger detection region that includes the adjacent (at distance $2d$) points of the sent signal point. The radii of the circles are $2a = 4d/\sqrt{3}$

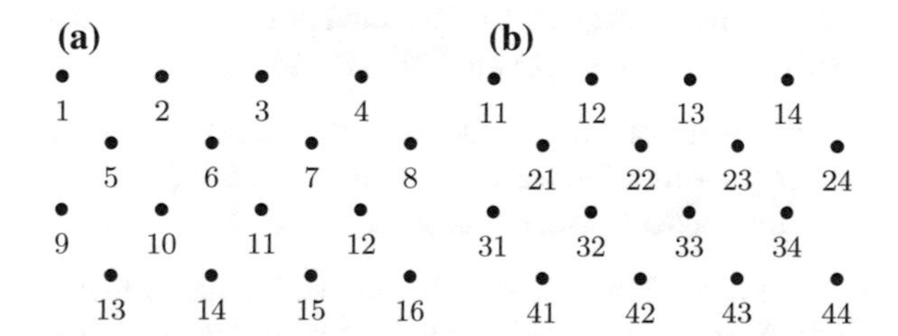

Fig. 4 Coded 16-TQAM

and $5a/2 = 5d/\sqrt{3}$, respectively. Therefore the expectation for the probability of symbol error P_c is bounded by

$$e^{-\frac{100}{7L^2-4}\text{SNR}_s} < P_c < e^{-\frac{64}{7L^2-4}\text{SNR}_s}. \quad (4)$$

In partial, for 16-TQAM we have

$$e^{-\frac{25}{27}\text{SNR}_s} < P_c < e^{-\frac{16}{27}\text{SNR}_s}. \quad (5)$$

But such a way of reasoning is wrong because it does not take in account the parameters of the code—the length of the codeword and how many errors per a codeword it can correct. Hence one have to bear in mind that one or more symbols will be incorrect decoded if the number of error detection per a codeword is larger than the code capacity. For example, if two successive signal points are received outside the circle with radius d they cannot be correctly decoded by a single error correcting code of length two (as the code used for simulations). Thus a symbol error appears. The probability for this even is a square of the error probability in uncoded case. Hence the probability for symbol error P_c in coded case has to satisfy the inequality

$$e^{-\frac{32}{7L^2-4}\text{SNR}_s} < P_c < e^{-\frac{24}{7L^2-4}\text{SNR}_s}. \quad (6)$$

In partial for 16-TQAM we have:

$$e^{-\frac{8}{27}\text{SNR}_s} < P_c < e^{-\frac{2}{9}\text{SNR}_s}. \quad (7)$$

5 Simulations

To compare the obtained bounds with real performance we have developed a software that simulates communication based on the 2^k-TQAM in AWGN channel for $k = 4, 6, 8$.

The software generalizes a random sequence of bits, transforms it into a sequence of signal points and adds to them AWGN corresponding to the prescribed (input) SNR. Then realizes the procedure of detection (and decoding if necessary) and compare the obtained points with the sent ones to calculate the SER. Also, the software transforms the sequence of detected points into a sequence of bits and calculates the BER. Three modes of simulation are realized:

- 'St'—uncoded case without any fixed rules of mapping k-bits sets to points;
- 'grey'—uncoded case with Grey mapping (Fig. 1 for $k = 4$);
- 'intc'—coded with a integer code (for $k = 4$ is used mapping given in Fig. 4b).

We carried out the simulations with pseudo-random sequences of length $\approx 10^9$ bits. As it was expected, the SERs for both uncoded cases, non-Grey (mode 'St') and

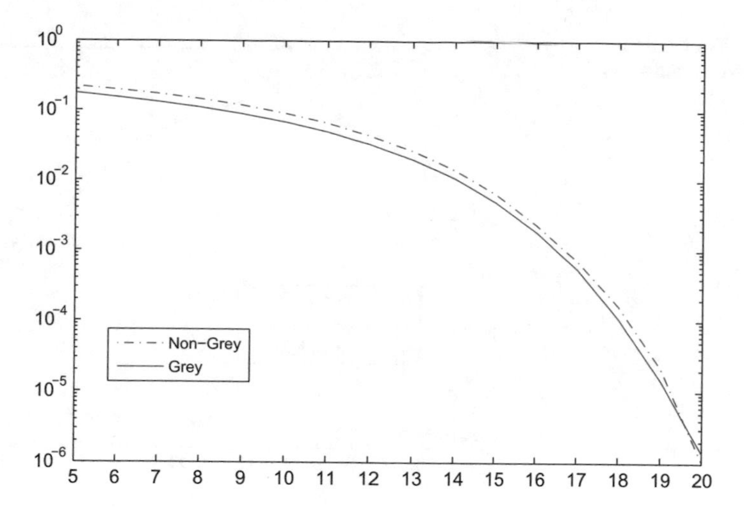

Fig. 5 BER of uncoded 16-TQAM—Grey and non-Grey mapping

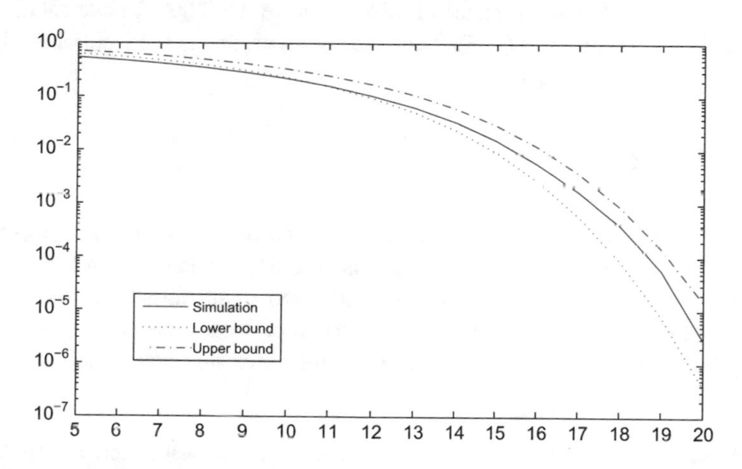

Fig. 6 SER of uncoded 16-TQAM—simulation result

Grey mapping, are almost the same, but the Grey mapping gives smaller BER. The bit error rates for uncoded 16-TQAM with and without Grey mapping are given in Fig. 5.

Figure 6 presents the comparison of the obtained by simulation SER with lower and upper bounds for uncoded case given by inequality (3). For small values of SNR all three graphs are very closed, the simulation even gives better result, but for the values that of practical interest the simulation graph is between the others as inequality (3) requires. The explanation is that for small SNR the deviation from (3) generated by contour points of the constellation is larger and it reflects on the mean SER.

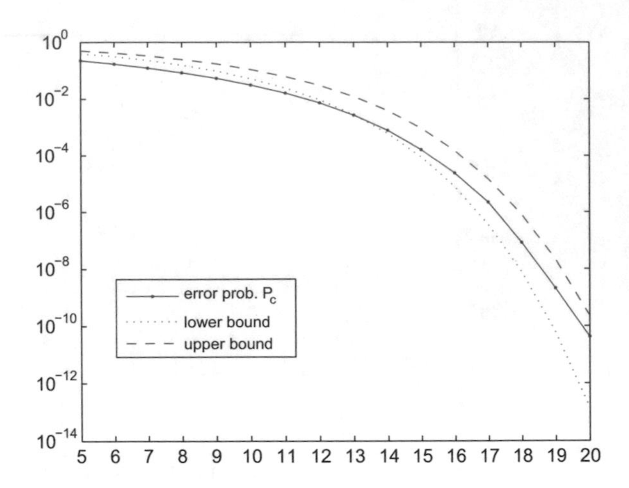

Fig. 7 SER of coded 16-TQAM—simulation result

Figure 7 demonstrates the result of simulation of 16-TQAM coded with single error integer code with $\mathbf{H} = (1, 2)$. The simulation confirms the inequality (7).

6 Conclusion

We derive a lower and an upper bounds for symbol error probability for TQAM. Our considerations and simulations show that this bounds give an acceptable for practice estimation of symbol error probability both in uncoded and coded with integer codes cases. The graphs show that in the all cases the real performance follows the behavior of the upper bound—its graph is just shifted in the better direction graph of the upper bound.

Acknowledgements This work was partially supported by the National Science Fund of Bulgaria under Grant DFNI-I02/8.

References

1. Huber, K.: Codes over Gaussian integers. IEEE Trans. Inf. Theory **40**(1), 207–216 (1994)
2. Haykin, S.: Digital Communications. Wiley, NY (1988)
3. Kostadinov, H., Morita, H., Manev, N.: Integer codes correcting single errors of specific types $(\pm e_1, \pm e_2, \ldots, \pm e_s)$. IEICE Trans. Fundam. **E86-A**(7), 1843–1849 (2003)
4. Kostadinov, H., Morita, H., Manev, N.: Derivation on bit error probability of coded QAM using integer codes. IEICE Trans. Fundam. **E87-A**(12), 3397–3403 (2004)
5. Kostadinov, H., Morita, H., Iijima, N., Han Vinck, A.J., Manev, N.: Soft decoding of integer codes and their application to coded modulation. IEICE Trans Fundam. **E39-A**(7), 1363–1370 (2010)

6. Kostadinov, H., Morita, H., Manev, N.L.: On (–1,+1) error correctable integer codes. IEICE Trans. Fundam. **E93-A**(12), 2758–2761 (2010)
7. Park, S.-J.: Triangular quadrature amplitude modulation. IEEE Commun. Lett. **11**(4), 292–294 (2007)
8. Park, S.-J.: Performance analysis of triangular quadrature amplitude modulation in AWGN channel. IEEE Commun. Lett. **16**, 765–768 (2012)
9. Varshamov, R., Tenengolz, G.: One asymmetrical error-correctable codes. Automatika i Telematika **26**(2), 288–292 (1965) (in Russian)

Construction of Multistage Scenario Tree for Insurance Activity

Tsvetanka Kovacheva

Abstract In the paper a stochastic asset model can be used for long term financial planning and observations in insurance. The scenario model is developed for the case when the large number of scenarios is generated and represents the uncertainty of stochastic parameters. The paper presents the construction the multistage scenario tree using the clustering-based approach. It is implemented on sampled data of nominal interest rate according to accepted stochastic model.

Keywords Scenario generation · Stochastic model · Uncertainty · Scenarios · Multistage scenario tree · Cluster analysis

1 Introduction

The insurance business has two basic parts: *underwriting activity*—collection of premiums for accepting risk for the others, that on average cover future claims and *investment activity*—the investment of those premiums and free reserves. The investment activity is similar to other financial institutions like pension banks, funds and corresponds to the asset side of the company.

The both activities are influenced by the uncertainty (randomness) from environment—risk, which has stochastic behavior. The insurance company cannot influence the behavior of risk factors. It can make the decisions to reduce the influence of these risks. Risk and insurance are associated with each other. The risk is concerned with a chance of loss. Some risks that confront the insurance company are: the risks of claims size and claims frequency, reinsurance risk, interest rate risk, inflation risk and others. The risks can be considered as uncontrollable factors, which are random variables. These variables are not under the control of the insurance company. Variables whose values are under control of the company named decision variables (*strategy of insurance*).

T. Kovacheva (✉)
Department of Mathematics, Technical University,
Studentska Str. 1, 9010 Varna, Bulgaria
e-mail: tsveta_kovacheva@tu-varna.bg

K. Georgiev et al. (eds.), *Advanced Computing in Industrial Mathematics*,
Studies in Computational Intelligence 681, DOI 10.1007/978-3-319-49544-6_8

Different scenarios are generated for the decision making in some situations of the insurance company activities. The scenarios are related with variation of risk factors over time. The scenario generator includes stochastic models for the risk factors affecting the company.

The most commonly used models for decision of long term financial planning and observations problems are CIR (Cox, Ingersoll and Ross) and HMT (Hibbert, Mowbray and Turbull) [1–3]. They are extensions of the famous short-rate Vasicek model [4]. The models involve decision stages and uncertainty.

The basic feature of these models is that a stochastic process describes the uncertain environment. The stochastic processes are estimated from historical data and calibrated using some prior information. The large number of interrelated risk factors (inflation, interest rate, etc.) is taken into account to understand their potential behavior over the planning period. Some parameters are completely unknown at the current point at time, when decision must be taken.

The large numbers of scenarios are simulated by using a random scenario generator and form the initial scenario fan. The next step is approximation the generated scenario fan into multistage scenario tree using a modified Ward clustering method. The constructed multistage scenarios tree is evaluated by calculation some statistical characteristics of the nominal interest rate.

The structure of the paper is following: Sect. 2 presents some basic notions and terminology for the paper. Section 3 gives the basic methods and steps of the scenario tree generation. Section 4 illustrates the construction of two-stage scenario tree, its advantages and disadvantages, and the necessity of construction multistage scenario tree. The modified Ward clustering method is proposed to create the scenario tree. Section 5 presents the model for long-term financial planning purposes. The simulation results on the base of the chosen model using the Statistical Analysis System (SAS) are given. Section 6 concludes the paper.

2 Basic Notions and Terminology

Insurance activity is most commonly represented with multivariate stochastic processes.

Assume that a discrete-time continuous space stochastic data process is given

$$\xi(\omega) = \{\xi_t(\omega)\}_{t=0}^{T} \tag{1}$$

The process is defined on probability space $(\Omega, \mathcal{F}, \mathrm{P})$ with parameter $t \in \{0, 1, \ldots, T\}$ (time) and state space $\Re^d$, Ω is a simple space, $\mathcal{F}$—a σ-algebra (σ-field)—a set of random events of Ω, $P : \mathcal{F} \to [0, 1]$—a probability measure. The function $\xi : \Omega \to \Re^d$ is said to be $\mathcal{F}$-measurable. For every $\omega \in \Omega$ the sequence $\{\xi_0(\omega), \xi_1(\omega), \ldots, \xi_T(\omega)\}$ is named *a data path* (trajectory, realization, scenario). The information structure is given by a filtration $\Im = (\mathcal{F}_0, \mathcal{F}_1, \ldots, \mathcal{F}_{\mathrm{T}})$ of an increasing sequence of σ-algebras, i.e. $\mathcal{F}_{\mathrm{t}} \subseteq \mathcal{F}_{\mathrm{t+1}}$ and $\sigma(\xi) \subseteq \Im$. The σ-algebra $\mathcal{F}_{\mathrm{t}}$ describes the available information at time t. The distribution of the process may be the result of an estimation based on historical data.

The decision process is

$$x = \{x_t\}_{t=0}^{T} \tag{2}$$

where—x_t—recourse decision at stage t, $x_t : \Omega \to \Re^n$ and $\mathcal{F}_t \subseteq \mathcal{F}$.

The stage is a moment in time, when decisions are taken. The time period is the interval between two time points. In each of the stages, the decisions are limited by constrains that may depend on the previous decisions and observations. Stages do not necessarily refer to time periods t, they correspond to steps in the decision process.

The sequence of decisions and observations is

$$x_0, (\xi_0, x_1), (\xi_1, x_2), \ldots, (\xi_{T-1}, x_T) \tag{3}$$

The process (2) is *nonanticipative* [5], i.e. every decision x_t taken at any stage $t > 1$ of the process depends only on the past observations and decisions.

To represent the process (1) as a tree with finite number of realizations, it is necessary to approximate it as close as possible by a discrete stochastic process

$$\xi^l(\omega) = \left\{\xi_t^l(\omega)\right\}_{t=0}^{T}, l = \overline{1, L} \tag{4}$$

A tree process (4) generates a filtration $\Im = (\mathcal{F}_0, \mathcal{F}_1, \ldots, \mathcal{F}_T)$ by $\sigma\left(\xi^l(\omega)\right) \subseteq \Im$.

The tree process (4) takes only finitely many values and $p_l = P\left(\xi^l\right)$, $p_l \geq 0$, $\sum_{l=1}^{L} p_l = 1$.

3 Scenario Tree Generation

Scenario tree generation methods are powerful decision-making tools when decisions have to be made under uncertainty. A major focus of scenario generation is to create a tree-structure of scenarios [6–10, 18].

The main scenario generation methods are:

- *sampling*: the method takes values from the distribution function, thus providing scenario values (Monte Carlo or random sampling, importance sampling, bootstrap, internal sampling, conditional sampling, stratified sampling or Markov chain, Monte Carlo sampling).
- *statistical methods*: determine some statistical characteristics of the given data which are used to determine the best fit theoretical distribution and generate the scenarios using this distribution (statistical moments or property matching, principal component analysis, regression and its variants).

- *simulation*: this involves the simulation of the process by inputting random numbers into its equation (stochastic process simulation, error correction model, vector auto-regressive).
- other methods: these methods include a variety of methods (artificial neural networks, clustering, scenario reduction, hybrid methods).

A scenario generation procedure involves some or all of the following *steps*:

Step 1. Choosing the model describing the stochastic parameters (for example econometric models and time series—autoregressive models, moving average models, autoregressive moving average models, Bayesian VAR, reduced rank regression, etc. and diffusion processes—Brownian motion, etc.).

Step 2. Collection historical observations of parameter values.

Step 3. Estimation/calibration of parameters with historical data.

Step 4. Generation of data paths according to the chosen model. Using statistical approximation (property matching, moment matching, non parametric methods) or sampling (random sampling, stratified sampling, bootstrapping) data paths can be generated performing discretization of the distribution over the time horizon T.

Step 5. Construction a scenario tree with the desired properties on the base of sampled data paths (scenarios) and constrains.

4 Construction of Multistage Scenario Tree

4.1 Construction of Two-Stage Scenario Tree

Two-stage scenario tree (Fig. 1) is constructed from the scenarios of the process (4). In *the first stage* all scenarios coincide at the time period $t = 0$ in the initial root node, i.e. $\xi_0^1 = \xi_0^2 = \cdots = \xi_0^L$, since the values of the random parameters are known with certainty during the first time period. In *the second stage* the tree has L branches from the initial root node, which are extended from the nodes at the time period $t = 1$ to the nodes at the time period $t = T$. The tree consists of $N = 1 + TL$ nodes. If such tree is used as input in multistage stochastic program, the model is a two-stage problem. The tree process is defined on the finite probability space $\Omega = (\omega = 1, \ldots, \omega = L)$. The corresponding filtration is $\Im = (\mathcal{F}_0, \mathcal{F}_1, ..., \mathcal{F}_{\mathrm{T}})$, such that the sequence of σ-algebras is

$$\mathcal{F}_0 \quad : \{\omega = 1, \ldots, \omega = L\} \text{ and } \mathcal{F}_1 = \mathcal{F}_2 = \cdots = \mathcal{F}_T : \{\omega = 1\}, \{\omega = 2\}, \ldots, \{\omega = L\} \quad (5)$$

The properties of the two-stage multiperiod program are the following [6]:

- decisions at all time periods are made at once and no further information is expected.
- hedging against all considered unrelated scenarios of possible developments is assumed.

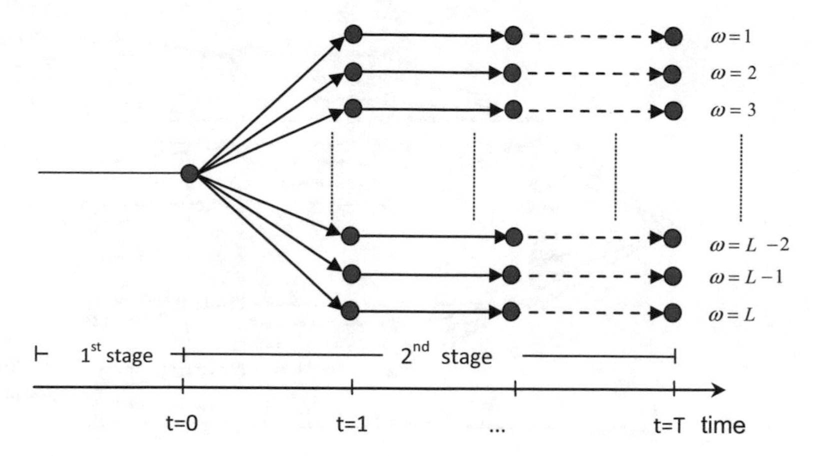

Fig. 1 Two-stage scenario tree

- except for the first stage no nonanticipative constrains appear.

Depending on the considered problem, such properties can be considered as disadvantages. This entails the construction of multistage trees for multistage models.

4.2 *Construction of Multistage Scenario Tree*

Multistage scenario tree can be obtained using the following methods: optimal discretization, barycentric approximation, sequential clustering [6]. In the result is received the structure shown in the Fig. 2.

The multistage scenario tree starts also from the initial root node at the time period $t = 0$. In the next time periods the structure of the scenario tree is described by branching into a finite number of scenarios K. Generally nodes of the tree at the time period t correspond to possible values of ξ_t^l that may occur. Each of them except for the root node is connected to a unique node at level $(t-1)$, called the *ancestor node or predecessor* and is also connected to nodes at level $(t+1)$, called *children nodes or successors*. The branching factor is

$$b_t(x) = \#\left(y : P\left(\xi_{t+1}^l = y \,\middle|\, \xi_t^l = x\right) > 0,\ t = \overline{0, T-1}\right) \tag{6}$$

where the nodes (x, t) and $(y, t+1)$ are connected. Typically, the branching factors are chosen beforehand and independent of x.

A scenario tree is a rooted tree where all the leaves are at depth T. Its structure is determined by the vector $(1, b_1, \ldots, b_t, \ldots, b_T)$, where b_t denotes the number of successors per node in stage t. The vector of the number of tree nodes in each stage is $(n_0, \ldots, n_t, \ldots, n_T)$. If the branching factor is the same for a given period, then:

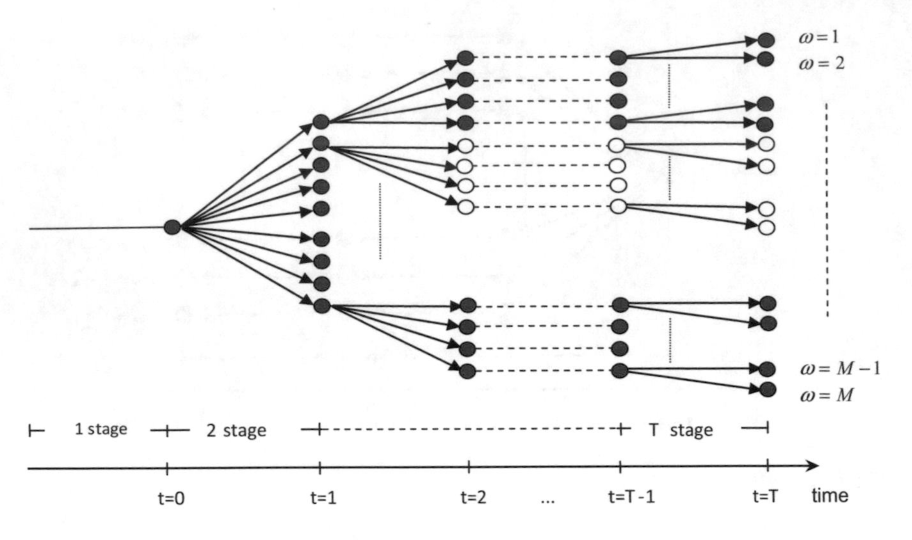

Fig. 2 Construction of multistage scenario tree

- *the total number of tree nodes* is

$$n_{tr} = 1 + \sum_{t=1}^{T} n_t \text{ or } n_{tr} = 1 + b_1 + b_1 b_2 + \cdots + b_1 b_2 \ldots b_T \tag{7}$$

where—n_t—number of nodes in stage t ($n_0 = 1$).

- *the total number of scenarios* is

$$S = \prod_{t=1}^{T} b_t \tag{8}$$

The tree process is defined on the finite probability space $\Omega = (\omega = 1, \omega = 2, \ldots, \omega = 32)$. The corresponding filtration is $\Im = (\mathcal{F}_0, \mathcal{F}_1, \ldots, \mathcal{F}_\mathrm{T})$, such that the sequence of σ-algebras is

$$\begin{aligned}
&\mathcal{F}_0 : \{\omega = 1, \ldots, \omega = M\} \\
&\mathcal{F}_1 : \{\omega = 1, \omega = 2, \omega = 3, \omega = 4\}, \ldots, \{\omega = M-3, \omega = M-2, \omega = M-1, \omega = M\} \\
&\ldots \\
&\mathcal{F}_{\mathrm{T}-1} : \{\omega = 1, \omega = 2\}, \{\omega = 3, \omega = 4\}, \ldots, \{\omega = M-1, \omega = M\} \\
&\mathcal{F}_\mathrm{T} : \{\omega = 1\}, \{\omega = 2\}, \ldots, \{\omega = M\}
\end{aligned} \tag{9}$$

The probability of scenario l or its *path probabilities* assigned to the terminal (leaf) nodes is

$$p_l = P\left(\xi^l\right) = P\left(\xi_0^l\right)\prod_{t=1}^{T} P\left(\xi_t^l \,\middle|\, \xi_0^l, \xi_1^l, \ldots, \xi_{t-1}^l\right), l = \overline{1, M} \tag{10}$$

The nonantipativity constraints are included in an implicit form.

The sum of probabilities for all scenarios in each stage of the tree is equal 1. As the depth of the scenario tree increases, the total probability for a deep node decreases, so it may be useful to simplify the problem.

The multistage scenario tree differs from the two-stage scenario tree by its robustness, stability of solutions (similar sub-scenarios result in similar decisions). It reflects interstage dependence and decreases the number of nodes.

The scenario trees are used in financial management models in life insurance and pension funds. The usual financial scenario processes (asset prices, interest rates, etc.) have to be extended for mortality risk [11]. Mortality appears as an independent risk factor. In [11] is given example the stochastic management of individual life insurance contracts for combine both risk factors.

4.3 Hierarchical Clustering Method of Construction the Multistage Scenario Tree

There are many techniques to create the scenario tree from the scenario fan as conditional/importance sampling, ad hoc/expert cutting and pasting, moment fitting, minimization of the distances of probability distributions, clustering, discretization schemes used instead of simulation or sampling [7].

In the paper the scenario tree is constructed from the scenario fan by clustering method [12–14]. The modified Ward (minimum variance) method is used.

The method attempts to minimize the error of sum of squares (ESS) of the clusters C_t^k that can be formed at each time period t. The objective function is

$$ESS_t = \sum_l \left(\xi_t^l\right)^2 - \left(\sum_l \xi_t^l\right)^2 \tag{11}$$

which is calculated for all scenarios belong to the clusters from the previous time period. The cluster centers $\overline{\xi}_t^k$ are computed. They correspond to the nodes of the scenario tree with respective probability and dispersion. The mean and dispersion of the cluster C_t^k at any time period t are

$$\overline{\xi}_t^k = M\left[\xi_t^k\right], \xi_t^k \in C_t^k \tag{12}$$

$$Var\left(\xi_t^k\right) = M\left[\left(\xi_t^k - \overline{\xi}_t^k\right)^2\right], \xi_t^k \in C_t^k \tag{13}$$

The probabilities of the centers $\overline{\xi}_t^k$ are the sum of probabilities of the individual scenarios $\xi_t^k \in C_t^k$.

In each time period t the total mean and variance are evaluated on the base of the cluster means and variances.

5 Simulation Results

The 2-factor model of Hibbert, Mowbray and Turbull (HMT) for long-term financial planning purposes model is used. It presents an original continuous-time mean-reverting stochastic process from Ornstein-Uhlenbeck type for the short-term interest rate [3, 15, 16, 19]. The process is defined by the system

$$\left| \begin{array}{l} dr_1(t) = \alpha_{r1}\left(r_2(t) - r_1(t)\right)dt + \sigma_{r1} dZ_{r1}(t) \\ dr_2(t) = \alpha_{r2}\left(\mu_r - r_{r2}(t)\right)dt + \sigma_{r2} dZ_{r2}(t) \end{array} \right. \tag{14}$$

where—$r_1(t)$, $r_2(t)$—real short-rate and mean reversion level for the real short-rate at time t; α_{r1}, α_{r2}—autoregressive parameters; σ_{r1}, σ_{r2}—annualized volatilities (standard deviations); μ_r—mean reversion levels for $r_2(t)$; $dZ_{r1}(t)$, $dZ_{r2}(t)$—shocks to the both real rate processes which is distributed normally; b_{r1}, b_{r2}-lower bounds for the real rates.

The analytic expressions in (14) can be used to derive the real interest rate $r_1(t)$ and $r_2(t)$ at any time. They are the continuous-time equivalents of two first-order autoregressive time-series processes. The real interest rates have a normal distribution. The inconvenience of the model is the possibility to obtain the negative real rates. The two processes $Z_1(t)$ and $Z_2(t)$ are Brownian motions with instantaneous correlation ρ. The Eqs. (14) allow calculating an expected path for future short-rates that is naturally related to current forward rates. The price of a zero-coupon bond at time t that pays one unit in real terms (i.e. protected from inflation) at time T is given by the following pricing equation

$$P_r(t, T) = e^{[A(T-t) - B_1(T-t)r_1(t) - B_2(T-t)r_2(t)]} \tag{15}$$

where—$A(s)$, $B_1(s)$ and $B_2(s)$ are the functions of the parameters for real interest rate movements [3, 15].

After obtained prices for real discount bonds, it is then possible to calculate also the continuously compounded yield at time t for maturity T

$$R_r(t, T) = -\log\left(P_r(t, T)\right)/(T - t) \tag{16}$$

The continuous process (14) can be approximated by the discrete one using longer intervals, such as monthly (requirement of long-term insurance problems). The discrete model is

$$\left| \begin{array}{l} \Delta r_{1t} = \alpha_{k1}(r_{2t} - r_{1t})\Delta t + \sigma_{r1}\varepsilon_{1t} \\ \Delta r_{2t} = \alpha_{k1}(\mu_r - r_{2t})\Delta t + \sigma_{r2}\varepsilon_{2t} \end{array} \right. \quad (17)$$

From $\Delta r_{1t} = r_{1,t+1} - r_{1t}$ and $\Delta r_{2t} = r_{2,t+1} - r_{2t}$, then

$$\left| \begin{array}{l} r_{1,t+1} = (1 - \alpha_{k1}\Delta t)r_{1t} + \alpha_{k1}r_{2t}\Delta t + \sigma_{r1}\varepsilon_{1t} \\ r_{2,t+1} = \alpha_{k2}\mu_r\Delta t + (1 - \alpha_{k2}\Delta t)r_{2t} + \sigma_{r2}\varepsilon_{2t} \end{array} \right. \quad (18)$$

In the discrete case the standard Brownian motion is presented by ε_{1t} and ε_{2t}. The obtained discrete samples from the system (18) represent the scenarios for the uncertain variables over the planning time period.

The simulation algorithm is [11]:

- generate $\varepsilon_i \sim N(0, 1)$ and multiply it by $\sqrt{\Delta t}$.
- simulate the real interest rates $r_2(t)$ using the second equation of (15) and $r_1(t)$ using the first equation of (15).
- determine maturity time T for discount bonds.
- use analytical expressions (15) for $P_q(t, T)$ and $R_r(t, T)$ at any term.
- calculate discount bond return at time t for maturity T using (16).

The parameters must be determined by calibration [3].

The same structure model for the short-term inflation rates is used, as (14) but with different parameters [3, 15, 16]. The equations governing the path of the short-term inflation rates are

$$\left| \begin{array}{l} dq_1(t) = \alpha_{q1}(q_2(t) - q_1(t))\,dt + \sigma_{q1}dZ_{q1}(t) \\ dq_2(t) = \alpha_{q2}\left(\mu_q - q_{r2}(t)\right)dt + \sigma_{q2}dZ_{q2}(t) \end{array} \right. \quad (19)$$

where the parameters are similar to those of the interest rates.

The pricing equation of $P_q(t, T)$ and equivalent yield of $R_q(t, T)$ are similar to Eqs. (15) and (16).

The discrete inflation model of (17) is

$$\left| \begin{array}{l} q_{1,t+1} = (1 - q_{k1}\Delta t)r_{1t} + \alpha_{k1}q_{2t}\Delta t + \sigma_{q1}\varepsilon_{1t} \\ q_{2,t+1} = \alpha_{q2}\mu_q\Delta t + (1 - \alpha_{q2}\Delta t)q_{2t} + \sigma_{q2}\varepsilon_{2t} \end{array} \right. \quad (20)$$

Analogically, the obtained discrete samples from the system (20) represent the scenarios for the uncertain variables over the planning time period. Discretization and simulation algorithms are similar to those of the real interest rate.

The Fisher equation allows to combined the real interest rate and the inflation expectations term structures and obtain the price of the nominal interest rate term structure

$$P_{nom}(t, T) = P_r(t, T)\,P_q(t, T) \quad (21)$$

The algorithm and calculations were implemented in the Statistical Analysis System (SAS) [17].

At first, the model is constrained in a way that guarantees the negative rates do not appear. Therefore the following initial conditions are assumed: the inflation level is 2.5 %, the mean reversion level for the real short-rate is 2.83 %, the current 3-month T-bill norm is 5 % and the current 10-year T-bond yield is 5.58 % [3].

Using the HMT stochastic asset model the representative set of $L = 1000$ scenarios for the nominal interest rate during time horizon of $T = 10$ years or 120 months with one month time increment is simulated. Its graphical representation is shown in Fig. 3. The initial root node is $\xi_0^1 = \cdots = \xi_0^L = 5\,\%$. The fan consists a 120,000 nodes.

For *estimating the optimal number of clusters* in the data is used the following statistics: cubic clustering criterion (CCC) and t^2 (PST2) statistics (Fig. 4).

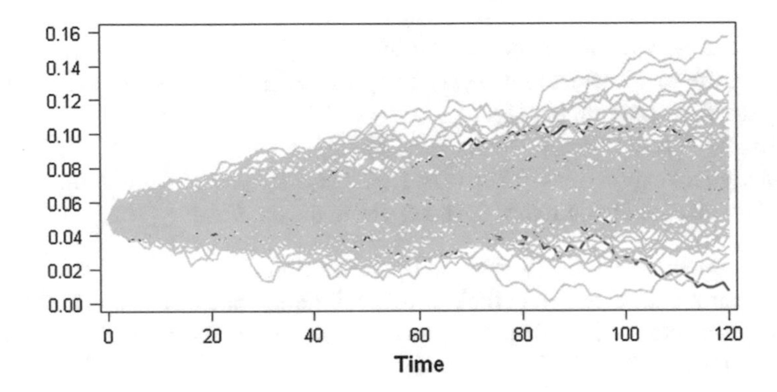

Fig. 3 Initial scenario fan

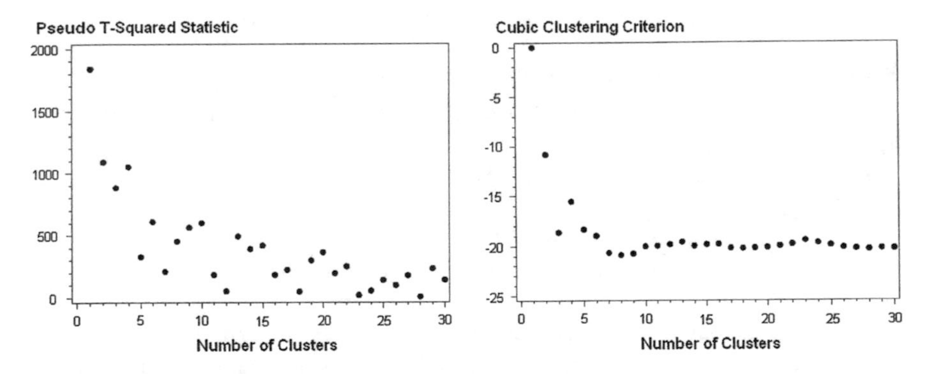

Fig. 4 Estimation of the number of clusters

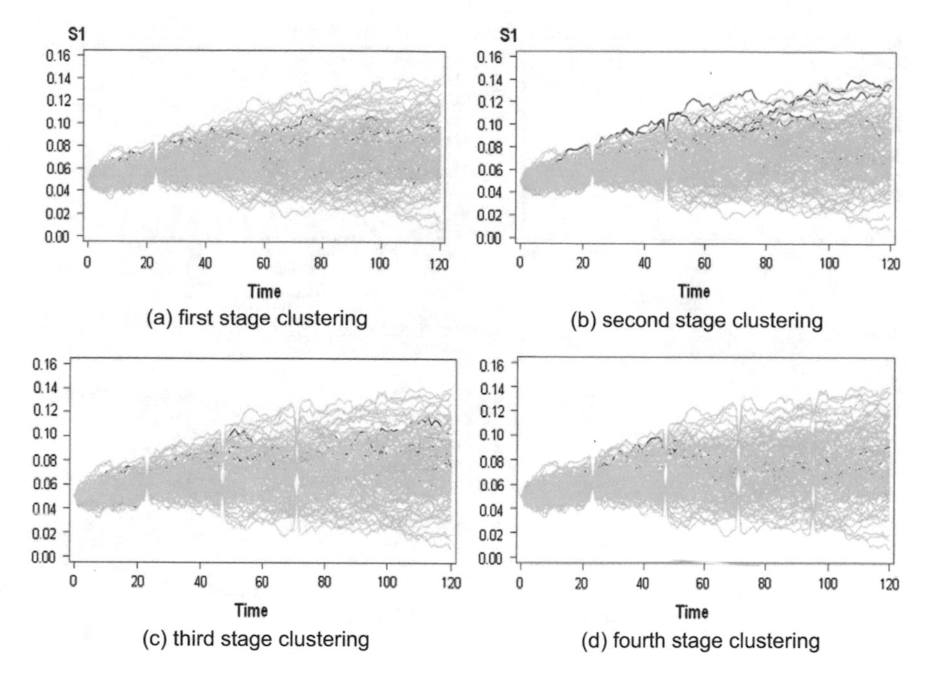

Fig. 5 Multistage scenario trees in consistently clustering

From the PST2 graph the first minimums indicate that the good numbers of clusters are 2 and 3. The peaks in the CCC graph show that the possible numbers of clusters are 2, 4 and 6. Considered together, these statistics suggest that the data can be clustered into two clusters. The clustering is made with $K = 2$ (branching factor).

By clustering the generated scenarios with Ward method scenario fan is transformed to multistage scenario tree also with K=2 at each stage, which is confirmed also by the results in both statistics—PST2 and CCC. Consecutively, the scenario fan is transformed to the following five multistage scenario trees (STs) with different decision moments: 2-stage ST at $t = 10$ (Fig. 3); 3-stage ST at $t = 2, 10$ (Fig. 5a); 4-stage ST at $t = 2, 4, 10$ (Fig. 5b); 5-stage ST at $t = 2, 4, 6, 10$ (Fig. 5c); 6-stage ST at $t = 2, 4, 6, 8, 10$ (Fig. 5d).

The dimensions of received STs are less then the scenario fan size. For example 6-stage tree with 2 branches consists of 63 nodes and 32 scenarios according formulae (6) and (7). At Fig. 6 are given graphically the branching structures of STs.

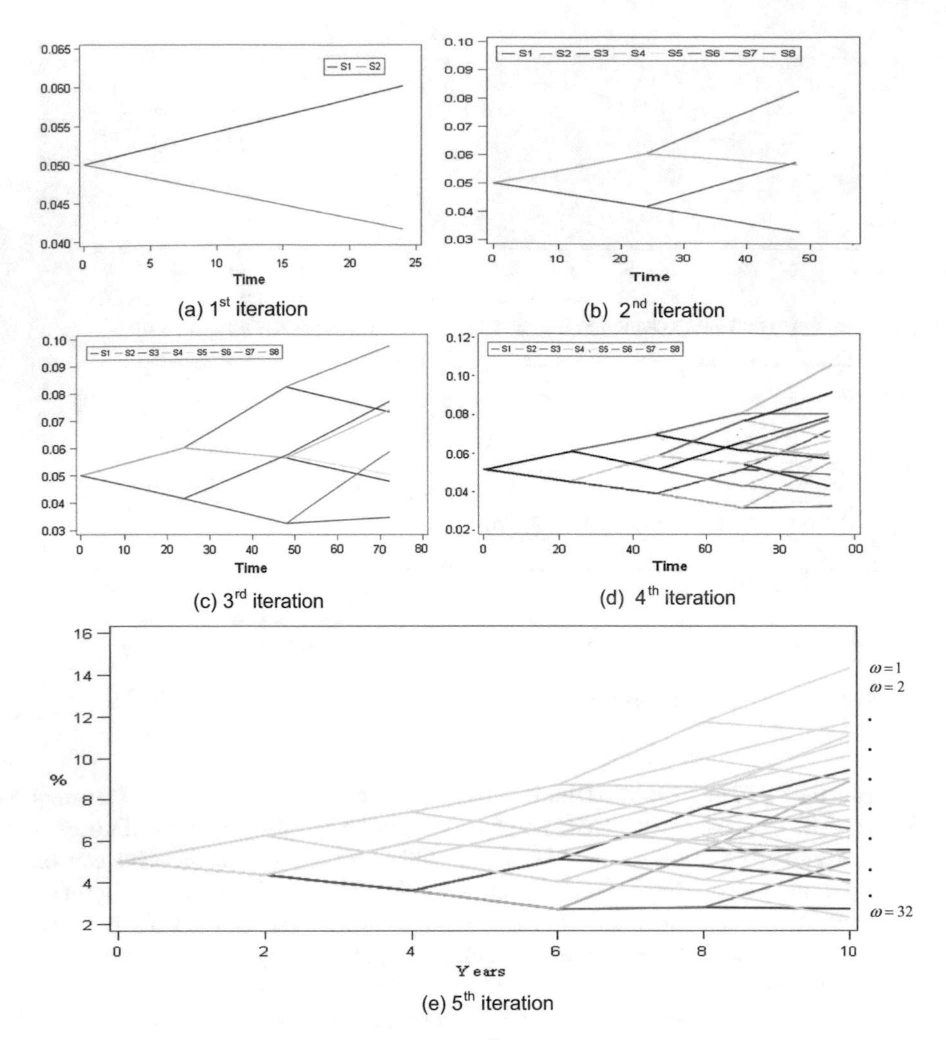

Fig. 6 Construction of multistage scenario tree

The tree process in Fig. 6 is defined on the finite probability space $\Omega = (\omega = 1, \omega = 2, \ldots, \omega = 32)$. The corresponding filtration is $\Im = (\mathcal{F}_0, \mathcal{F}_1, \ldots, \mathcal{F}_5)$, such that the sequence of σ-algebras is

$\mathcal{F}_0 : \{\omega = 1, \omega = 2, \ldots, \omega = 32\}$; $\mathcal{F}_1 : \{\omega = 1, \ldots, \omega = 16\}, \{\omega = 17, \ldots, \omega = 32\}$;

$\mathcal{F}_2 : \{\omega = 1, \ldots, \omega = 8\}, \ldots, \{\omega = 25, \ldots, \omega = 32\}$; $\mathcal{F}_3 : \{\omega = 1, \ldots, \omega = 4\}, \ldots, \{\omega = 29, \ldots, \omega = 32\}$;

$\mathcal{F}_4 : \{\omega = 1, \omega = 2\}, \{\omega = 3, \omega = 4\}, \ldots, \{\omega = 31, \omega = 32\}$;

$\mathcal{F}_5 : \{\omega = 1\}, \{\omega = 2\}, \ldots, \{\omega = 32\}$

The generated STs are evaluated by calculated some statistical characteristics of the nominal interest rate. The mean value (12) and the dispersion (13) at five determined time moments ($t = 2, 4, 6, 8$ and 10 years) of the time are calculated. For the scenario fan they are given in Table 1 and for the STs—in Table 2. The comparison of the statistical characteristics of the scenario fan and STs from the tables shows that they are retained.

6 Conclusions

In this paper a procedure based on scenario simulation and clustering technique for construction the multistage scenario tree is presented. The large set of scenarios of nominal interest rate is generated to represent the uncertainty in a sensible way account: the goal of the model, its structure and the available information.

Tables 1 and 2 show that constructed scenario tree according to the described method with optimal number of clusters $K = 2$ presents sufficiently precise initial scenario fan.

The size of the constructed scenario trees is much smaller than scenario fan. This approach decreases the calculating time.

The proposed model can be incorporated into a variety of insurance applications. The following investigations are planned in the future:

- comparison of the considered model with others, as CIR model;
- construction of the event trees describing the occurrence of events (in case of life-insurance death or survival events) using the life (mortality) tables in models for managing individual or portfolios contracts for life insurance and pension funds.

Table 1 Characteristics of scenario fan

Statistical characteristics/decision moments	Time moments, in years				
	$t = 2$	$t = 4$	$t = 6$	$t = 8$	$t = 10$
Mean	0.05365	0.05777	0.06264	0.06844	0.07224
Dispersion	0.01177	0.01706	0.02091	0.02348	0.02600

Table 2 Characteristics of multistage scenario trees

		Decision moments, in Years					
		Characteristic	$t = 2$	$t = 4$	$t = 6$	$t = 8$	$t = 10$
Multistage scenario trees	2-stage tree	Mean	–	–	–	–	0.07224
		Dispersion					0.02600
	3-stage tree	Mean	0.05411	–	–	–	0.07541
		Dispersion	0.01174				0.02564
	4-stage tree	Mean	0.05411	0.05956	–	–	0.06894
		Dispersion	0.01174	0.01698			0.02452
	5-stage tree	Mean	0.05411	0.05956	0.06048	–	0.07039
		Dispersion	0.01174	0.01698	0.01851		0.02444
	6-stage tree	Mean	0.05411	0.05956	0.06048	0.06778	0.07253
		Dispersion	0.01174	0.01698	0.01851	0.02113	0.02336

References

1. Brigo, D., Mercurio, F.: Interest Rate Model-Theory and Practice: With Smile. Inflation and Credit, Springer Finance (2001)
2. Cox, J.C., Ingersoll, J.E., Ross, S.A.: A theory of the term structure of interest rates. Econometrica **53**, 385–407 (1985). doi:10.2307/1911242
3. Hibbert, J., Mowbray, P., Turnbull, C.: A stochastic asset model and calibration for long-term financial planning purposes. Technical Report, Barrie & Hibbert Limited, pp. 1–76 (2001)
4. Vasicek, O.: An equilibrium characterization of the term structure. J. Financ. Econ. **5**, 177–188 (1977)
5. Dupacova, J., Consigli, G., Wallace, S.W.: Scenarios for multistage stochastic programs. Ann. Oper. Res. **100**, 25–53 (2000)
6. Domenica, N., Birbilis, G., Mitra, G., Valente, P.: Stochastic programming and scenario generation within a simulation framework: an information system perspective, pp. 1–41. Carisma Center, Brunel University, Uxbridge, Middlesex (2003)
7. Dupacova, J., Hurt, J., Stepan, J.: Stochastic Modeling in Economics and Finance. Kluwer Academic Publishers (2002)
8. Kaut, M.: Scenario Generation for Stochastic Programming Introduction and Selected Methods, pp. 1–61. SINTEF Technology and Society (2011)
9. Mitra, S.: Scenario generation for stochastic programming. OptiRisk Syst. 1–33 (2008)
10. Mitra, S., Di Domenica, N.: A review of scenario generation methods. Int. J. Comput. Sci. Math. **3**(3), 226–244 (2015). ISSN: 1752-5055
11. Pflug, G., Romisch, W.: Modeling, Measuring and Managing Risk. World Scientific Publishing Co. Pte. Ltd. (2007)
12. Jain, A., Murty, M.N., Flynn, P.J.: Data clustering: a review. ACM Comput. Surv. **31**, 3 (1999)
13. Kaufmann, L., Rousseeuw, P.: Finding Groups in Data. An Introduction to Cluster Analysis. Wiley-Interscience, Canada (1990)
14. Kovacheva, Ts.: Cluster analysis. Inf. Technol. Control. Sofia. **3**, 24–29 (2004)
15. Kovacheva, Ts.: Risk factors analysis affecting the insurance company's activities. "Applied Mathematics and Mechanics", Proceeding, Ulyanovsk State Technical University, Ulyanovsk, pp. 252–258 (2011). ISBN: 978-5-9795-0904-4
16. Neftci, S.: An Introduction to the Mathematics of Financial Derivatives. Elsevier (2000)
17. Statistical Analysis with SAS/STAT® Software. http://www.sas.com/technologies/analytics/statistics/stat/index.html
18. Heitsch, H., Römisch, W.: Scenario tree modelling for multistage stochastic programs. DFG Research Center Matheon "Mathematics for key technologies" (2005)
19. Pranevicius, H., Sutiene, K.: Scenario generation for multistage stochastic programs, pp. 1–6. KTU, Kaunas (2008)

Microphone Array for Non-contact Monitoring of Rolling Bearings

Volodymyr Kudriashov, Vladislav Ivanov, Kiril Alexiev and Petia Koprinkova-Hristova

Abstract A non-contact approach for detection of lubrication loss in ball bearings is described in the paper. An acoustic camera consisting of array of microphones and camera is used for measuring bearing noise. It is found that the lubrication loss increases the obtained sound pressure from 3 to 33 dB, in the frequency range 10 Hz–20 kHz. Automatic detection of the lubrication loss may be done by a thresholding technique.

Keywords Acoustic camera · Multi-sensor system · Acoustic noise source localization · Non-contact monitoring · Direction of arrival

1 Introduction

Acoustic and vibration measurement methods are used for monitoring of rolling element bearings. Review on these methods shows that emphasis is on vibration measurement methods [16]. The vibration measurement in frequency domain enables localization of particular defect [16]. However, the acoustic emission (AE) measurements are suitable to detect defects earlier than they appear in the vibration acceleration spectra [19, 20]. The AE measurements can detect growth of subsurface crack whereas the vibration monitoring can detect surface defects only [16]. The

V. Kudriashov (✉) · K. Alexiev · P. Koprinkova-Hristova
Mathematical Methods for Sensor Information Processing Department,
Institute of Information and Communication Technologies, Bulgarian Academy of Sciences, Acad. G. Bonchev Str., bl.25 A, 1113 Sofia, Bulgaria
e-mail: Kudriashov.Vladimir@gmail.com

K. Alexiev
e-mail: Alexiev@bas.bg

P. Koprinkova-Hristova
e-mail: PKoprinkova@bas.bg

V. Ivanov
Technical University of Sofia, 8 Kl. Ohridski Blvd., 1756 Sofia, Bulgaria
e-mail: VVI@tu-sofia.bg

K. Georgiev et al. (eds.), *Advanced Computing in Industrial Mathematics*,
Studies in Computational Intelligence 681, DOI 10.1007/978-3-319-49544-6_9

high quality of ultrasonic AE measurements is mainly determined by a difference in the frequency range between AE signals (above about 50 kHz) and interfering signals in a vibrational frequency range (below about 50 kHz) [12, 13, 16]. Both the amplitudes and the quantity of AE counts that exceed chosen threshold level are used to detect defects in the bearings [15]. The threshold level depends on operating conditions [11]. The ultrasonic AE sensor is usually placed on the bearing housing. Mounting surface roughness and contact grease affect losses of incoming AE [12]. The above-described methods require the contact between the sensor and the object under investigation. The latter may lead to additional costs due to production line stops, required to assure the contact.

An array of microphones may be used for non-contact bearing monitoring [1]. The microphone array portability is useful to avoid the production stops. The microphone signal contains descriptors of examined bearing. Even close placement of microphone to the bearing do not assure the absence of penetration of unwanted emissions. These emissions are received from various spatial positions. Their impact affects seriously the quality of measurements. However, the desired and the unwanted signals may income from different spatial positions. The array of microphones may be focalized to known position of the desired source of acoustic emission. It enables to separate desired and unwanted signals using the spatial difference between their sources. The quality of the separation depends on applied signal processing methods, equipment performance, and experimental conditions. We consider condition monitoring based upon processing of sound pressure measured near the bearing, in the frequency range below 20 kHz.

2 Experimental Setup

The work exploits acoustic camera manufactured by Brüel & Kjaer (Sound and Vibration Measurement A/S). The acoustic camera comprises microphone array and optic camera. The acoustic camera generates acoustic images of sound pressure in cross range—elevation coordinates, for noise source localization. The optic image is overlaid onto the acoustic image to increase the quality of acoustic noise source identification. The microphone array of the acoustic camera could be focalized to the spatial position of the bearing. Portability of the acoustic camera enhances workflow, additionally. Thus, the usage of the acoustic camera enables to increase the quality of bearing monitoring.

The acoustic camera has 18 channels. The channel consists of a microphone (type 4958), an input module (type 3053-B-120 or 3050-B-060), and an acquisition software (Pulse LabShop). The microphone frequency band is in range 10 Hz–20 kHz. The microphone array diameter is $\sim$0.32 m. The acoustic camera uses beamforming and acoustic holography to generate acoustic images. The microphone array beamwidth is $\sim 6°$, for frequency 10 kHz. The acoustic camera enables export of measured data for post-processing.

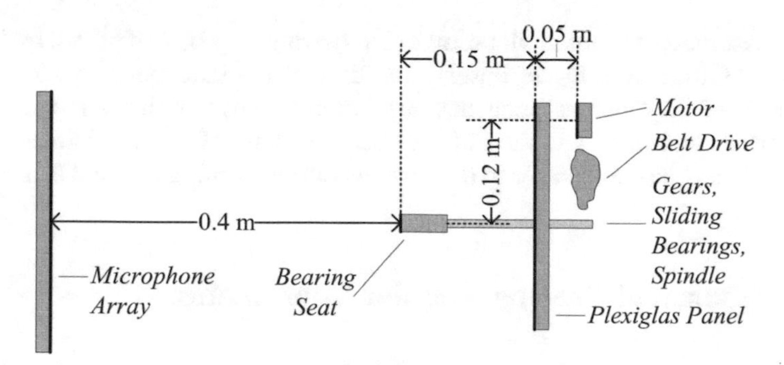

Fig. 1 Scheme of the experimental setup: *top view*

The microphone array was placed near to the bearing diagnosis equipment (model: MS-730A, NSK Ltd, Fig. 1). The belt drive passes rotation of the alternating current motor to gears. The gears pass the rotation to the spindle. A bearing seat is a mounting place for the inner ring of the bearing, on the ending of the spindle. The outer ring of the bearing is loaded with pressing screw. The equipment is inside a room. Distances between the microphone array and bearing seat, motor and others are given in Fig. 1.

The experiments were carried out in a room, where multipath propagation of acoustic signals of both desired and interfering sources exist. The interfering sources are: the equipment motor and gears, walking and speaking people inside the room and urban noises from outdoor. Sound pressure of the interfering signals is comparable to the acoustic emission of some bearing samples, at the output of single microphone channel.

The fundamental fault frequencies that the bearing generates when balls pass over a surface anomaly are a function of the bearing geometry and relative speed between its inner and outer rings [7]. All measured bearings have the same geometry. Their manufacturers are different: SKF, NSK Ltd., KBS, and HF. The number of rolling elements (balls) is nine. The ball diameter is ~7.938 mm. The pitch diameter is ~39.04 mm. The contact angle of these bearings is zero. The latter allows the radial load of the outer race of the bearing with the pressing screw. The spindle rotational speed slightly varies from 1820 to 1830 rpm. The calculated values of the fundamental fault frequencies are 12–166 Hz. The fundamental fault frequencies are proportional to rotational speed of the spindle (Fig. 1). The speed can be controlled by the frequency of oscillations of alternating current, which feeds the motor. The 2.2 kW frequency inverter manufactured by Elmark (type: EL-ZVF9) is applied to vary the frequency. The frequency increasing from 50 to 65 Hz enables obtainment of the spindle rotational speed ~2334 rpm. The corresponding fundamental fault frequencies are 15–211 Hz. Further increasing of the feeding current is not possible because it affects the stability of the rotational speed, even at a reduced pressure of the screw.

The frequencies are under 0.5 kHz, for normal rotational speeds [16]. The beamwidth of the microphone array at such low-frequency range is wider than 120°.

Non-contact measurement of vibration of the bearing may be committed by bistatic radiometer [10]. Increasing of observation time of the radiometer to 10 s enables obtainment of the Doppler frequency resolution 0.1 Hz. The latter is suitable for remote measurement of vibration in the spatial volume of interest. Dimensions of the volume may be comparable to dimensions of some rolling element bearings.

3 Assessment of Bearing Angular Coordinates

The received acoustic signal contains desired and interfering components. Spatial diversity between emission source of interest and another interfering source(s) may be used, to suppress the impact of the latter. The microphone array is focused on a specified spatial position (volume) [4]. The focalization enables both maintaining of signals from the volume and partial suppression of signals, incoming from unwanted directions. The suppression is limited by directional properties of the employed microphone array. Thus, the focalization of microphone array enables to improve the quality of received signal in comparison with a single microphone.

The focalization is described in [4]. It requires compensation of propagation delays of received signals. The delays are determined by the propagation model of incoming signals and the distances between microphones and measured bearing. The delayed signals could be summed or processed in another way. We assume, that the noises of microphones and corresponding acquisition channels are non-correlated. We assume also that input signals are stationary. To improve the quality of obtained results all autocorrelation terms of a cross-spectral matrix are replaced by zeros [4].

The focalization is carried out in a frequency domain. The delays depend on Euclidean distances between the bearing seat and microphones of the array (Fig. 1). The frequency resolution of the developed software is 4 Hz. The frequency resolution is limited by Fourier transform size at the particular configuration of the software.

The software allows generation of acoustic images and estimation of the focalized spectra of the random input signals at a finite acquisition time. The difference of the focalized spectra estimates (FSE) from original one depends on the signal features, environmental conditions and a quality of the measurements of spatial coordinates of the source. The quality of measurement of angular coordinates of the source with the software depends on several factors noted below.

3.1 Preliminary Calculations

The angles of arrival are measured as coordinates of the peak of the acoustic image. The quality of the measurement may be described by its accuracy. The accuracy of measurement is defined as the root mean square errors (RMSE) and it depends on the signal-to-noise ratio (SNR), on interfering signal and on equipment quality, etc. High SNR and absence of interferences are required for high quality measurements.

Cramer-Rao Lower Bound defines the fundamental limitation for attainable RMSE of measurement in case of background noise, at high SNR [14]. Thus, the attainable RMSE of measurement of direction of arrival is written as [2]:

$$\sigma = \frac{180}{\pi} \frac{\nu_s}{f_c \, L_{ef} \sqrt{SNR}} \quad (1)$$

where σ is the RMSE of the direction of arrival measurement; ν_s is the incoming wave propagation velocity, at measurements conditions; f_c is the center frequency at acoustic image generation; L_{ef} is the effective size of the microphone array for the specified direction of arrival; *SNR* is the signal-to-noise ratio [2]. In the case of low SNR, the attainable RMSE of measurement in noise background may be defined by Ziv-Zakai, Barankin and other lower bounds [3, 17, 18].

The RMSE (1) of measurement of direction of arrival is ~6°, at SNR = 20 dB (q = 10), f_c =1 kHz and L_{ef} = 0.32 m. The RMSE value is 10 times less than a theoretical beamwidth of the microphone array, at far range. The footprint of the RMSE value is comparable to the pitch diameter of observed bearings. The pitch diameter is not less than ±3 RMSE values, for $f_c \geq 6$ kHz.

3.2 Generation of Acoustic Image

NSK bearing sample with unknown defect was used to evaluate angular coordinates of the focalization position because it delivers powerful acoustic emission. A prior defined range enabled generation of the acoustic image, in azimuth-elevation coordinates. The acoustic image was generated for center frequency 10 kHz, bandwidth 1 kHz and averaging time equal to 0.25 s (Fig. 2). The peak contains sound pressure ~30.5 dB to 1 Pa, denoted hereinafter as dB. The threshold level of the image is 12 dB. The image peak is placed at azimuth 0° and elevation 1°. In the case of small amount of data, the angular coordinates could be obtained by approach, described in [5]. Both modified imaging approach and threshold level are suitable to obtain all 3D spatial coordinates of the bearing [2, 6, 10].

At the above-mentioned conditions, the dimensions of the resolution cell of the focalized microphone array depend mainly on the acoustic frequency. At low frequencies, the resolution cell increases. At center frequency 1 kHz, the footprint of the microphone array is ~0.5 m, at range 0.4 m. Thus, the interfering signal sources in Fig. 2 become to get inside the resolution cell. That is why the FSE cannot differ significantly from non-focalized one at frequencies below ~1 kHz.

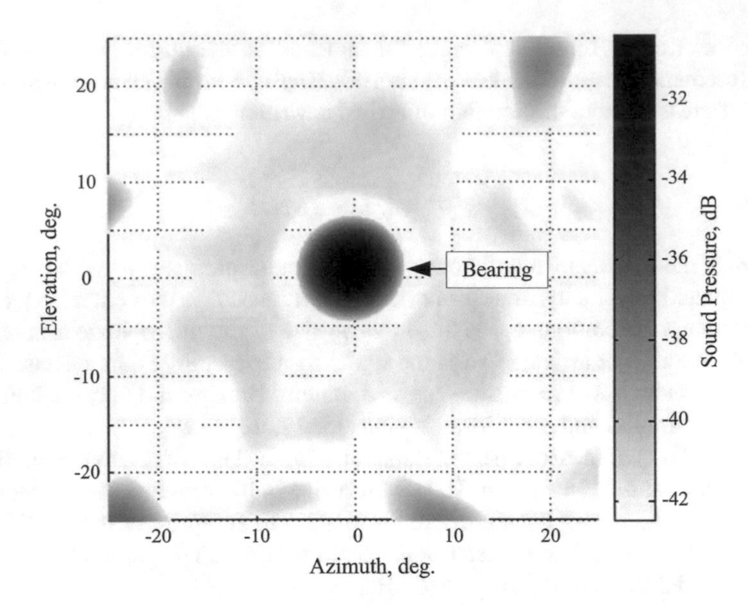

Fig. 2 Acoustic image for the experimental setup with NSK bearing sample

3.3 Interfering Background

The interfering background is generated by unwanted sources of acoustic emission, mainly. The background level was measured without bearing installed on the spindle (Fig. 1). Hereinafter, the measured angular coordinates are used to obtain FSE with frequency resolution 4 Hz. The observation time is 0.25 s. The signals update rate is 1 s. At the current experiment, 100 FSE were averaged, to decrease the variation of the sound pressure values. The result of the above is considered as the interfering background level.

The background FSE is obtained for two frequencies of the motor feeding frequency: 65 and 50 Hz. As well, the background FSE is obtained without the motor feeding frequency inverter (lowest curve, Fig. 3). The obtained FSE shows that the frequency inverter generates interfering sound emission at frequencies 4, 4.7, 7.6–8.9, 11–13 kHz (both upper curves, Fig. 3). Two of the background FSE curves are shifted up and down on the Fig. 3 using auxiliary pedestals of ± 15 dB, i.e. the FSE of measurements carried out at motor feeding frequency 65 Hz is moved upwards while the FSE of measurements carried out with switched off motor feeding frequency inverter is shifted downwards.

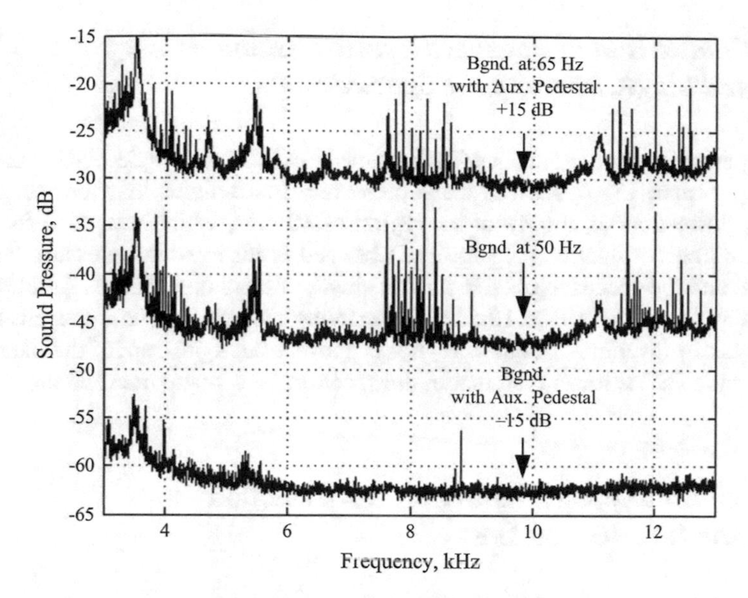

Fig. 3 Averaged realizations of the focalized spectra estimates for background signal

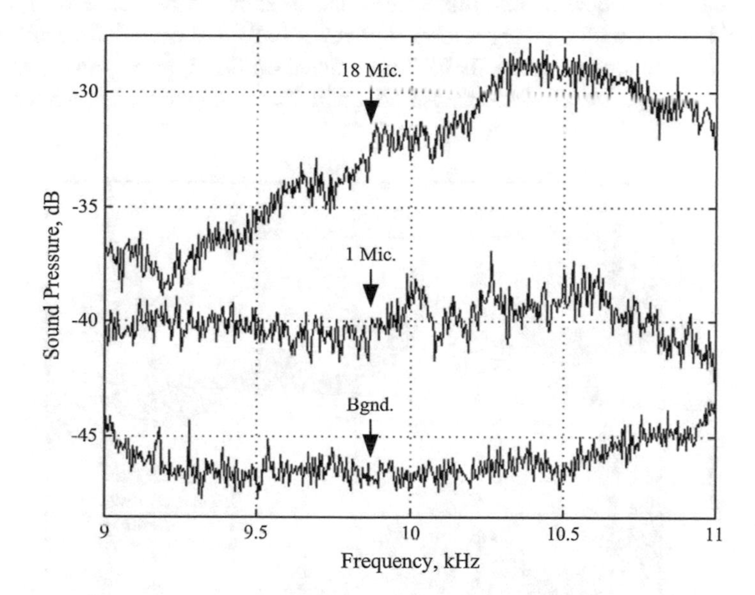

Fig. 4 Realizations of the spectra estimates for NSK bearing sample with unknown defect

3.4 Comparison of Focalized Spectra Estimates and Single Microphone Spectrum

Test experiment was carried out with the same NSK bearing sample. FSE of the bearing (upper curve, Fig. 4) showed sound pressure ~14 dB higher than the background spectra (lowest curve, Fig. 4), at the region of 10 kHz, which was used for angular coordinates evaluation (Fig. 2). The obtained profit in sound pressure between the FSE and the spectrum obtained using single microphone channel (intermediate curve, Fig. 4) is up to 9 dB, at 10.5 kHz. Interfering acoustic signals are received from the frequency inverter at frequencies, noted above. Sound pressure of the interfering peaks in the FSE is lower than that in the spectrum of a single microphone.

4 Focalized Spectra Estimates for Detection of the Lubrication Loss

Insufficient lubrication may cause fails of a bearing or defects of its surfaces as well. As an initial step, we simulated lubrication loss to show its influence on FSE. We used SKF bearing with a priory known defects of balls and cage. The bearing was degreased using a special spray. Its FSE is depicted on Fig. 5, upper curve. Then the bearing was finely lubricated and measured again. The obtained FSE is ~3 dB lower,

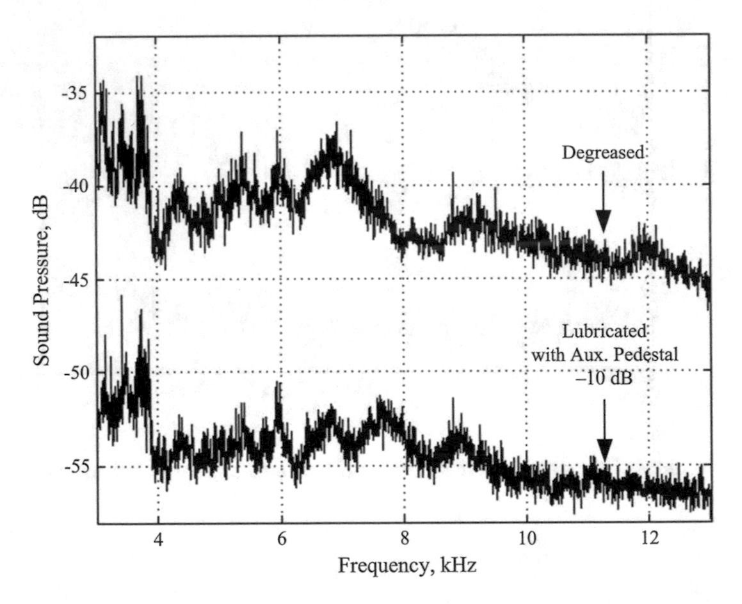

Fig. 5 Focalized spectra estimates in case of presence and absence of artificial lubrication loss, for SKF bearing with cage and balls defects

comparable to the previous one. The latter FSE is shown with auxiliary pedestal −10 dB (lower curve, Fig. 5), to assure separation from the former one. The shape of both FSE is similar at the frequency range under ~6.5 and 9–11 kHz.

Two KBS bearings were used to analyze their FSE (Fig. 6). Both bearings have unknown surface defects. Additionally, one of them has lubrication loss. Their FSE were obtained in the frequency range from 4 to 7 kHz, for feeding current's frequency of 50 Hz. The FSE for the bearing with lubrication loss (upper curve, Fig. 6) is about 25 dB higher than FSE of bearing without the loss (lower curve, Fig. 6). The difference between the depicted spectra is less than ±6 dB, after accounting the noted 25 dB level. For the feeding current's frequency 65 Hz, the difference between the obtained FSE is up to 7.6 dB, after accounting of the noted 25 dB level. In the frequency range 10 Hz–20 kHz, the maximum difference between these FSE is more than 33 dB, for both feeding current's frequencies.

The experiment was repeated with bearings, manufactured by HF. Prior information on defects in their surfaces is absent. One of them has lubrication loss. The difference between that FSE is up to 18 dB, for the frequency range 5–16 kHz (Fig. 7). The change of the feeding current's frequency from 50 to 65 Hz increases the difference, but its maximal value is about 18 dB, for the above frequency range. In the frequency range 10 Hz–20 kHz, the mean difference between the obtained FSEs is ~9.6 dB (~6.7 dB) for the feeding current's frequency 65 Hz (50 Hz). The experimental results showed that the lubrication loss assures FSE difference not less than 3 dB.

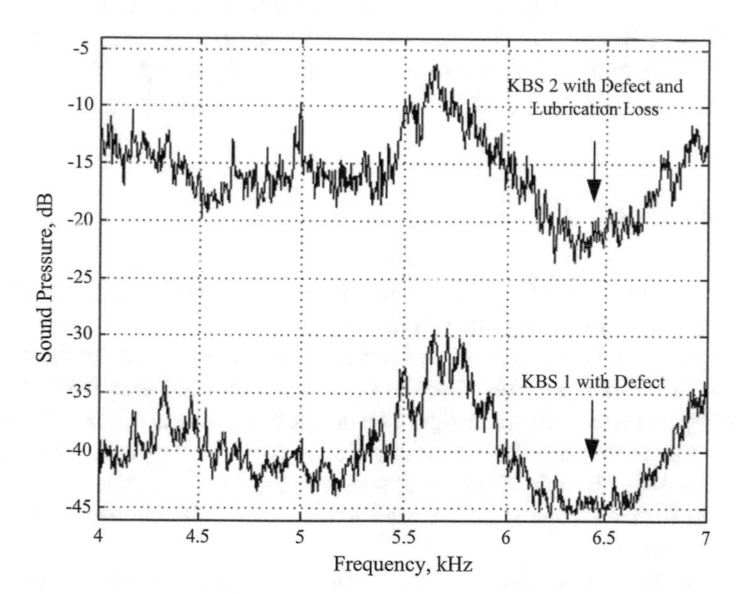

Fig. 6 Focalized spectra estimates in case of lubrication loss, for KBS bearings with unknown surface defects

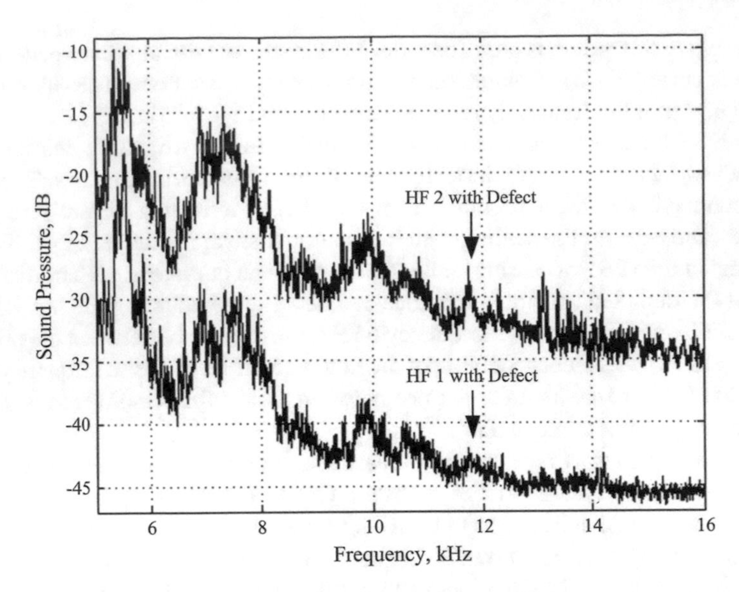

Fig. 7 Focalized spectra estimates in case of lubrication loss, for HF bearings with unknown surface defects and feeding current frequency 50 Hz

In order to analyze results in automatic mode and process high enough amount of data, collected by the acoustic camera, a specialized application has to be developed to distinguish good bearing from bad one (damaged bearing) [16]. A review of different methods shows a high success rate of methods that use fuzzy logic [9] and/or neural networks [8].

5 Conclusions

Acoustic emissions of bearings were used for remote monitoring of the condition of the bearings, in the frequency range 10 Hz–20 kHz.

The experimental setup consisting from microphone array and the optic camera was applied to monitor bearings. Angular coordinates of the bearing seat were determined using generated acoustic image. The developed software enables the acoustic image generation and focalized spectra estimation with a frequency resolution of 4 Hz. Fundamental fault frequencies of used bearings were calculated. The focalized spectra for interfering signals were estimated, for different assemblies of the experimental setup.

The lubrication loss leads to increase in the focalized spectra from 3 dB to more than 25 dB in the frequency range 4–6 kHz (Figs. 5, 6, 7). For the tested KBS bearings, the maximum increase in the focalized spectra is higher than 33 dB in the frequency range 10 Hz–20 kHz. The automated detection of the lubrication loss may be easily realized using simple thresholding technique to differentiate the obtained focalized spectra from the reference one, at appropriate environmental conditions.

Acknowledgements This work was partly supported by the project AComIn, grant 316087, funded by the FP7 Capacity Program (Research Potential of Convergence Regions).

References

1. Akcay, H., Germen, E.: Identification of acoustic spectra for fault detection in induction motors. In: AFRICON 2013, pp. 1–5. IEEE. doi:10.1109/AFRCON.2013.6757650
2. Barton, D.K., Leonov, S.A. (eds.): Radar Technology Encyclopedia (Electronic Edition). Artech House Inc., Boston (1998). ISBN 0-89006-893-3
3. Bell, K.L., Steinberg, Y., Ephraim, Y., Van Trees, H.L.: Extended Ziv-Zakai lower bound for vector parameter estimation. IEEE Trans. Inf. Theory **43**(2), 624–637 (1997). doi:10.1109/18.556118
4. Christensen, J.J., Hald, J.: Beamforming. Brüel & Kjær Tech. Rev. **1**, 1–48 (2004)
5. Chyrka, I.: Fast direction-of-arrival estimation for single source. In: International Conference on Telecommunications and Remote Sensing (ICTRS-2015), pp. 54–58. SCITEPRESS, Bulgaria (2015). ISBN 978-989-758-152-6
6. Doukovska, L.A.: Adaptive Hough detector threshold analysis in presence of randomly arriving impulse interference. In: 4th Microwave and Radar Week MRW-2010—11th International Radar Symposium, IRS 2010—Conference Proceedings, pp. 142–145. IEEE (2010). ISSN 2155-5754. ISBN 978-995569018-4
7. Graney, B.P., Starry, K.: Rolling element bearing analysis. Mater. Eval. **70**(1), 78–85 (2012)
8. Koprinkova-Hristova, P., Chyrka, Iu., Kudriashov, V., Alexiev, K., Ivanov, V., Nedyalkov, P.: Smart feature extraction from acoustic camera multi-sensor measurements (presented at The International Conference ADVANCED COMPUTING FOR INNOVATION ACOMIN'2015, to appear in). In: Kacprzyk, J. (ed.) Innovative Approaches and Solutions in Advanced Intelligent Systems. SCI. Springer, Heidelberg (2016)
9. Liu, T.I., Singonahalli, J.H., Iyer, N.R.: Detection of roller bearing defects using expert system and fuzzy logic. Mech. Syst. Signal Process. **10**(5), 595–614 (1996). doi:10.1006/mssp.1996.0041
10. Lukin, K., Kudriashov, V.V., Vyplavin, P., Palamarchuk, V.: Coherent imaging in the range-azimuth plane using a bistatic radiometer based on antennas with beam synthesizing. Aerosp. Electr. Syst. Mag. **29**(7), 16–22 (2014). doi:10.1109/MAES.2014.130142
11. Morhain, A., Mba, D.: Bearing defect diagnosis and acoustic emission. Proc Inst Mech Eng Part J: J Eng Tribol 217(4):257–272 (2003). ISSN 1350-6501
12. Oh, H., Azarian, M.H., Pecht, M.: Estimation of fan bearing degradation using acoustic emission analysis and Mahalanobis distance. In: Applied Systems Health Management Conference 2011: Enabling Sustainable Systems. MFPT 2011, pp. 1–12. Virginia Beach VA, USA (2011)
13. Pinner, T., Obando, H.S., Moeser, G., Burger, W.: Monitoring lathe tool's wear condition by acoustic emission technology. In: Third International Conference on Condition Monitoring of Machinery in Non-Stationary Operations (CMMNO 2013), pp. 183–193. Springer, Berlin (2014) doi:10.1007/978-3-642-39348-8_15. ISSN 2195-4356
14. Seidman, L.P.: Performance limitations and error for parameter estimation. Proc. IEEE **58**(5), 644–652 (1970). doi:10.1109/PROC.1970.7720
15. Sikorski, W. (ed.): Acoustic Emission—Research and Applications. InTech, Rijeka (2013)
16. Tandon, N., Choudhury, A.: A review of vibration and acoustic measurement methods for the detection of defects in rolling element bearings. Tribol. Int. **32**(8), 469–480 (1999). doi:10.1016/S0301-679X(99)00077-8

17. Weinstein, E., Weiss, A.J.: Fundamental limitations in passive time-delay estimation-part I: narrow-band systems. IEEE Trans. Acoust. Speech Signal Process. **31**(2), 472–486 (1983). doi:10.1109/TASSP.1983.1164061
18. Weinstein, E., Weiss, A.J.: Fundamental limitations in passive time-delay estimation-part II: wide-band systems. IEEE Trans. Acoust. Speech Signal Process. **32**(5), 1064–1078 (1984). doi:10.1109/TASSP.1984.1164429
19. Yoshioka, T., Fujiwara, T.: New acoustic emission source locating system for the study of rolling contact fatigue. Wear **81**(1), 183–186 (1982). doi:10.1016/0043-1648(82)90314-3
20. Yoshioka, T., Fujiwara, T.: Application of acoustic emission technique to detection of rolling bearing failure. Am. Soc. Mech. Eng. **14**, 55–76 (1984)

Geothermal Effects for BOD Removal in Horizontal Subsurface Flow Constructed Wetlands: A Numerical Approach

Konstantinos Liolios, Vassilios Tsihrintzis, Krassimir Georgiev and Ivan Georgiev

Abstract A simplified numerical approach is presented for the simultaneous groundwater flow, geothermal energy (heat) transport and contaminant transport and removal in shallow unconfined aquifers. Emphasis is given to Biochemical Oxygen Demand (BOD) removal in Horizontal Subsurface Flow Constructed Wetlands (HSF CW), under non-isothermal conditions. The system of the governing non-linear partial differential equations is treated numerically by using the family computer code Visual MODFLOW. In a numerical example, where BOD is injected in entering geothermal water, the so-resulted computational results are compared with available experimental data.

Keywords Computational environmental engineering · Geothermal problems · Groundwater flow · Constructed wetlands · Heat and mass transport through porous media · Biochemical oxygen demand

1 Introduction

Geothermal energy (heat) and mass transport through porous media are governing concepts concerning the operational modelling of geothermal reservoirs [1–7]. Such

K. Liolios (✉) · K. Georgiev · I. Georgiev
Institute of Information and Communication Technologies,
Bulgarian Academy of Sciences, Sofia, Bulgaria
e-mail: kostisliolios@gmail.com

K. Georgiev
e-mail: georgiev@parallel.bas.bg

V. Tsihrintzis
Department of Infrastructure and Surveying Engineering,
School of Rural and Surveying Engineering,
National Technical University of Athens, Athens, Greece

I. Georgiev
Institute of Mathematics and Informatics,
Bulgarian Academy of Sciences, Sofia, Bulgaria

K. Georgiev et al. (eds.), *Advanced Computing in Industrial Mathematics*,
Studies in Computational Intelligence 681, DOI 10.1007/978-3-319-49544-6_10

geothermal energy storage tools can in some cases be used for feeding with warm geothermal water adjacent aquifers of CW [8, 9] and so to influence their operation. CW are recently a good alternative solution for small settlements in order to treat municipal wastewater. As the use of these systems is currently becoming very popular in many countries, it seems necessary to find their optimal design characteristics in order to maximize their removal efficiency and to decrease their area and construction cost [9–11].

The traditional numerical simulation of CW operation is based on the concepts of groundwater flow and the contaminant transport and removal through porous media, under the assumption of isothermal conditions. Thus, an averaged constant temperature level is commonly assumed during the time of operation required by the hydraulic residence time (*HRT*), which usually ranges from one to four weeks (7–28 days) according to contaminant type, weather and local conditions, etc. [8, 9].

In reality, temperature is a variable quantity during *HRT* and influences significantly, among others, the water viscosity, the hydraulic conductivity coefficients and the decay of contaminant in CW [8, 9, 12]. This holds especially for the case of HSF CW, which are adjacent to systems of aquifers for geothermal energy storage (ATES). Such a case is when geothermal water is feeding greenhouses [4, 5, 13], where the warm entering water into the HSF CW is mixed with injected municipal wastewater for reuse reasons. So, it seems necessary to investigate the temperature variability and to take into account the temperature dependence of the decay coefficients, in order to compute in a more realistic way the contaminant removal and the performance of CW.

The present paper deals numerically with a non-isothermal system, in which both thermal energy (heat) and mass (of a solute) are transported through a porous medium. The case of horizontal groundwater flow in shallow aquifers is considered, taking into account practical aspects from the environmental engineering point of view. Such is the case of rectangular and thin HSF CW, where the vertical depth is much smaller in comparison to other two dimensions (length and width). This geometric peculiarity and the developing small groundwater velocities (usually about 10–50 cm/day) allow on the one hand the assumption of non-variable water density to be accepted and on the one hand the Soret and Dufour effects [1, 14, 15] to be ignored. Indeed, cross-diffusion (Soret and Dufour) effects can be neglected since they are unlikely to be significant, because density gradients in shallow natural geothermal systems are small in general [16].

A simplified numerical simulation of this BOD transport and removal in HSF CW is presented under temperature variability. For the computational realization of the approach, the temperature dependence of the decay coefficients and fluid viscosity are needed. Besides the groundwater flow, the solute concentration fields and the temperature distribution are computed. The family computer code Visual MODFLOW [17] is used, where finite-difference techniques are applied for the spatial and temporal discretization of the governing equations. Finally, in a numerical example, computational results are compared with available experimental data [18].

2 The Mathematical Formulation of the Problem

Simultaneous heat and solute transport in saturated groundwater flow systems is considered and modeled taking into account [1, 2, 19–23]. In [1], a general and mathematically strict formulation for 3-dimensional groundwater flow is given, which requires the solution of a system of nonlinear partial differential equations with totally eight (8) unknown space-time functions. These functions are the hydraulic pressure, the solute concentration, the three velocity components, the temperature, the variable fluid density and the viscosity.

Herein the emphasis is given to the problem of modeling the operation of HSF CW under geothermal effects. Practical considerations from the environmental engineering point of view, as well as some aspects presented and discussed in significant research efforts, e.g. [21, 23], are taken into account in order to simplify the problem formulation and its numerical solution and to reduce the number of the unknown space-time functions

The 3-dimensional groundwater flow equation, under the assumption of non-variable water density and using tensorial notation $(i, j = 1, 2, 3)$, following [19, 20] is written as:

$$\frac{\partial}{\partial x}(K_{ij}(T)\frac{\partial h}{\partial x_j}) + q_s = S_s\frac{\partial h}{\partial t} \tag{1}$$

where $K_{ij}(T)$ is a component of the (temperature dependent) hydraulic conductivity tensor, in [LT^{-1}]; h is the hydraulic head, in [L]; q_s is the volumetric flow rate per unit area of aquifer, representing fluid sources (positive) and sinks (negative) when precipitation and evapotranspiration effects are taken in to account, respectively, in [T^{-1}]; S_s is the specific storage of the porous materials, in [L^{-1}]; and $T = T(x, y, z; t)$ is the temperature, in [°C].

Equation (1), accompanied by suitable boundary and initial conditions, provides as solution the hydraulic head: $h = h(x_i; t)$. It is reminded that when the aquifer is considered as a shallow one, and this is usually the case of HSF CW, the non-linear Boussinesq groundwater flow equation, see e.g. [1], can be used alternatively, instead of Eq. (1):

$$\frac{\partial}{\partial x}(K_x(T)h\frac{\partial h}{\partial x}) + \frac{\partial}{\partial y}(K_y(T)h\frac{\partial h}{\partial y}) + q_u = S_y\frac{\partial h}{\partial t} \tag{2}$$

Here, q_u is the source/sink term, in [LT^{-1}]; and S_y is the specific yield [dimensionless]. As concerns the dependence of the hydraulic conductivity tensor with temperature, this can be presented in the form:

$$K_{ij}(T) = K_{ij}(T_0)\frac{\mu(T_0)}{\mu(T)} \tag{3}$$

where $K_{ij}(T_0)$, in [LT^{-1}], and $\mu(T_0)$, in [ML^{-1}T^{-1}], are the hydraulic conductivity and the viscosity, respectively, in a reference temperature T_0 (usually $T_0 = 20\,°C$);

and $\mu(T)$ is the dynamic viscosity in temperature T, in [$ML^{-1}T^{-1}$]. The function $\mu(T)$ is a non-linear function and is estimated by experiments or given by formulas which are available in most hydrogeology books, e.g. [23].

The Darcy velocity field q_i, in [LT^{-1}], is computed through the Darcy relationship:

$$q_i = -K_{ij}(T)\frac{\partial h}{\partial x_j} \tag{4}$$

Further, the partial differential equation which describes the development and transport of a contaminant with adsorption in 3-dimensional, transient groundwater flow systems, can be written:

$$\varepsilon R_d \frac{\partial C}{\partial t} = \frac{\partial}{\partial x_i}(\varepsilon D_{Sij}\frac{\partial C}{\partial x_j}) - \frac{\partial}{\partial x_i}(q_i C) + q_s C_s + \Sigma \mathrm{R_n} \tag{5}$$

where ε is the porosity (assumed constant) of the subsurface medium [dimensionless]; R_d is the delay factor [dimensionless]; C is the dissolved concentration of solute, in [ML^{-3}]; D_{Sij} is the solute hydrodynamic dispersion coefficient tensor, in [L^2T^{-1}]; C_s is the concentration of the source or sink flux, in [ML^{-3}]; and ΣR_n is the chemical reaction term, in [$ML^{-3}T^{-1}$]. The seepage or linear pore water velocity v_i, in [LT^{-1}], is related to the specific discharge or Darcy flux through the relationship: $q_i = v_i\varepsilon$. As concerns the solute dispersion coefficient tensor D_{Sij}, it is dependent on the molecular diffusion and on the mechanical dispersion [1, 19] based on the Darcy velocity through the corresponding longitudinal and horizontal dispersivities a_L and a_T, respectively, both in [L].

The last term in Eq. (5), for the usual linear reaction case, is given by the formula:

$$\Sigma \mathrm{R_n} = -\lambda \varepsilon R_d C \tag{6}$$

where λ is the first-order removal coefficient, in [T^{-1}]. For usual solute removal in HSF CW, it is considered that λ depends mainly on the temperature T. Indeed, based on experimental results, temperature effects have often been summarized [8–10] by the use of the modified Arrhenius equation [24]:

$$\lambda_\mathrm{T} = \lambda(\mathrm{T}) = \lambda_{20} \cdot \theta_T^{(\mathrm{T}-20)} \tag{7}$$

Here, λ_T is the rate constant at temperature T, in [T^{-1}]; λ_{20} is the rate constant at temperature $T = 20\,°\mathrm{C}$, in [T^{-1}]; and θ is the temperature factor [dimensionless]. Based on experimental results, for BOD is usually taken $\theta = 1.06$ [8] and $\lambda_{20} = 0.22\,\mathrm{d}^{-1}$ [24]. Formulas similar to (7) can be given for other solutes.

The heat-transport equation, derived from the conservation principle of enthalpy for the aquifer medium (fluid and solid porous medium), is written [23, 25–27]:

$$\rho_m c_m \frac{\partial T}{\partial t} = \frac{\partial}{\partial x_i}(D_{Tij}\frac{\partial T}{\partial x_j}) - \rho_w c_w \frac{\partial}{\partial x_i}(q_i T) + \Sigma \mathrm{R_T} \tag{8}$$

D_{Tij} is the thermal hydrodynamic dispersion coefficient tensor, in [MLT^{-3}K^{-1}]; ρ_w is the fluid (water) density, in [ML^{-3}]; c_w is the heat capacity of water, in [L^{2}T^{-2}K^{-1}]; and the term $\rho_w c_w$ is the volumetric heat capacity of the water, in [ML^{-1}T^{-2}K^{-1}]. The last term ΣR_T concerns temperature sources/sinks, in [ML^{-1}T^{-3}]. The term $\rho_m c_m$ is the volumetric heat capacity of the aquifer medium (fluid and solid porous medium), in [ML^{-1}T^{-2}K^{-1}], and is given by the relation [23]:

$$\rho_m c_m = \varepsilon \rho_w c_w + (1 - \varepsilon)\rho_s c_s \tag{9}$$

where ρ_s is the dry solid density, in [ML^{-3}]; and c_s is the heat capacity of the solid, in [MLT^{-3}K^{-1}]. As concerns D_{Tij}, it is dependent on the isotropic conductivity coefficient k_m of the porous medium (water plus solid), in [MLT^{-3}K^{-1}], and on the Darcy velocity through the corresponding thermal dispersitivities b_L and b_T, both in [L]. The coefficient k_m is given by the relation:

$$k_m = \varepsilon k_w + (1 - \varepsilon)k_s \tag{10}$$

where k_w and k_s are the conductivity coefficients of the water and the solid, respectively, both in [MLT^{-3}K^{-1}].

The non-linear system of the above partial differential equations (1)–(10), combined with appropriate initial and boundary conditions, describe the 3-dimensional flow of groundwater, the heat transport and the transport and removal of contaminants in a porous medium. The functions $\lambda(T)$ and $\mu(T)$ are considered as known from experimental results. Thus, the unknowns of the problem are the following six space-time functions: The hydraulic head: $h = h(x, y, z; t)$; the three Darcy velocity components: $q_i = q_i(x, y, z; t)$; the temperature: $T = T(x, y, z; t)$; and the concentration: $C = C(x, y, z; t)$. Although the practical considerations from the environmental point of view have simplified the problem, it remains difficult to be solved analytically, and for this reason numerical methods are required.

3 The Numerical Treatment of the Problem

For the numerical solution of the problem described in the previous paragraph, the Finite Difference Method (FDM) is chosen, among the well-known [19, 20] available numerical methods (Finite Element Method—FEM, Finite Volume Method—FVM, etc.). The reason is that the HSF CW have a rectangular scheme, which allows the use of orthogonal grid systems for the space discretization. Moreover, FDM is the basis for the computer code family MODFLOW [17], which is widely used for the simulation of groundwater flow and mass transport, see e.g. [10, 11]. Thus, in the present study, the MODFLOW code, accompanied by the effective computer package MT3DMS, is used. MODFLOW code family has been already applied successfully by considering heat as a tracer, according to [25]. As concerns the Finite Volume Method (FVM), mathematical aspects for its application in flows through porous media have been presented in [28, 29].

Other effective codes, e.g. HYDRUS [30, 31], HST3D [32] and SUTRA [33], have been also used successfully for heat transfer and variable-density groundwater problems, see [19]. The SEAWAT code [34], which is a coupled version of MODFLOW and MT3DMS, has been also applied. This code uses the following equation to represent dynamic viscosity as a function of temperature $\mu(T)$, in relevance to Eq. (3):

$$\mu(T) = A_1 \cdot A_2^{\left(\frac{A_3}{T+A_4}\right)} \quad (11)$$

The values of the constants are [34]: $A_1 = 2.394 \times 10^{-5}$, $A_2 = 10$, $A_3 = 248.37$ and $A_4 = 133.15$. Equation (11) in SEAWAT code has been adopted from the SUTRA code [33], where, for solute transport, viscosity is taken to be constant. For example, at temperature $T = 20\,°\mathrm{C}$, $\mu(T)$ is equal with $\mu(20) = 1 \times 10^{-3}\,\mathrm{kg/m/s}$.

An evaluation of MODFLOW/MT3DMS for heat transport simulation of closed geothermal systems has been presented in [35]. Concerning a temperature range between 0 and 40 °C of the geothermal water, the reported temperature influences on several physical parameters (such as density and viscosity of water), on the thermal conductivity and the heat capacity of the porous medium, are shown in Fig. 1. The behavior of hydraulic conductivity in regard to temperature shown in Fig. 1a, displays an approximately linear trend in a range from 0 to 40 °C. Similar approximately linear trends, ascending or descending, show the other quantities except water density, which shows a small variation of 0.7 % for the temperature variation.

Due to the above approximately linear trends of quantities, which are dependent on temperature, the non-linear characteristics involved in Eqs. (1)–(4) can be overcome by a heuristic approach applied in the time-discretization procedure. So, denoting the time-step by Δt, in each current time moment t the old values of $K_{ij}(T)$, as well as of q_i, known from the previous time-moment $(t - \Delta t)$ are used in Eqs. (1), (5) and (8).

As concerns the functions $\lambda(T)$ and their dependence on temperature for HSF CW, these have been determined and reported by Liolios et al. in [10], for a sufficient number of pilot HSF CW and for the most common wastewater contaminants.

4 Numerical Example and Representative Results

The herein proposed numerical approach is applied and calibrated by using available experimental results of five pilot-scale HSF CW. These units were operated for two years in the facilities of the Laboratory of Ecological Engineering and Technology, in Democritus University of Thrace, Xanthi, Greece. The experiment purpose was to investigate the effect of temperature, *HRT*, vegetation type and porous media material and grain size on the performance of HSF CW treating wastewater containing usual solutes.

The above pilot-scale units are rectangular tanks made of steel, with dimensions $L = 3$ m long, 0.75 m wide and 1 m deep, whereas the depth of the porous material is $d = 0.45$ m. As the ratio d/L is small (0.15), these CW can be considered as shallow

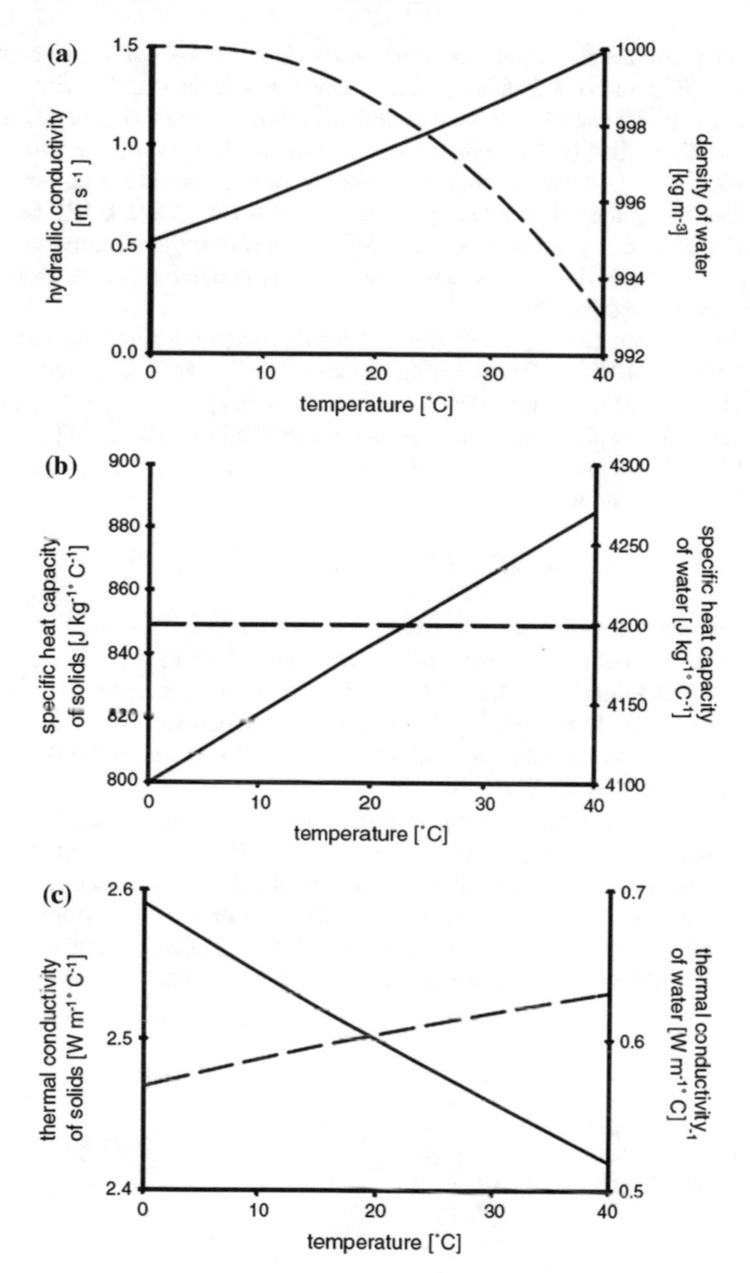

Fig. 1 Thermal dependencies of: **a** hydraulic conductivity and density of the water; **b** thermal conductivity of solids and water; and **c** specific heat capacity of solids and water. *Left axes* are associated with the *solid lines*, while *right axes* are associated with the *dashed lines*. By Hecht-Méndez et al. [35]

aquifers. Three of the above five units contained medium gravel (MG), one contained fine gravel (FG) and one cobbles (CO). Concerning the three units with medium gravel, one (MG-R) was planted with common reeds (*Phragmites australis*) and one (MG-C) with cattails (*Typha latifolia*), while one (MG-Z) was kept unplanted. The other two units (FG-R and CO-R) were planted with common reeds. For further details concerning the experimental procedure, results etc., see [10, 18]. Hereafter, the simulation of the operation of the unit MG-R and some representative computed results concerning BOD transport and removal under geothermal water feeding are next presented and discussed.

The MG-R tank feeding is geothermal water containing input concentration: $C_{in} = 360\,\text{mg BOD/Liter}$. The entering volume rate Q_{in}, in [L/day], corresponds to a hydraulic residence time: $HRT = 14\,\text{days}$ (2 weeks). The $\lambda(T)$ functions for BOD removal in the above tanks have been presented in [10]. The $\lambda(T)$ function for the MG-R tank is shown in Fig. 2 and concerns the values $\lambda(T)$ computed by the Arrhenius-type equation:

$$\lambda_{MG-R}(T) = 0.2115 \times (1.043)^{(T-20)}, \tag{12}$$

Further, the input-data for the MODFLOW code are: $S_s = 10^{-5}\,\text{m}^{-1}$; $S_y = 0.37$; $\varepsilon = 0.37$; Diffusion coefficient: $D = 3.6 \times 10^{-5}\,\text{m}^2/\text{s}$; $a_L = 0.027\,\text{m}$; $a_L/a_T = 0.015$; Initial hydraulic head: $h = 0.45\,\text{m}$; Geothermal water density: $\rho_w = 1000\,\text{kgr/m}^3$; $c_w = 4200\,\text{Joule/kg/}^\circ\text{C}$; Dry solid (MG) density: $\rho_s = 2650\,\text{kgr/m}^3$; Mean (at 20 °C): $c_s = 840\,\text{Joule/kg/}^\circ\text{C}$; Mean (at 20 °C): $k_w = 0.60\,\text{W/m/}^\circ\text{C}$; Mean (at 20 °C): $k_s = 2.5\,\text{W/m/}^\circ\text{C}$.

Some representative results are presented in next Table 1 and Fig. 3: For various temperatures of the entering geothermal water ($T_{in} = 10, 20, 30, 40$ and $50\,^\circ\text{C}$), the corresponding computed averaged concentrations at characteristic sections ($C_{1/3}$ for distance $L/3 = 1\,\text{m}$ from the inlet, $C_{2/3}$ for distance $2L/3 = 2\,\text{m}$ from the inlet, and C_{out} for the outlet) are presented in Table 1. These results are after three weeks (21 days) of simulation, when a steady-state has been reached, whereas the temper-

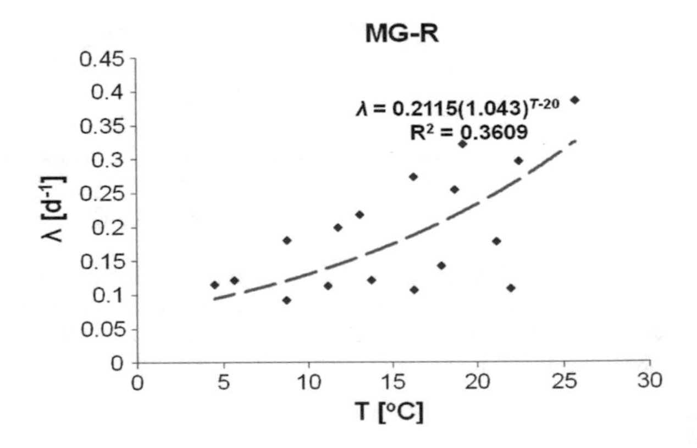

Fig. 2 Function $\lambda(T)$ concerning the dependence of the decay coefficient on temperature for the MG-R tank. By Liolios et al. [10]

Table 1 Computed concentrations [mg BOD/L] along the pilot unit MG-R, after 21 days of simulation and for output environment temperature: $T_{out} = 5\,°C$

Geothermal water inlet temperature (°C)	$C_{1/3}$ (mg/L)	$C_{2/3}$ (mg/L)	C_{out} (mg/L)
10	197.30	116.14	82.57
20	175.87	94.72	61.14
30	139.06	65.81	38.24
40	115.28	38.89	20.66
50	93.26	24.46	14.75

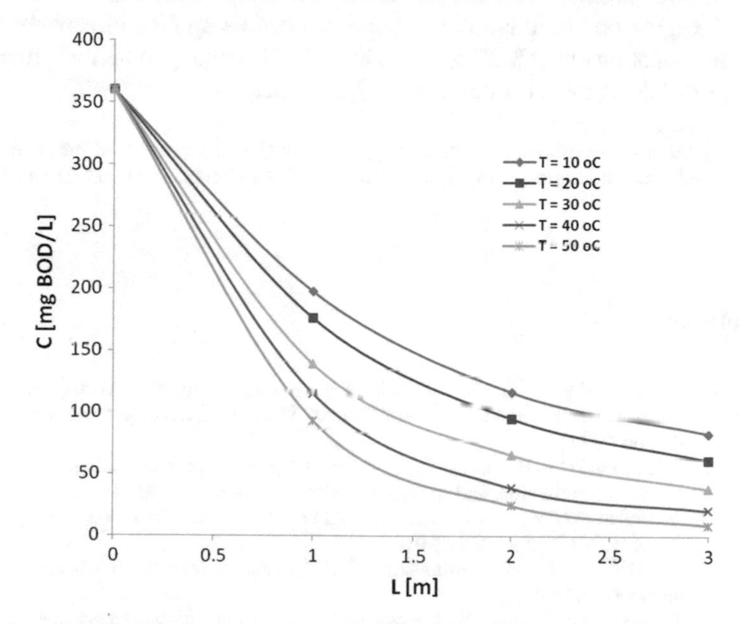

Fig. 3 Concentrations profiles along the tank MG-R after 21 days, for geothermal water inlet temperature: $T_{in} = 10, 20, 30, 40$ and $50\,°C$ and for output environment temperature: $T_{out} = 5\,°C$

ature in the output environment remains: $T_{out} = 5\,°C$. In Fig. 3, the corresponding concentrations profiles along the tank MG-R are presented.

When the concentration values of the above Table 1 and Fig. 3 are averaged, then they are in a good agreement with the experimental measured values and concerning average values for all the experiments temperatures [18]. This proves the effectiveness of the proposed numerical approach.

5 Concluding Remarks

A simplified numerical approach has been presented for the simultaneous groundwater flow, geothermal energy (heat) transport and contaminant transport and removal in shallow unconfined aquifers. This approach proposes an efficient numerical treatment for the system of non-linear partial differential equations, governing the simulation of contaminant transport and removal in HSF CW under temperature variability. In a numerical example from environmental engineering, a good agreement between computed and experimental results concerning contaminant concentrations has been proved. Thus, the proposed numerical approach can be applied effectively for the numerical simulation of the BOD removal in HSF CW under geothermal effects, and for the optimal design and operation of such facilities.

Acknowledgements The research is partly supported by the FP7 project AComIn: Advanced Computing for Innovation, grant 316087 and Bulgarian NSF Grants, DMU 03-62 and DFNI I-01/5.

References

1. Bear, J.: Dynamics of Fluids in Porous Media. Dover Publications, New York (1988)
2. Bear, J., Bachmat, Y.: Introduction to Modeling of Transport Phenomena in Porous Media. Klouwer, Boston (1990)
3. Molson, J.W., Frind, E.O., Palmer, C.D.: Thermal energy storage in an unconfined aquifer: 2. Model development, validation, and application. Water Resour. Res. **28**(10), 2857–2867 (1992)
4. Lee, K.S.: A review on concepts, applications, and models of aquifer thermal energy storage systems. Energies **3**(6), 1320–1334 (2010)
5. Lund, J.W., Boyd, T.L.: Direct utilization of geothermal energy 2015 worldwide review. Geothermics **60**, 66–93 (2016)
6. Burow, K.R., Constantz, J., Fujii, R.: Heat as a tracer to estimate dissolved organic carbon flux from a restored wetland. Groundwater **43**(4), 545–556 (2005)
7. Vafai, K.: Handbook of Porous Media, 2nd edn. CRC Press, Boca Raton (2005)
8. Kadlec, R.H., Wallace, S.: Treatment Wetlands, 2nd edn. CRC Press, Boca Raton (2009)
9. Vymazal, J., Kröpfelová, L.: Wastewater Treatment in Constructed Wetlands with Horizontal Sub-Surface Flow. Springer, Berlin (2008)
10. Liolios, K.A., Moutsopoulos, K.N., Tsihrintzis, V.A.: Modeling of flow and BOD fate in horizontal subsurface flow constructed wetlands. Chem. Env. J. **200–202**, 681–693 (2012)
11. Liolios, K., Tsihrintzis, V., Moutsopoulos, K., Georgiev, I., Georgiev, K.: A computational approach for remediation procedures in horizontal subsurface flow constructed wetlands. In: Lirkov, I., Margenov, S., Wasniewski, J. (eds.) LNCS, vol. 7116, pp. 299–306. Springer, Berlin (2012)
12. Ma, R., Zheng, C.: Effects of density and viscosity in modeling heat as a groundwater tracer. Groundwater **48**(3), 380–389 (2010)
13. Twidell, J., Weir, T.: Renewable Energy Resources, 3rd edn. Routledge, New York (2015)
14. Postelnicu, A.: Influence of chemical reaction on heat and mass transfer by natural convection from vertical surfaces in porous media considering Soret and Dufour effects. Heat Mass Trans. **43**(6), 595–602 (2007)
15. Moorthy, M.B.K., Kannan, T., Senthilvadivu, K.: Soret and Dufour effects on natural convection heat and mass transfer flow past a horizontal surface in a porous medium with variable viscosity. WSEAs Trans. Heat Mass Trans. **3**(8), 121–129 (2013)

16. McKibbin, R.: Modeling heat and mass transport processes in geothermal systems. In: Vafai, K. (ed.) Handbook of Porous Media. Taylor and Francis, New York (2005)
17. Waterloo Hydrogeologic Inc.: Visual MODFLOW v. 4.2. Users Manual. U.S. Geological Survey, Virginia, USA (2006)
18. Akratos, C.S., Tsihrintzis, V.A.: Effect of temperature, HRT, vegetation and porous media on removal efficiency of pilot-scale horizontal subsurface flow constructed wetlands. Ecol. Eng. J. **29**(2), 173–191 (2007)
19. Zheng, C., Bennett, G.D.: Applied Contaminant Transport Modelling, 2nd edn. Wiley, New York (2002)
20. Bear, J., Cheng, A.D.: Modeling Groundwater Flow and Contaminant Transport. Springer, Berlin (2010)
21. Dagan, G.: Some aspects of heat and mass transfer in porous media. Fund. Trans. Phen. Por. Med. 55–64 (1972)
22. Nield, D.A., Bejan, A.: Convection in Porous Media, 3rd edn. Springer, New York (2006)
23. De Marsily, G.: Quantitative Hydrogeology. Academic Press, London (1986)
24. Tanner, C.C., Clayton, J.S., Upsdell, M.P.: Effect of loading rate and planting on treatment of daily farm wastewaters in constructed wetlands-I. Removal of oxygen demand, suspended solids and faecal coliforms. Water Res. **29**, 17–26 (1995)
25. Anderson, M.P., Woessner, W.W.: Applied Groundwater Modeling: Simulation of Flow and Advective Transport. Academic Press, London (2002)
26. Anderson, M.P.: Heat as a ground water tracer. Groundwater **43**(6), 951–968 (2005)
27. Domenico, P.A., Schwartz, F.W.: Physical and Chemical Hydrogeology, 2nd edn. Wiley, New York (1998)
28. Lazarov, R.D., Mishev, I.D., Vassilevski, P.S.: Finite volume methods for convection-diffusion problems. SIAM J. Numer. Anal. **33**(1), 31–55 (1996)
29. Ewing, R., Lazarov, R., Lin, Y.: Finite volume element approximations of nonlocal reactive flows in porous media. Num. Meth. Part. Dif. Eqs. **16**(3), 285–311 (2000)
30. Simunek, J., Jacques, D., Langergraber, G., Bradford, S.A., Šejna, M., van Genuchten, M.T.: Numerical modeling of contaminant transport using HYDRUS and its specialized modules. J. Indian Inst. Sci. **93**(2), 265–284 (2013)
31. Langergraber, G., Giraldi, D., Mena, J., Meyer, D., Peña, M., Toscano, A., Brovelli, A., Korkusuz, E.A.: Recent developments in numerical modelling of subsurface flow constructed wetlands. Sci. Tot. Environ. **407**(13), 3931–3943 (2009)
32. Kipp, K.L.: HST3D; A Computer Code for Simulation of Heat and Solute Transport in Three-Dimensional Ground-Water Glow Systems (No. 86-4095). U.S. Geological Survey, Denver, Colorado (1987)
33. Voss, C.I.: A Finite-Element Simulation Model for Saturated-Unsaturated, Fluid-Density-Dependent Ground-Water Flow with Energy Transport or Chemically-Reactive Single-Species Solute Transport (No. 84-4369). U.S. Geological Survey, Denver, Colorado (1984)
34. Langevin, C.D., Thorne Jr., D.T., Dausman, A.M., Sukop, M.C., Guo, W.: SEAWAT v. 4: A Computer Program for Simulation of Multi-Species Solute and Heat Transport (No. 6-A22). U.S. Geological Survey, Florida Integrated Science Center (2008)
35. Hecht-Méndez, J., Molina-Giraldo, N., Blum, P., Bayer, P.: Evaluating MT3DMS for heat transport simulation of closed geothermal systems. Groundwater **48**(5), 741–756 (2010)

Further Results of Mean-Value Type in $\mathbb{C}$ and $\mathbb{R}$

Lubomir Markov

Abstract We prove an extension of Pompeiu's Mean Value Theorem to holomorphic functions in the spirit of the Evard-Jafari Theorem, a (new?) mean value theorem in $\mathbb{R}$, and an extension of the latter in $\mathbb{C}$.

Keywords Evard-Jafari Theorem · Complex Mean Value Theorem · Flett's Theorem · Pompeiu's Mean Value Theorem · Davitt-Powers-Riedel-Sahoo Theorem

2010 Mathematics Subject Classification: 30C15 · 26A24 · 30C99

The purpose of the present paper is to further the investigation initiated in our recent work [4] and to establish several theorems of mean-value type that are believed to be new. As in [4], for two complex numbers $A = a_1 + ia_2$ and $B = b_1 + ib_2$ we denote by (A, B) the open segment connecting A and B, and by $[A, B]$ the closed such segment. The line through A and B shall be denoted by $\overleftrightarrow{A, B}$. Throughout, we consider real or complex functions which are differentiable or analytic, respectively, on some domain. Recall the following results:

Proposition 1 (Two Complex Mean Value Theorems, see [2, 4]) *Let $f(z)$ be holomorphic on the open convex set $D_f \subseteq \mathbb{C}$ and let $A, B \in D_f$. Then $\exists\, z_1, z_2 \in (A, B)$, such that*

This paper was presented at the international conference "BGSIAM'15" (10th Annual Meeting of the Bulgarian Section of SIAM), 21–22 December 2015, Sofia, Bulgaria.

L. Markov (✉)
Department of Mathematics and CS, Barry University,
11300 NE Second Avenue, Miami Shores, FL 33161, USA
e-mail: lmarkov@barry.edu

K. Georgiev et al. (eds.), *Advanced Computing in Industrial Mathematics*,
Studies in Computational Intelligence 681, DOI 10.1007/978-3-319-49544-6_11

$$\Re\{f'(z_1)\} = \Re\left\{\frac{f(B) - f(A)}{B - A}\right\},$$
$$\Im\{f'(z_2)\} = \Im\left\{\frac{f(B) - f(A)}{B - A}\right\},$$

and $\exists\, w_1, w_2 \in (A, B)$, *such that*

$$\Re\{(B - A)f'(w_1)\} = \Re\{f(B) - f(A)\},$$
$$\Im\{(B - A)f'(w_2)\} = \Im\{f(B) - f(A)\}.$$

The case $f(B) = f(A)$ reduces the first set of equations in Proposition 1 to the Evard-Jafari Theorem [2], which is an extension of Rolle's Theorem to the complex domain.

Proposition 2 (Generalized Flett's Mean Value Theorem, see [1]) *Suppose that I is an open interval, $f : I \subseteq \mathbb{R} \to \mathbb{R}$ is differentiable, and $[a, b] \subset I$. Then $\exists\, \xi \in (a, b)$ such that*

$$f'(\xi) = \frac{f(\xi) - f(a)}{\xi - a} + \frac{1}{2}\frac{f'(b) - f'(a)}{b - a}(\xi - a).$$

When $f'(b) = f'(a)$, Proposition 2 reduces to the now classical theorem established by Flett [3].

Proposition 3 (Theorem 2 in [4]) *Let $f(z) = u(z) + iv(z)$ be holomorphic on the open convex set $D_f \subseteq \mathbb{C}$ and let $A = a_1 + ia_2 \in D_f,\ B = b_1 + ib_2 \in D_f$. Then $\exists\, z_1, z_2 \in (A, B)$ such that*

$$\Re\{f'(z_1)\} = \Re\left\{\frac{f(z_1) - f(A)}{z_1 - A} + \frac{1}{2}\frac{f'(B) - f'(A)}{B - A}(z_1 - A)\right\},$$
$$\Im\{f'(z_2)\} = \Im\left\{\frac{f(z_2) - f(A)}{z_2 - A} + \frac{1}{2}\frac{f'(B) - f'(A)}{B - A}(z_2 - A)\right\}.$$

Proposition 3 is an improved version (see [4]) of the

Davitt-Powers-Riedel-Sahoo Mean Value Theorem [1] For $\alpha, \beta \in \mathbb{C}$, define $\langle \alpha, \beta \rangle = \Re(\alpha\bar{\beta})$. Let $D_f \subset \mathbb{C}$ be open and convex, $f : D_f \to \mathbb{C}$ be holomorphic, $A, B \in D_f$. Then $\exists\, z_1, z_2 \in (A, B)$ such that

$$\Re[f'(z_1)] = \frac{\langle B - A, f(z_1) - f(A)\rangle}{\langle B - A, z_1 - A\rangle} + \frac{1}{2}\frac{\Re[f'(B) - f'(A)]}{B - A}(z_1 - A),$$
$$\Im[f'(z_2)] = \frac{\langle B - A, -i[f(z_2) - f(A)]\rangle}{\langle B - A, z_2 - A\rangle} + \frac{1}{2}\frac{\Im[f'(B) - f'(A)]}{B - A}(z_2 - A).$$

Proposition 4 (Pompeiu's Mean Value Theorem, see [5, 6]) *Suppose $I \not\ni 0$ is an open interval, $f : I \subset \mathbb{R} \to \mathbb{R}$ is differentiable, and $[a, b] \subset I$. Then $\exists \xi \in (a, b)$ such that*

$$f(\xi) - \xi f'(\xi) = \frac{bf(a) - af(b)}{b - a}. \tag{1}$$

Propositions 2 and 4 have simple geometric interpretations. For example, Pompeiu's Theorem says that the tangent at $(\xi, f(\xi))$ has the same y-intercept as the secant through $(a, f(a))$ and $(b, f(b))$. Surprisingly, Pompeiu omits the condition $0 \notin [a, b]$ in the original paper [5]. It isn't clear whether he considered it an obvious restriction, as the theorem is plainly false without it. A correct version can be found in [6].

Our first result is, to the best of our knowledge, a new theorem of the differential calculus:

Theorem 1 *Suppose $I \not\ni 0$ is an open interval, $f : I \subset \mathbb{R} \to \mathbb{R}$ is differentiable, and $[a, b] \subset I$. Then $\exists \xi \in (a, b)$ such that*

$$f(\xi) - \xi f'(\xi) = \frac{bf(\xi) - \xi f(b)}{b - \xi} + \frac{a}{2} \cdot \frac{f(a) - af'(a) - f(b) + bf'(b)}{b - a} \cdot \frac{b - \xi}{\xi},$$

or equivalently

$$f'(\xi) = \frac{f(b) - f(\xi)}{b - \xi} + \frac{a}{2} \cdot \frac{f(b) - bf'(b) - f(a) + af'(a)}{b - a} \cdot \frac{b - \xi}{\xi^2}.$$

One geometric interpretation of this statement is as follows: for a smooth function whose domain does not contain 0, if the tangents at $(a, f(a))$ and $(b, f(b))$ have the same y-intercept, there is an interior point ξ in (a, b) such that the tangent at $(\xi, f(\xi))$ passes through $(b, f(b))$.

Proof of Theorem 1 (**cf. [6, p. 84]**)

Define $\Phi(t) = tf(\frac{1}{t})$ on the interval $J = \{1/t : t \in I\} \supset \left[\frac{1}{b}, \frac{1}{a}\right] \overset{\text{def}}{=} [\beta, \alpha]$. Clearly $\Phi(t)$ is differentiable, and so Proposition 2 implies $\exists \eta \in (\beta, \alpha)$ such that

$$\Phi'(\eta) = \frac{\Phi(\eta) - \Phi(\beta)}{\eta - \beta} + \frac{1}{2}\frac{\Phi'(\alpha) - \Phi'(\beta)}{\alpha - \beta}(\eta - \beta),$$

or

$$f\left(\frac{1}{\eta}\right) - \frac{1}{\eta}f'\left(\frac{1}{\eta}\right) = \frac{\eta f\left(\frac{1}{\eta}\right) - \beta f\left(\frac{1}{\beta}\right)}{\eta - \beta} + \frac{1}{2}\frac{f\left(\frac{1}{\alpha}\right) - \frac{1}{\alpha}f'\left(\frac{1}{\alpha}\right) - f\left(\frac{1}{\beta}\right) + \frac{1}{\beta}f'\left(\frac{1}{\beta}\right)}{\alpha - \beta}(\eta - \beta).$$

Put $\dfrac{1}{\eta} = \xi$; $\eta \in (\beta, \alpha) \Leftrightarrow \xi \in (a, b)$. We obtain

$$f(\xi)-\xi f'(\xi) = \frac{\frac{1}{\xi}f(\xi)-\frac{1}{b}f(b)}{\frac{1}{\xi}-\frac{1}{b}} + \frac{1}{2}\,\frac{f(a)-af'(a)-f(b)+bf'(b)}{\frac{1}{a}-\frac{1}{b}}\left(\frac{1}{\xi}-\frac{1}{b}\right),$$

and after some simple algebraic manipulations both forms of the result follow. □

Neither Pompeiu's Theorem nor Theorem 1 hold true for complex functions of a complex variable. In the case of Pompeiu, take $f(z)=ze^{1/z}-z$ on the segment $\left[\frac{i}{4\pi},\frac{i}{2\pi}\right]$. The right-hand side of (1) is

$$\frac{\frac{i}{2\pi}\left(\frac{i}{4\pi}(e^{-4\pi i}-1)\right)-\frac{i}{4\pi}\left(\frac{i}{2\pi}(e^{-2\pi i}-1)\right)}{\frac{i}{2\pi}-\frac{i}{4\pi}}=0,$$

whereas

$$f(z)-zf'(z)=ze^{1/z}-z-z\Big(e^{1/z}-\frac{1}{z}e^{1/z}-1\Big)=e^{1/z}\neq 0,\ \forall z\in\mathbb{C}-\{0\}.$$

A counterexample for Theorem 1 is provided by $f(z)=ze^{1/z}-1$ on the segment $\left[\frac{i}{4\pi},\frac{i}{2\pi}\right]$. We have $f(z)-zf'(z)=ze^{1/z}-1-z\Big(e^{1/z}-\frac{1}{z}e^{1/z}\Big)=e^{1/z}-1$, and consequently $f(z)-zf'(z)=0$ for $z=\frac{i}{2k\pi}$, $\forall k\in\mathbb{Z}-\{0\}$. Thus it must be shown that the equation

$$f(z)-zf'(z)=\frac{\frac{i}{2\pi}f(z)-zf\left(\frac{i}{2\pi}\right)}{\frac{i}{2\pi}-z}$$

has no solution in $\left(\frac{i}{4\pi},\frac{i}{2\pi}\right)$. The last equation becomes

$$\Big(e^{1/z}-1\Big)\Big(\frac{i}{2\pi}-z\Big)=\frac{i}{2\pi}\Big(ze^{1/z}-1\Big)-z\Big(\frac{i}{2\pi}-1\Big),$$

or equivalently

$$-e^{1/z}-2\pi ize^{1/z}+ze^{1/z}=z. \tag{2}$$

On the imaginary axis we have $z=iy$ and (2) reduces to

$$\big[\cos(1/y)-i\sin(1/y)\big]\big(2\pi y-1+iy\big)=iy.$$

Expand this relation, divide by y, set $\frac{1}{y}=\theta$ and obtain

$$2\pi\cos\theta-2\pi i\sin\theta-\theta\cos\theta+i\theta\sin\theta+i\cos\theta+\sin\theta=i.$$

Comparing the real and imaginary parts and manipulating, we get the two equations

$$(\theta - 2\pi)\cos\theta - \sin\theta = 0, \quad \big[(\theta - 2\pi)\cos(\theta/2) - \sin(\theta/2)\big]\sin(\theta/2) = 0.$$

The only solution to this system is $\theta = 2\pi \notin (2\pi, 4\pi)$. Thus, Theorem 1 fails in the complex domain.

The two theorems allow, however, a version for complex functions in the spirit of Evard-Jafari [2], i.e., in terms of real and imaginary parts. We have:

Theorem 2 *Suppose f is holomorphic on an open convex set $D_f \subset \mathbb{C}$, $D_f \not\ni 0$. Let $[A, B] \subset D_f$ and $\overleftrightarrow{A, B} \ni 0$. Then $\exists\, z_1, z_2 \in (A, B)$ such that*

$$\Re\{f(z_1) - z_1 f'(z_1)\} = \Re\left\{\frac{Bf(A) - Af(B)}{B - A}\right\},$$

$$\Im\{f(z_2) - z_2 f'(z_2)\} = \Im\left\{\frac{Bf(A) - Af(B)}{B - A}\right\}.$$

Theorem 3 *Suppose f is holomorphic on an open convex set $D_f \subset \mathbb{C}$, $D_f \not\ni 0$. Let $[A, B] \subset D_f$ and $\overleftrightarrow{A, B} \ni 0$. Then $\exists\, z_1, z_2 \in (A, B)$ such that*

$$\Re\{f(z_1) - z_1 f'(z_1)\} = \Re\left\{\frac{Bf(z_1) - z_1 f(B)}{B - z_1} + \frac{A}{2} \cdot \frac{f(A) - Af'(A) - f(B) + Bf'(B)}{B - A} \cdot \frac{B - z_1}{z_1}\right\},$$

$$\Im\{f(z_2) - z_2 f'(z_2)\} = \Im\left\{\frac{Bf(z_2) - z_2 f(B)}{B - z_2} + \frac{A}{2} \cdot \frac{f(A) - Af'(A) - f(B) + Bf'(B)}{B - A} \cdot \frac{B - z_2}{z_2}\right\}.$$

Proof of Theorems 2 and 3

Consider the domain $D^* = \left\{\frac{1}{z} : z \in D_f\right\}$ and define $f^*(s) = sf(\frac{1}{s})$, $s \in D^*$. Clearly $D_f \supset [A, B] \longrightarrow \left[\frac{1}{B}, \frac{1}{A}\right] \subset D^*$, $\left[\frac{1}{B}, \frac{1}{A}\right] \not\ni 0$, $\overleftrightarrow{\frac{1}{B}, \frac{1}{A}} \ni 0$, and $f^*(s)$ is holomorphic on D^*.

(i) To prove Theorem 2, apply the (first) complex Mean Value Theorem: $\exists\, s_1, s_2 \in \left(\frac{1}{B}, \frac{1}{A}\right)$ such that

$$\Re\{(f^*)'(s_1)\} = \Re\left\{\frac{f^*(\frac{1}{A}) - f^*(\frac{1}{B})}{\frac{1}{A} - \frac{1}{B}}\right\},$$

$$\Im\{(f^*)'(s_2)\} = \Im\left\{\frac{f^*(\frac{1}{A}) - f^*(\frac{1}{B})}{\frac{1}{A} - \frac{1}{B}}\right\},$$

i.e.,

$$\Re\left\{f\left(\frac{1}{s_1}\right)-\frac{1}{s_1}f'\left(\frac{1}{s_1}\right)\right\}=\Re\left\{\frac{\frac{1}{A}f(A)-\frac{1}{B}f(B)}{\frac{1}{A}-\frac{1}{B}}\right\},$$
$$\Im\left\{f\left(\frac{1}{s_2}\right)-\frac{1}{s_2}f'\left(\frac{1}{s_2}\right)\right\}=\Im\left\{\frac{\frac{1}{A}f(A)-\frac{1}{B}f(B)}{\frac{1}{A}-\frac{1}{B}}\right\}.$$

Put $z_1=\frac{1}{s_1}$, $z_2=\frac{1}{s_2}$; we have $s_j\in\left(\frac{1}{B},\frac{1}{A}\right)\Leftrightarrow z_j\in(A,B)$, $j=1,2$. Simplifying the quantities on the right-hand sides establishes the result.

(ii) To prove Theorem 3, apply Proposition 3: $\exists\, s_1,s_2\in\left(\frac{1}{B},\frac{1}{A}\right)$ such that

$$\Re\left\{\left(f^*\right)'(s_1)\right\}=\Re\left\{\frac{f^*(s_1)-f^*(\frac{1}{B})}{s_1-\frac{1}{B}}+\frac{1}{2}\frac{\left(f^*\right)'(\frac{1}{A})-\left(f^*\right)'(\frac{1}{B})}{\frac{1}{A}-\frac{1}{B}}\left(s_1-\frac{1}{B}\right)\right\},$$
$$\Im\left\{\left(f^*\right)'(s_2)\right\}=\Im\left\{\frac{f^*(s_2)-f^*(\frac{1}{B})}{s_2-\frac{1}{B}}+\frac{1}{2}\frac{\left(f^*\right)'(\frac{1}{A})-\left(f^*\right)'(\frac{1}{B})}{\frac{1}{A}-\frac{1}{B}}\left(s_2-\frac{1}{B}\right)\right\},$$

or

$$\Re\left\{f\left(\frac{1}{s_1}\right)-\frac{1}{s_1}f'\left(\frac{1}{s_1}\right)\right\}=\Re\left\{\frac{s_1f(\frac{1}{s_1})-\frac{1}{B}f(B)}{s_1-\frac{1}{B}}+\frac{1}{2}\frac{f(A)-Af'(A)-f(B)+Bf'(B)}{\frac{1}{A}-\frac{1}{B}}\left(s_1-\frac{1}{B}\right)\right\},$$
$$\Im\left\{f\left(\frac{1}{s_2}\right)-\frac{1}{s_2}f'\left(\frac{1}{s_2}\right)\right\}=\Im\left\{\frac{s_2f(\frac{1}{s_2})-\frac{1}{B}f(B)}{s_2-\frac{1}{B}}+\frac{1}{2}\frac{f(A)-Af'(A)-f(B)+Bf'(B)}{\frac{1}{A}-\frac{1}{B}}\left(s_2-\frac{1}{B}\right)\right\}.$$

Putting $z_1=\frac{1}{s_1}$, $z_2=\frac{1}{s_2}$ as above and simplifying concludes the proof. □

Applying Theorem 2 to the function $f(z)=ze^{1/z}-z$ on the segment $\left[\frac{i}{4\pi},\frac{i}{2\pi}\right]$, we have the equations

$$\Re\{f(z)-zf'(z)\}=\Re\{e^{1/z}\}=0,\quad \Im\{f(z)-zf'(z)\}=\Im\{e^{1/z}\}=0,$$

which upon setting $z=iy$ become

$$\cos\left(\frac{1}{y}\right)=0,\quad \sin\left(\frac{1}{y}\right)=0.$$

Their respective solutions are $\frac{1}{y_1}=\frac{5\pi}{2}\left(\text{or } \frac{7\pi}{2}\right)\in(2\pi,4\pi)$, $\frac{1}{y_2}=3\pi\in(2\pi,4\pi)$.

Applying Theorem 3 to the function $f(z)=ze^{1/z}-1$ on the segment $\left[\frac{i}{4\pi},\frac{i}{2\pi}\right]$, we have the two already derived equations

$$(\theta - 2\pi)\cos\theta - \sin\theta = 0, \quad \big[(\theta - 2\pi)\cos(\theta/2) - \sin(\theta/2)\big]\sin(\theta/2) = 0,$$

where $\theta = 1/y$. The first equation turns into $(\theta - 2\pi) = \tan\theta$ and has a solution $\frac{1}{y_1} = \theta_1 \approx 10.7765947651 \in (2\pi, 4\pi)$. Similarly the second equation reduces to $(\theta - 2\pi) = \tan(\theta/2)$ and has a solution $\frac{1}{y_2} = \theta_2 \approx 8.6143076776 \in (2\pi, 4\pi)$.

References

1. Davitt, R., Powers, R., Riedel, T., Sahoo, P.: Flett's mean value theorem for holomorphic functions. Math. Mag. **72**, 304–307 (1999)
2. Evard, J.-C., Jafari, F.: A complex Rolle's Theorem. Amer. Math. Monthly **99**, 858–861 (1992)
3. Flett, T.M.: A mean value theorem. Math. Gazette **42**, 38–39 (1958)
4. Markov, L.: Mean value theorems for analytic functions. Serdica Math. J. **41**, 471–480 (2015)
5. Pompeiu, D.: Sur une proposition analogue au théorème des accroissements finis. Mathematica (Cluj) **22**, 143–146 (1946)
6. Sahoo, P., Riedel, T.: Mean Value Theorems and Functional Equations. World Scientific Publishing Singapore (1998)

Computer Aided Modeling of Ultrasonic Surface Waves Propagation in Materials with Gradient of Properties

Todor Partalin and Yonka Ivanova

Abstract The present work deals with modeling of generating, propagation and receiving of the Rayleigh impulse in material with gradient of properties. The spectrum of an ultrasonic signal, having passed through a material, is determined by the spectrum of the exciting electrical signal, frequency characteristics of the transmitting and receiving transducers and by material characteristics. The dispersion of the Rayleigh wave is obtained as a result of the simulation done by spectral decomposition of impulse and processing components considering wave penetration and wave velocity changes caused by gradient of mechanical characteristics. Simulated ultrasonic waveforms allow evaluation of the time delay effect induced by stress gradient. Application of ultrasonic wave for stress analyses is possible on the basis of spectral analysis and phase comparisons of ultrasonic impulse.

Keywords Ultrasonic surface Rayleigh waves · Stress analyses · Stress gradient · Rayleigh wave dispersion · Spectral decomposition · Spectral analysis

1 Introduction

The propagation of ultrasonic waves in material with internal stresses is influenced by the change of some physical and mechanical properties such as bulk density and elastic modulus due to the deformations created by loads. Changes depend on the type of wave (longitudinal, transverse and surface) and modify basic parameters such as velocity, polarization and phase [1–6]. Research related to the propagation

T. Partalin
Faculty of Mathematics and Informatics, Sofia Univeristy "St.Kliment Ohridski", Sofia, Bulgaria
e-mail: topart@fmi.uni-sofia.bg

Y. Ivanova (✉)
Faculty of Physics, Sofia University "St.Kliment Ohridski", Sofia, Bulgaria
e-mail: yonka@imbm.bas.bg

Y. Ivanova
Institute of Mechanics, Bulgarian Academy of Sciences, Sofia, Bulgaria

K. Georgiev et al. (eds.), *Advanced Computing in Industrial Mathematics*,
Studies in Computational Intelligence 681, DOI 10.1007/978-3-319-49544-6_12

of surface ultrasonic waves in medium with mechanical stress provide opportunities for their use in determining the physical and mechanical properties of the surface and subsurface layers of materials and stress analysis in metal structures [4–8].

The aim of the current work is to model the evolution of ultrasonic Rayleigh impulse in half-plane with gradient of mechanical stress by spectral decomposition and composing an impulse taking into account the time delay of spectral components. It is assumed that the material is homogeneous and the applied stress is in the elastic region.

2 Formulation

The model task refers to the bending of a beam (Fig. 1a) [9]. In the case of bending (Fig. 1a), the longitudinal lines (layers) on the upper side are elongated and on the lower side are shortened. The surface in which longitudinal lines do not change in length is the neutral surface, its intersection with the cross-sectional plane is neutral

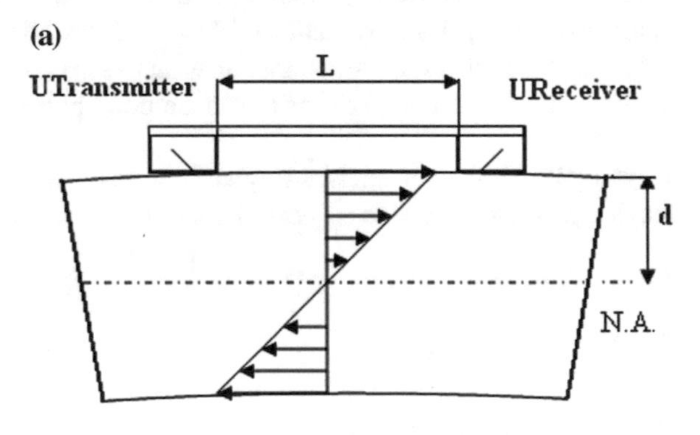

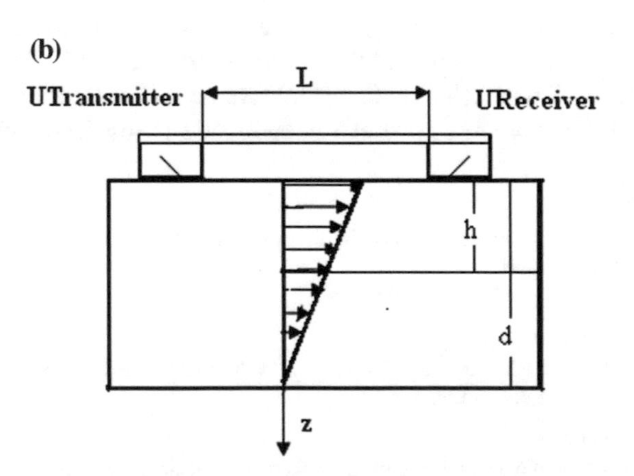

Fig. 1 **a** Bending of the beam. **b** Idealized schema

axis (N.A.). The maximum tensile and compressive stresses ($\sigma_{\max}$) occur at the points located farthest from the N.A.

In order to evaluate the acoustoelastic effect we model the dispersion of ultrasonic Rayleigh wave caused by stress gradient. The stresses and deformations are sufficiently small and therefore all the terms in the approximations of second and higher order should be neglected [1, 2]. The curvature of medium surface is ignored. The idealized schema is shown in Fig. 1b.

Electrical impulse excites through a transmitting transducer an ultrasonic surface wave that passes the distance L to the receiver where it is converted into electrical signal. The acoustoelastic effect will be assessed by analyzing the shape, velocity and spectrum of ultrasonic pulse.

3 Linearized Model to Present the Conditions for Ultrasonic Wave Propagation

We follow the most common conventional concepts of homogeneous, isotropic material, composed of atoms. In the case of uniaxial tensile the material volume changes and can be presented by the expression [1–3, 9]:

$$\frac{\Delta V}{V_o} = \frac{V - V_o}{V_o} \approx (1 + \varepsilon)(1 - \nu.\varepsilon)^2 - 1 \approx (1 - 2.\nu)\varepsilon \tag{1}$$

where V_o is the initial volume, before applying the mechanical load, σ is mechanical stress, ε is the relative deformation and ν is Poisson's ratio. So, the volume density of the material subjected to loading $\rho(\varepsilon)$ is presented with (2).

$$\rho(\varepsilon) = \frac{\rho_0}{1 + \varepsilon.(1 - 2.\nu)} \tag{2}$$

where, ρ_o is the initial density before applying a stress.

This approach represents the change of the density in a point. The velocity of longitudinal ultrasonic wave according to [2] and considering the density change (2) is:

$$c_l^2(\varepsilon) = \frac{E}{\rho(\varepsilon)} \cdot \frac{(1 - \nu)}{(1 + \nu) \cdot (1 - 2.\nu)} = \frac{E}{\rho_0} \cdot \frac{(1 - \nu)}{(1 + \nu) \cdot (1 - 2.\nu)} . (1 + \varepsilon.(1 - 2.\nu)) \tag{3}$$

Obviously, the linear approximations and considerations at such an elementary level do not suggest differences in the velocities in different directions. However, from the physical point of view, it is clear that the average distance between atoms in stressed material is not isotropic value and the change in density is due to the tension in the loading direction. This leads to a diminution of the velocity in this direction and different velocities in the other directions depending on the geometry of the loaded sample.

Materials seem to behave differently in tension and compression. Such bimodularity has been observed, for example, in rocks, composites [10, 11] as well as alloys such as cast iron, steels, copper alloys, aluminums etc. [2]. The ratio of Young's Modulus at the tension (E^+) and compression (E^-) (E^+/E^-) varies between 0.75 for cast iron and 0.95–0.98 for different kinds of steels [2].

Figure 2 illustrates simple potential energy curves as a function of interatomic spacing [10]. The energy E, shape and depth of the curve defines physical properties. The curves indicate the strength of the bond based on the depth of the potential well. The minima in the energy curve determine the equilibrium interatomic spacing r_o. In

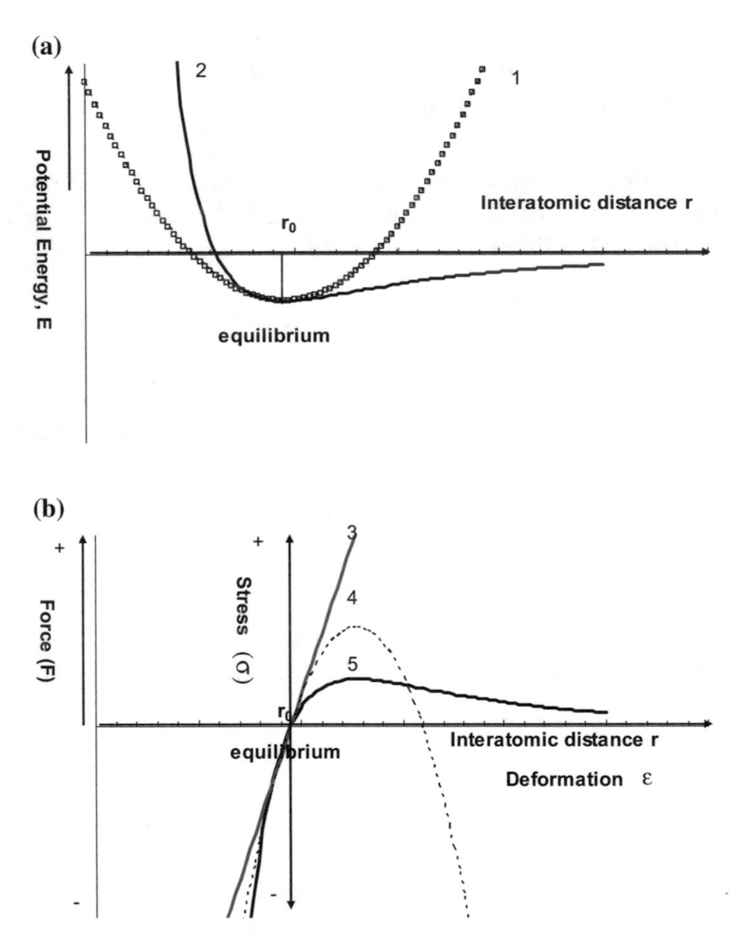

Fig. 2 **a** Potential energy curve E versus interatomic distance: *1*—Quadratic approximation of potential function; *2*—Empirical potential function (Lennard-Jones). **b** Force F versus interatomic distance; *3*—Hooke's law of elasticity $\sigma_o = E.\varepsilon_o$, *4*—$\sigma = \varepsilon.E_o(1 - \delta.\varepsilon)$; *5*—Derivative of potential function 2

the absence of external forces, the crystal has equilibrium spacing. The slope of the tangent to the force curve evaluated at $r = r_o$ is presented by Young's modulus (E_o).

$$\sigma_o = E_o.\varepsilon_o, \tag{4}$$

where E_o is an average Young's Modulus.

When an external force is applied, the interatomic spacing is altered. Both tension and compression increase the lattice potential energy. The stress can be expressed by the approximation (curve 4, Fig. 2b):

$$\sigma = \varepsilon.E(\varepsilon) \tag{5}$$

For our purposes, we can accept (6) as the simplest approximation that expresses the asymmetry of the potential energy of the interaction between the atoms of the material [10, 11].

$$\sigma = \varepsilon.E_o(1 - \delta.\varepsilon) \tag{6}$$

where δ is material coefficient

The coefficient δ can be estimated as follows [2]

$$E^- = E_o(1 + \delta.\varepsilon) \;\&\; E^+ = E_o(1 - \delta.\varepsilon) \Rightarrow \delta = \frac{E^- - E^+}{2.\varepsilon.E_o} \tag{7}$$

Taking into account (6) and neglecting the second order terms, the velocity of longitudinal ultrasonic wave propagating in the stressed media can be expressed by the relation [1, 2]:

$$c_l^2(\varepsilon) = \frac{E_0}{\rho}.(1 - \delta.\varepsilon) \approx \frac{E_0}{\rho_0}.\frac{(1-\nu)}{(1+\nu)\cdot(1-2.\nu)}.(1 + \varepsilon.((1-2\nu) - \delta)) \tag{8a}$$

$$c_l^2(\varepsilon) = c_0^2.(1 + \varepsilon.(1 - 2\nu - \delta)).\frac{(1-\nu)}{(1+\nu)\cdot(1-2.\nu)}, \tag{8b}$$

where c_0 is velocity of longitudinal wave in material without stresses ($c_0^2 = \frac{E_o}{\rho_o}$).

After some algebraic transformations we obtain the expression for velocity of ultrasonic wave

$$c_l(\varepsilon) \approx c_0.(1 + (1 - \delta - 2.\nu).\frac{\varepsilon}{2}) = c_0.(1 + k\varepsilon), \tag{9}$$

where k can be expressed by the relation

$$k = \frac{(1-\nu).(1-\delta-2.\nu)}{2.(1+\nu)\cdot(1-2.\nu)} \tag{10}$$

The coefficient k depends on the value δ and can take values between -0.5 and 0.5 hence it can be expected a negative growth rate of ultrasound velocity $c_l(\varepsilon)$ for some materials.

The change of velocity of the longitudinal wave depends on the Poisson's ratio ν as well. For carbon iron alloys (cast iron and steels) ν is about 0.23–0.30, and for non-ferrous alloys such as copper and zinc is between 0.32 and 0.42 [2]. The value of Poisson's ratio is different in tension and compression loading. The relative difference $(\nu^+ - \nu^-)/\nu^+$ is from 4 % for carbon steels to 20 % for cast iron. However the velocity changes caused by acoustoelastic effect are very small up to 0.1 % for steels and around 1.5 % for cast iron [2].

We will use the generally accepted expression for Rayleigh wave velocity c_R [7, 12], that gives a linear relation between wave velocity and deformation, respectively tension (11)

$$c_R(\varepsilon) = k_{Rt}.c_t = k_{Rt}k_{tl}.c_l(\varepsilon) = k_{Rt}k_{tl}.c_o^{\cdot}(1 + k.\varepsilon) = c_{Ro}.(1 + k_1.\sigma), \tag{11}$$

where $c_{Ro} = c_o.k_{Rt}.k_{tl}$ is Rayleigh wave velocity in an medium without stress, and coefficients k_{Rt} and k_{tl} are $k_{Rt} \approx 0.93$, $k_{tl} \approx 0.55$ for carbon steel with Poisson's ratio $\nu = 0.29$ [12], $k_1 = k/E_o$.

The velocity of the wave depends also on direction of propagation and deformation [2, 7]. As a result the dependence of the velocity of the surface Rayleigh wave on deformation is expressed by equation:

$$c_R(\varepsilon) \approx c_{Ro}.(1 + k.\varepsilon) \tag{12}$$

The presence of properties' gradient normal to material surface is significant because the difference in penetration of the mechanical perturbation into the medium depends on the frequency. Stress σ and its corresponding deformation ε are different in depth of material. Therefore the velocity $c_R(\varepsilon)$ will depend on penetration depth of the surface wave (h), which is in the range of the wavelength λ [7, 12]:

$$h \approx 1,7.\lambda \tag{13}$$

The mechanical stress varies linear from the surface to the N.A and depends on depth h (Fig. 1b).

$$\sigma(h) = \sigma_{\max}.\frac{d-h}{d} = \varepsilon_{\max}.E_0.\frac{d-h}{d} \tag{14}$$

The dependencies (11–14) lead to appearance of dispersion of the ultrasonic Rayleigh wave (15)

$$c_R(\lambda) = c_R\left[\varepsilon\left\{\sigma\left(h(\lambda)\right)\right\}\right]. \tag{15}$$

4 Simulations of Ultrasonic Rayleigh Impulse Propagation in Stressed Media

In order to determine the influence of tension gradient we perform a numerical modeling of ultrasonic Rayleigh impulse which is passed through a stressed material with stress gradient. The spectrum of an ultrasonic signal is determined by the spectrum of the exciting electrical signal, the frequency characteristics of the transmitting and receiving transducer and by the material characteristics.

The excitation of ultrasonic oscillations in the piezoelectric sensor is realized by means of shock loading (Fig. 3). The transmitter/receiver can to be taken as an electromechanical system which oscillates around the basic resonance frequency f_o [13–15]. An exemplary presentation of its amplitude-frequency characteristics (AFC) is shown in Fig. 4. The basic frequency of transducers is accepted to be $f_o = 4\,\text{MHz}$. The real AFC of the transducers are given by more complex formulas, considering the acoustic resistance of a piezoelectric sensor, attenuator, protector etc. [16, 17]. They differ from the used exemplary presentation, but the differences are not substantial for the model problem. Practically, the AFC are experimentally recorded according [15, 16]. The shape of the generated acoustic signal is obtained by applying an inverse Fourier transformation of the product:

$$S_a(f) = S_E(f).S_t(f) \tag{16}$$

Figure 4 show spectrum of electrical signal (S_E), amplitude-frequency characteristics of transmitter (S_t) and receiver (S_r) and spectrum of acoustic signal S_a.

The impact of the medium with stress gradient is modeled using narrowband filters (S_m) with which the impulse is decomposed into components with different frequency. AFC of filters ($F\#$) and their sum S_m are shown in Fig. 5.

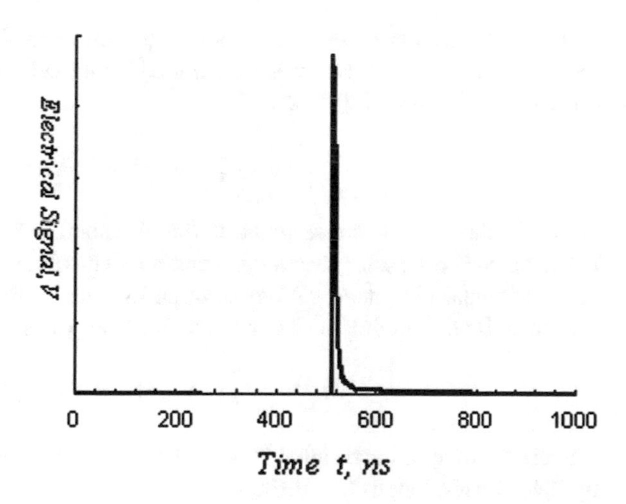

Fig. 3 Exciting electrical signal

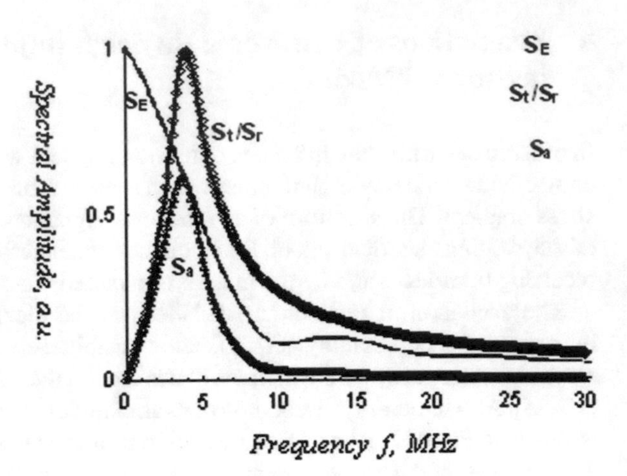

Fig. 4 Spectra of signals: S_E—spectrum of electrical signal, S_t, S_r—AFC of transmitting/receiving transducers, S_a—spectrum of acoustic signal

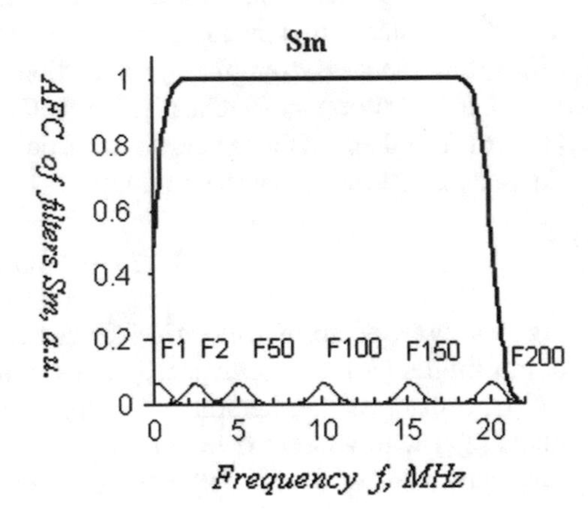

Fig. 5 Amplitude frequency characteristics of filters $F\#$ and their sum S_m

Each of signal components has a penetration depth $h_i \approx 1,7.\lambda_i$ referring to its frequency f_i and its velocity is determined by the deformation, respectively stress corresponding to depth (13) and (14)

$$c_R(\varepsilon) \cong c_R(\lambda_i) = c_R(f_i) \tag{17}$$

Acoustic components are summed in the ultrasonic receiving transducer according to their arrival time after passing the length L. The ultrasonic receiver transforms the acoustic signal into electrical one through its transfer function (S_r). The process is presented by the product of the transfer functions of the separate units

$$S(f) = S_E(f).S_t(f).S_m(f).S_r(f) \tag{18}$$

The arrival time of each signal is related with its penetration depth. It is determined by the acoustic length L and the velocity of ultrasonic Rayleigh wave $c_R(\lambda_i)$. The

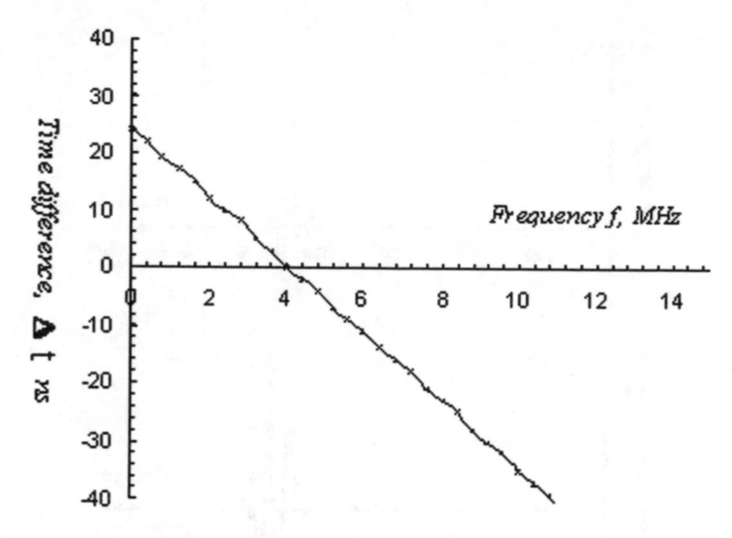

Fig. 6 Dependency of time difference ($\Delta\ t$) on frequency f

difference in the flight time of the wave in a material with (t_s) and without stress gradient (t_o) is

$$\Delta t(\lambda) \cong \Delta t(f) = t_o - t_s = \frac{L}{c_{Ro}} - \frac{L}{c_R(\varepsilon_{(\lambda\approx h)})} = \frac{L}{c_{Ro}}.(1 - \frac{1}{1 + k\varepsilon_{(\lambda\approx h)}}) \approx \frac{L}{c_{Ro}} k\varepsilon \tag{19}$$

where $c_R(\varepsilon_{(\lambda\approx h)})$ and c_{Ro} are the velocities of surface waves in media with and without stress gradient.

In the present study the dependency of time difference on frequency was chosen to be linear as shown in the Fig. 6. The influence of nonlinear gradient of properties can be modeled with nonlinear time delay.

5 Discussion

The results from simulation of ultrasonic Rayleigh impulse propagation through a stressed material are shown in Figs. 7 and 8. Figure 7 shows signals of ultrasonic Rayleigh impulse in material without stress gradient (a) and with stress gradient (b). The signal which is passed through the media with stress gradient in the Fig. 7b is shifted and its shape is changed. The arrival time difference between these signals is very small, less than 0.2 % of the flight time between ultrasonic transducers. The obtained signal and its spectrum depend on the length of the acoustic path, because the delay of the time is proportional to the base length L.

The change of amplitude spectra of signals is not significant. The greater is the influence of stress gradient on the shape of signal and its phase spectrum (20) as it

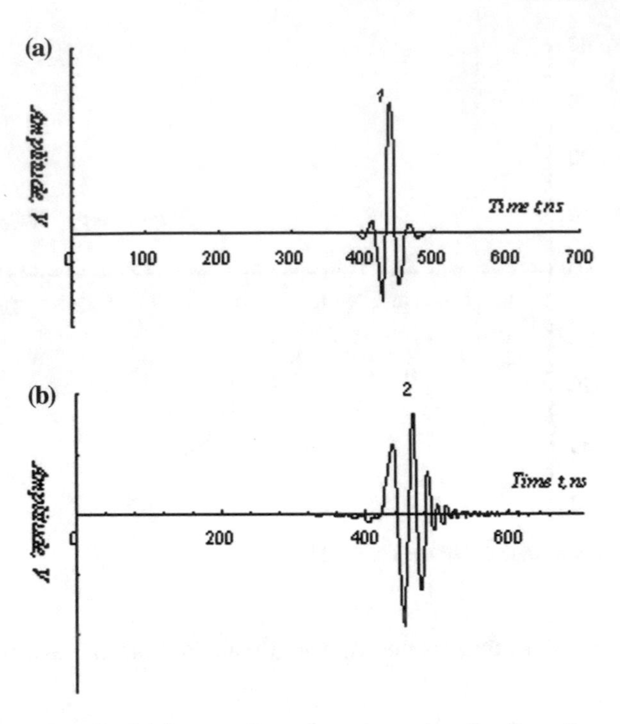

Fig. 7 Waveforms from Rayleigh waves *Curve 1*—unstressed media; *Curve 2*—media with stress gradient

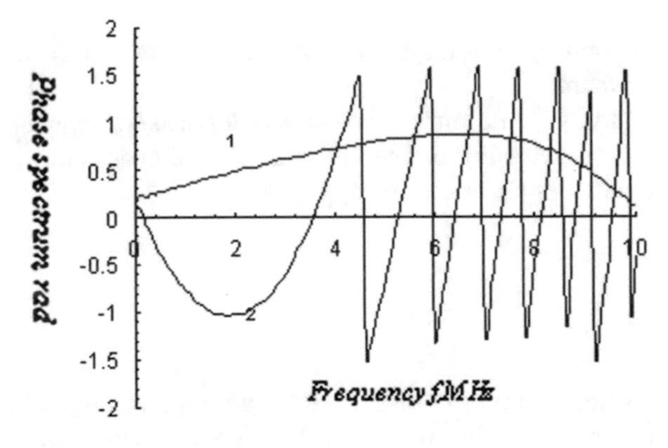

Fig. 8 Phase spectra of ultrasonic signals *Curve 1*—Phase spectrum of ultrasonic signal in unstressed media; *Curve 2*—Phase spectrum of ultrasonic signal in media with stress gradient

is shown in the Fig. 8.

$$\varphi(f) = arctg\left(\frac{\operatorname{Im} S(f)}{\operatorname{Re} S(f)}\right) \tag{20}$$

where $\operatorname{Im} S(f)$ and $\operatorname{Re} S(f)$ are imaginary and real part of the spectrum.

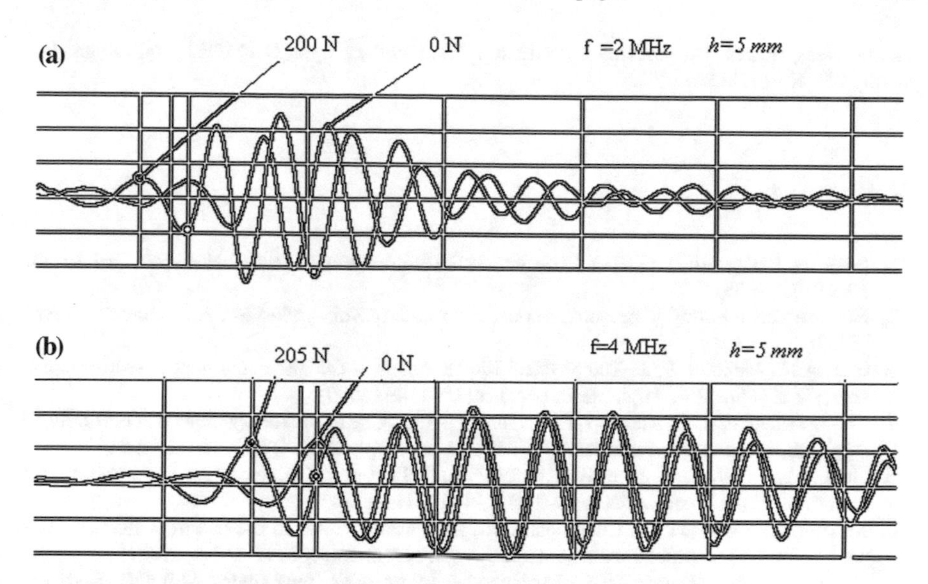

Fig. 9 Signals before and after bending load: thickness sample—5 mm, stress gradient $2\sigma/h = 20 \times 10^9$ Pa (*a*-frequency 2 MHz, *b*-frequency 4 MHz) [18]

Experiments on Rayleigh wave propagation in materials with stress gradient are presented in [18]. The signals from ultrasonic surface waves before and after bending of cantilever made of carbon steel are shown in Fig. 9a, b for 2 and 4 MHz. The value of stress gradient $(2\sigma/h)$ is 20×10^9 Pa. It can be noted the change of signals shapes and appearance of dispersion of waves caused by stress loading.

6 Conclusion

The propagation of Rayleigh wave in stressed medium is quite complex phenomenon. The presented simulation of the evolution of the wave pulse shows the importance of the transverse gradient of the mechanical stress. It remains only a simulation and not properly represents the physical process of wave motion. The performed numerical experiments show that the stress gradient affects the shape of signals and less the amplitude spectra. The biggest difference can be seen in the phase spectrum. This simulation is illustrative and concerns mechanical stress gradient. Nevertheless any normal to the surface gradient of mechanical parameters of medium such as density, modulus of elasticity and so one, caused by phase composition, temperature, chemical or heat treatment etc. make a similar impact. It allows trying assumptions and preliminary results with perspective to compare with planned experiment as well as to support the clarification of a phenomenological theory. The results may find application in material science and seismology. Application of the Rayleigh waves for stress and strain state analysis of the materials is possible on the basis of spectral analysis and phase and impulse shape comparisons.

Acknowledgements The research is partly supported by the project N 147/2015 with Sofia University "St. Kliment Ohridski".

References

1. Bach, F., Askegaard, V.: General stress-velocity expressions in acoustoelasticity. Exp. Mech. **19**, 69–75 (1979)
2. Nondestructive testing, Handbook, Acoustic tensometry, vol. 4, Moscow, Publ. House "Spectr" (2007)
3. Guz, A.N., Makhort, F.G.: The physical fundamentals of the ultrasonic nondestructive stress analysis of solids. Int. Appl. Mech. **36**(9), 1119–1149 (2000)
4. Chernoochenko, A.A., Makhort, F.G., Gushcha, O.I.: Use of the theory of acoustoelasticity of Rayleigh waves to determine stresses in solids. Int. Appl. Mech. **27**(1), 38–42 (1991)
5. Hirao, M., Fukuoka, H., Hori, K.: Acoustoelastic effect of Rayleigh surface wave in isotropic material. ASME J. Appl. Mech. **48**, 119–124 (1981)
6. Makhort, F., Gushcha, O., Chernoochenko, A.: Surface waves in the determination of near-surface stresses of structural elements. Int. Appl. Mech. **36**(8), 1047–1051 (2000)
7. Bobrenko, V.M., Vangeli, M.S., Kutsenko, A.I.: Acoustic Tensometry. Shtinitsa, Kishinev (1991)
8. Si-Chaib, M.O., Menad, S., Djelouah, H., Bocquet, M.: An ultrasound method for the acoustoelastic evaluation of simple bending stresses. NDT E Int. **34**, 521–529 (2001)
9. Hearn, E.J.: Mechanics of Materials. Elsevier (1997). ISBN: 978-0-7506-3265-2
10. Padmavathi, D.A.: Potential energy curves & material properties. Mater. Sci. Appl. **2**, 97–104 (2011)
11. Destrade, M., Murphy, J., Rashid, B.: Differences in tension and compression in the nonlinearly elastic bending of beams. Int. J. Struct. Changes Solids Mech. Appl. **1**(1), 73–81 (2009)
12. Viktorov I.A.: Physic basis application of ultrasonic Rayleigh and Lamb waves in technique, Mir, Science (1966)
13. Sharp, R. (ed.): Methods of Nondestructive Testing. Moskow, Mir (1972)
14. Kutzarov, S.: Passive LC and RC Diagram. Sofia, Technika (1980)
15. Transducers, U. (ed.): Kikuchi. Mir, Moskow (1972)
16. ASTM E 1065 A1, Measurement of frequency response
17. Merkulova, V.M., Tokarev, V.M.: Calculation of wide-band Piesotransducers used for ultrasonic and immersion defectoscopy. Defectoskopiya 4 (1972)
18. Ivanova, Y., Partalin T.: Investigation of stress-state in rolled sheets by ultrasonic techniques. Ultragarsas (Ultrasound) **66**(1) (2011)

InterCriteria Analysis of Simple Genetic Algorithms Performance

Tania Pencheva and Maria Angelova

Abstract Recently developed approach of InterCriteria Analysis is here applied aiming at an assessment of the performance of such a promising stochastic optimization technique as simple genetic algorithms. Considered algorithms, as representatives of the biologically-inspired ones, are chosen as an object of investigation since they are proven as quite successful in solving of many challenging problems in the field of complex dynamic systems optimization. In this investigation simple genetic algorithms are applied for the purposes of parameter identification of a fermentation process. Altogether six simple genetic algorithms are here considered, differ from each other in the execution order of main genetic operators, namely selection, crossover and mutation. The apparatuses of index matrices and intuitionistic fuzzy sets, underlying the InterCriteria Analysis, are implemented to assess the performance of simple genetic algorithms for the parameter identification of *Saccharomyces cerevisiae* fed-batch fermentation process. The obtained results after the InterCriteria Analysis application are thoroughly analysed towards the algorithms outcomes, such as convergence time and model accuracy.

1 Introduction

In general, modelling of fermentation processes is a challenging and rather difficult to be solved problem. Logically explanation of this fact is connected with a complex structure of the considered processes, usually described by a systems of non-linear differential equations with several specific growth rates. Thus, model parameter

---

T. Pencheva (✉) · M. Angelova
Institute of Biophysics and Biomedical Engineering, Bulgarian Academy of Sciences, 105 Acad. G. Bonchev Str., 1113 Sofia, Bulgaria
e-mail: tania.pencheva@biomed.bas.bg

M. Angelova
e-mail: maria.angelova@biomed.bas.bg

K. Georgiev et al. (eds.), *Advanced Computing in Industrial Mathematics*, Studies in Computational Intelligence 681, DOI 10.1007/978-3-319-49544-6_13

identification is of a key importance for modelling process. Genetic algorithms (GA), as representatives of biologically inspired metaheuristic techniques, have pointed more and more attention mainly due to the fact that they reach a good solution, even global optima, within reasonable computing time. Many applications demonstrate GA as suitable for solving complex problems [10, 11, 14], even for such a difficult task as parameter identification of fermentation process models [1, 3, 15]. Looking for any kind of relations between model parameters and optimization algorithm performance might lead to significant improvement of the parameter identification procedure. For that purpose, recently developed approach of InterCriteria Analysis (ICrA) [8] is going to be applied.

InterCriteria Analysis is an approach based on two fundamental concepts—of the Intuitionistic Fuzzy Sets [7] and the Index Matrices [4–6]. ICrA allows detecting of possible correlations between pairs of involved criteria and provides on this basis an additional information for the investigated objects. Following the main goal of ICrA development, namely ICrA to be successfully applied for decision making in different areas of science and practice, the idea this approach to be tested in the field of parameter identification of fermentation processes models is intuitively appeared. First applications of ICrA for that purposes are presented in [2, 12, 16]. In [2] ICrA is implemented for the purposes of a parameter identification of *S. cerevisiae* fermentation process. Authors discover relations between convergence time, model accuracy and model parameters, when two genetic algorithms parameters, namely rates of crossover and mutation, are investigated. In [16] model parameter identification of a non-linear model of an *E. coli* fed-batch fermentation process is considered using GA. Then ICrA is applied to explore the existing relations and dependencies of defined model parameters and GA outcomes—execution time and objective function value. Authors in [12] again consider an *E. coli* fed-batch fermentation process and apply ICrA in order to find correlation between the kinetic variables. Structural and parametric identification of the process are performed based on the obtained by ICrA evaluations of dependencies between biomass, substrate, oxygen and carbon dioxide.

Current research aims to demonstrate an another attempt for a successful application of ICrA for the purposes of model parameters identification of fermentation processes. ICrA is going to be applied for assessment of the performance of different simple genetic algorithms used for the purposes of parameter identification of a *S. cerevisiae* fermentation process. Altogether six simple genetic algorithms that differ from each other in the sequence of execution of the main genetic operators, namely selection, crossover and mutation, are applied. ICrA implementation, based on the results of a series of parameters identification procedures, is expected to reveal some relations between GA outcomes and model parameters themselves.

2 Problem Formulation

Model parameter identification of a considered here *S. cerevisiae* fed-batch fermentation process is performed based on the real data from a cultivation carried out in the *Institute of Technical Chemistry University of Hanover, Germany*. The experimental data set consists of on-line measurement of substrate (glucose) and off-line measurements of biomass and ethanol. The detailed description of the process conditions and experimental data could be found in [13].

According to the mass balance, mathematical model of a *S. cerevisiae* fed-batch fermentation process is commonly described by the following system of non-linear differential equations [13]:

$$\frac{dX}{dt} = (\mu_{2S}\frac{S}{S+k_S} + \mu_{2E}\frac{E}{E+k_E})X - \frac{F_{in}}{V}X \tag{1}$$

$$\frac{dS}{dt} = -\frac{\mu_{2S}}{Y_{S/X}}\frac{S}{S+k_S}X + \frac{F_{in}}{V}(S_{in} - S) \tag{2}$$

$$\frac{dE}{dt} = -\frac{\mu_{2E}}{Y_{E/X}}\frac{E}{E+k_E}X + \frac{F_{in}}{V}E \tag{3}$$

$$\frac{dV}{dt} = F_{in} \tag{4}$$

where X is the biomass concentration, [g/l]; S—substrate concentration, [g/l]; E—ethanol concentration, [g/l]; F_{in}—feeding rate, [l/h]; V—bioreactor volume, [l]; S_{in}—substrate concentration in the feeding solution, [g/l]; μ_{2S}, μ_{2E}—maximum values of the specific growth rates, [1/h]; k_S, k_E—saturation constants, [g/l]; $Y_{S/X}$, $Y_{E/X}$—yield coefficients, [-].

For the considered here model (Eqs. (1)–(4)), the following vector including six model parameters should be identified: $p = [\mu_{2S}, \mu_{2E}, k_S, k_E, Y_{S/X}, Y_{E/X}]$.

As an optimization criterion, mean square deviation between the model output and the experimental data for biomass, substrate and ethanol, obtained during the cultivation has been used:

$$J = \sum_{i=1}^{m}\left(X_{\exp}(i) - X_{\mathrm{mod}}(i)\right)^2 + \sum_{i=1}^{n}\left(S_{\exp}(i) - S_{\mathrm{mod}}(i)\right)^2 + \sum_{i=1}^{l}\left(E_{\exp}(i) - E_{\mathrm{mod}}(i)\right)^2 \to \min \tag{5}$$

where m, n and l are the experimental data dimensions; $X_{\exp}$, $S_{\exp}$ and $E_{\exp}$ are the available experimental data for biomass, substrate and ethanol; X_{mod}, S_{mod} and E_{mod}

are the model predictions for biomass, substrate and ethanol with a given model parameter vector.

3 Simple Genetic Algorithms for Parameter Identification

As mentioned above, genetic algorithms belong to the methods based on biological evolution and inspired by Darwins theory of survival of the fittest [11]. Frequently used as an alternative to the conventional optimization techniques, GA are successfully applied to solve complex problems in different research fields [11]. Standard Simple GA (SGA) initially presented by Goldberg [11] work with a population of coded parameter sets (called chromosomes), and search a global optimal solution using three main genetic operators in a sequence selection, crossover, mutation. This algorithm is here denoted as SGA-SCM, coming from the operators execution order selection, crossover, mutation. SGA-SCM starts with a creation of a randomly generated initial population. After that each solution is evaluated and assigned a fitness value. According to the fitness function, the most suitable solutions are selected. Then, crossover proceeds to form a new offspring. Mutation with determinate probability is then applied aiming to prevent falling of all solutions into a local optimum. GA terminate when some criterion, such as number of generations reached, or evolution time passed, or fitness threshold reached, or fitness convergence satisfied, etc., is fulfilled. For the purposes of this investigation, GA terminate when a certain number of generations is reached.

Following the main idea of GA to imitate the processes in nature, one can assume that the probability crossover to come first and then mutation is comparable to that both processes to occur in a reverse order; or selection to be performed after crossover and mutation, no matter of their order. Following this idea, many modifications of SGA-SCM, concerning the sequence of execution of main genetic operators have been developed aiming to improve the model accuracy and algorithms convergence time [1, 3]. Three modifications, namely SGA-SMC (following the mentioned above notation), SGA-CMS, and SGA-MCS have been proposed and basically investigated for parameter identification of a fed-batch fermentation process of *S. cerevisiae* in [3]. Another two modifications possible to occur—SGA-CSM and SGA-MSC, are proposed in [1] for the same purpose. The performance of mentioned above altogether six modifications of SGA-SCM, applied for the parameter identification of a *S. cerevisiae* fed-batch fermentation process, is going to be assessed by the promising approach of ICrA.

4 InterCriteria Analysis

InterCriteria analysis, based on the apparatuses of index matrices and intuitionistic fuzzy sets, is given in details in [8]. Here, for a completeness, ICrA is briefly presented, as exposed in [2].

An intuitionistic fuzzy pair (IFP) [9] is an ordered pair of real non-negative numbers $\langle a, b \rangle$, where $a, b \in [0, 1]$ and $a + b \leq 1$, that is used as an evaluation of some object or process. According to [9], the components (a and b) of IFP might be interpreted as degrees of "membership" and "non-membership" to a given set, degrees of "agreement" and "disagreement", degrees of "validity" and "non-validity", degrees of "correctness" and "non-correctness", etc.

The apparatus of index matrices (IM) is initially presented in [4] and discussed in more details in [5, 6]. For the purposes of ICrA application, the initial index set consists of the criteria (for rows) and objects (for columns) with the IM elements assumed to be real numbers. Further, an IM with index sets consisting of the criteria (for rows and for columns) with IFP elements determining the degrees of correspondence between the respective criteria is constructed, as it is doing to be briefly presented below.

Let the initial IM is presented in the form of Eq. (6), where, for every p, q, $(1 \leq p \leq m, 1 \leq q \leq n)$, C_p is a criterion, taking part in the evaluation; O_q—an object to be evaluated; a_{C_p,O_q}—a real number or another object, that is comparable about relation R with the other a-object, so that for each i, j, k: $R(a_{C_k,O_i}, a_{C_k,O_j})$ is defined.

$$A = \begin{array}{c|ccccccc} & O_1 & \dots & O_k & \dots & O_l & \dots & O_n \\ \hline C_1 & a_{C_1,O_1} & \dots & a_{C_1,O_k} & \dots & a_{C_1,O_l} & \dots & a_{C_1,O_n} \\ \vdots & \vdots & \ddots & \vdots & \ddots & \vdots & \ddots & \vdots \\ C_i & a_{C_i,O_1} & \dots & a_{C_i,O_k} & \dots & a_{C_i,O_l} & \dots & a_{C_i,O_n} \\ \vdots & \vdots & \ddots & \vdots & \ddots & \vdots & \ddots & \vdots \\ C_j & a_{C_j,O_1} & \dots & a_{C_j,O_k} & \dots & a_{C_j,O_l} & \dots & a_{C_j,O_n} \\ \vdots & \vdots & \ddots & \vdots & \ddots & \vdots & \ddots & \vdots \\ C_m & a_{C_m,O_1} & \dots & a_{C_m,O_k} & \dots & a_{C_m,O_l} & \dots & a_{C_m,O_n} \end{array}, \quad (6)$$

Let $\overline{R}$ be the dual relation of R in the sense that if R is satisfied, then $\overline{R}$ is not satisfied, and vice versa. For example, if "R" is the relation "$<$", then $\overline{R}$ is the relation "$>$", and vice versa. If $S^{\mu}_{k,l}$ is the number of cases in which $R(a_{C_k,O_i}, a_{C_k,O_j})$ and $R(a_{C_l,O_i}, a_{C_l,O_j})$ are simultaneously satisfied, while $S^{\nu}_{k,l}$ is the number of cases is which $R(a_{C_k,O_i}, a_{C_k,O_j})$ and $\overline{R}(a_{C_l,O_i}, a_{C_l,O_j})$ are simultaneously satisfied, it is obvious, that

$$S^{\mu}_{k,l} + S^{\nu}_{k,l} \leq \frac{n(n-1)}{2}.$$

Further, for every k, l, satisfying $1 \leq k < l \leq m$, and for $n \geq 2$,

$$\mu_{C_k,C_l} = 2\frac{S^{\mu}_{k,l}}{n(n-1)}, \quad \nu_{C_k,C_l} = 2\frac{S^{\nu}_{k,l}}{n(n-1)} \tag{7}$$

are defined. Therefore, $\langle \mu_{C_k,C_l}, \nu_{C_k,C_l} \rangle$ is an IFP. Next, the following IM is constructed:

$$\begin{array}{c|ccc} & C_1 & \dots & C_m \\ \hline C_1 & \langle \mu_{C_1,C_1}, \nu_{C_1,C_1} \rangle & \dots & \langle \mu_{C_1,C_m}, \nu_{C_1,C_m} \rangle \\ \vdots & \vdots & \ddots & \vdots \\ C_m & \langle \mu_{C_m,C_1}, \nu_{C_m,C_1} \rangle & \dots & \langle \mu_{C_m,C_m}, \nu_{C_m,C_m} \rangle \end{array}, \tag{8}$$

that determines the degrees of correspondence between criteria C_1, ..., C_m.

The sum $\mu_{C_k,C_l} + \nu_{C_k,C_l}$ is not always equal to 1. The difference

$$\pi_{C_k,C_l} = 1 - \mu_{C_k,C_l} - \nu_{C_k,C_l} \tag{9}$$

is considered as a degree of "uncertainty".

5 Numerical Results and Discussion

All identification procedures have been performed on a Computer system with Intel(R) Core(TM) i5-3450 CPU 3.1 GHz, 8 GB Memory, Windows 7 (64-bit) operating system.

To estimate the model parameters (vector p) of the considered model (Eqs. 3 and 4), six SGA with different execution order of main genetic operators selection, crossover and mutation, have been consequently applied. In current investigation, GA operators and parameters are tuned according to [3]. Due to the stochastic nature of genetic algorithms, 30 independent runs for each of the applied here six modifications of SGA have been performed.

Altogether eight criteria are considered: C_1—objective function value J; C_2—convergence time T; C_3 and C_4—specific growth rates, respectively μ_{2S} and μ_{2E}; C_5 and C_6—saturation constants, respectively k_S and k_E; C_7 and C_8—yield coefficients, respectively $Y_{S/X}$ and $Y_{E/X}$. Concerning the objects—six objects are here investigated, corresponding to SGA modifications: O_1 is the SGA-SCM; O_2—SGA-SMC; O_3—SGA-CMS; O_4—SGA-CSM; O_5—SGA-MSC; and O_6—SGA-MCS.

Equations (10)–(12) present, respectively, the average values (Eq. 10), the best estimates (Eq. 11), and the worst ones (Eq. 12), of the values of objective function J, the algorithm convergence time T, [s], as well as of the model parameters. The objective functions (criterion C_1) is given with more digits after the decimal point in order to be distinguishable.

$$A_1(average) = \begin{array}{c|cccccc} & O_1 & O_2 & O_3 & O_4 & O_5 & O_6 \\ \hline C_1 & 0.022065 & 0.022103 & 0.022212 & 0.022069 & 0.022217 & 0.022144 \\ C_2 & 215.85 & 198.06 & 218.32 & 232.75 & 203.55 & 216.47 \\ C_3 & 0.99 & 0.98 & 0.92 & 0.96 & 0.98 & 0.99 \\ C_4 & 0.15 & 0.14 & 0.11 & 0.15 & 0.11 & 0.12 \\ C_5 & 0.14 & 0.13 & 0.11 & 0.14 & 0.13 & 0.13 \\ C_6 & 0.80 & 0.80 & 0.80 & 0.80 & 0.80 & 0.80 \\ C_7 & 0.40 & 0.40 & 0.42 & 0.40 & 0.42 & 0.41 \\ C_8 & 2.02 & 1.81 & 1.41 & 2.02 & 1.41 & 1.64 \end{array} \quad (10)$$

$$A_2(best) = \begin{array}{c|cccccc} & O_1 & O_2 & O_3 & O_4 & O_5 & O_6 \\ \hline C_1 & 0.022059 & 0.022096 & 0.022134 & 0.022138 & 0.022061 & 0.022134 \\ C_2 & 347.45 & 278.82 & 411.44 & 217.78 & 279.13 & 264.30 \\ C_3 & 1.00 & 0.98 & 0.99 & 0.91 & 0.99 & 1.00 \\ C_4 & 0.15 & 0.14 & 0.12 & 0.12 & 0.15 & 0.12 \\ C_5 & 0.14 & 0.13 & 0.13 & 0.11 & 0.14 & 0.13 \\ C_6 & 0.80 & 0.80 & 0.80 & 0.80 & 0.80 & 0.80 \\ C_7 & 0.40 & 0.40 & 0.41 & 0.41 & 0.40 & 0.41 \\ C_8 & 2.02 & 1.82 & 1.65 & 1.66 & 2.03 & 1.65 \end{array} \quad (11)$$

$$A_3(worst) = \begin{array}{c|cccccc} & O_1 & O_2 & O_3 & O_4 & O_5 & O_6 \\ \hline C_1 & 0.02207 & 0.02212 & 0.02218 & 0.02229 & 0.02208 & 0.02220 \\ C_2 & 14.55 & 14.15 & 14.74 & 13.65 & 13.87 & 16.15 \\ C_3 & 1.00 & 0.99 & 0.98 & 0.96 & 0.94 & 0.90 \\ C_4 & 0.15 & 0.13 & 0.12 & 0.08 & 0.15 & 0.12 \\ C_5 & 0.14 & 0.13 & 0.14 & 0.11 & 0.13 & 0.12 \\ C_6 & 0.80 & 0.80 & 0.80 & 0.80 & 0.80 & 0.80 \\ C_7 & 0.40 & 0.41 & 0.41 & 0.43 & 0.40 & 0.41 \\ C_8 & 2.02 & 1.79 & 1.58 & 1.10 & 2.06 & 1.57 \end{array} \quad (12)$$

Looking at the obtained results, there is much less than 1 % difference concerning the objective function value J between the best among the best results ($J = 0.022059$) and the worst among the worst results ($J = 0.02229$), while the convergence time T increases more than 25 times. Such a small deviation of J proves all of the considered here six modifications of SGA as equally reliable and it is of user choice to make a compromise between the model accuracy and convergence time.

Based on three constructed *IM*s (($A_1(average)$), ($A_2(best)$) and ($A_3(worst)$)), ICrA is then implemented. Six *IM*s (IM_1 to IM_6) that determine the degrees of "agreement" (μ_{C_k,C_l}) and "disagreement" (ν_{C_k,C_l}) between criteria for three considered cases (average, best and worst) are obtained.

$$IM_1 = \begin{array}{c|cccccccc} & C_1 & C_2 & C_3 & C_4 & C_5 & C_6 & C_7 & C_8 \\ \hline C_1 & \mathbf{1} & 0.47 & 0.40 & 0.07 & 0.13 & 0.13 & 1.00 & 0.07 \\ C_2 & 0.47 & \mathbf{1} & 0.40 & 0.47 & 0.53 & 0.40 & 0.47 & 0.60 \\ C_3 & 0.40 & 0.40 & \mathbf{1} & 0.67 & 0.73 & 0.60 & 0.40 & 0.53 \\ C_4 & 0.07 & 0.47 & 0.67 & \mathbf{1} & 0.93 & 0.93 & 0.07 & 0.87 \\ C_5 & 0.13 & 0.53 & 0.73 & 0.93 & \mathbf{1} & 0.87 & 0.13 & 0.80 \\ C_6 & 0.13 & 0.40 & 0.60 & 0.93 & 0.87 & \mathbf{1} & 0.13 & 0.80 \\ C_7 & 1 & 0.47 & 0.40 & 0.07 & 0.13 & 0.13 & \mathbf{1} & 0.07 \\ C_8 & 0.07 & 0.60 & 0.53 & 0.87 & 0.80 & 0.80 & 0.07 & \mathbf{1} \end{array} \tag{13}$$

$$IM_2 = \begin{array}{c|cccccccc} & C_1 & C_2 & C_3 & C_4 & C_5 & C_6 & C_7 & C_8 \\ \hline C_1 & \mathbf{0} & 0.53 & 0.60 & 0.93 & 0.87 & 0.87 & 0.00 & 0.93 \\ C_2 & 0.53 & \mathbf{0} & 0.60 & 0.53 & 0.47 & 0.60 & 0.53 & 0.40 \\ C_3 & 0.60 & 0.60 & \mathbf{0} & 0.33 & 0.27 & 0.40 & 0.60 & 0.47 \\ C_4 & 0.93 & 0.53 & 0.33 & \mathbf{0} & 0.07 & 0.07 & 0.93 & 0.13 \\ C_5 & 0.87 & 0.47 & 0.27 & 0.07 & \mathbf{0} & 0.13 & 0.87 & 0.20 \\ C_6 & 0.87 & 0.60 & 0.40 & 0.07 & 0.13 & \mathbf{0} & 0.87 & 0.20 \\ C_7 & 0 & 0.53 & 0.60 & 0.93 & 0.87 & 0.87 & \mathbf{0} & 0.93 \\ C_8 & 0.93 & 0.40 & 0.47 & 0.13 & 0.20 & 0.20 & 0.93 & \mathbf{0} \end{array} \tag{14}$$

$$IM_3 = \begin{array}{c|cccccccc} & C_1 & C_2 & C_3 & C_4 & C_5 & C_6 & C_7 & C_8 \\ \hline C_1 & \mathbf{1} & 0.27 & 0.27 & 0.13 & 0.13 & 0.47 & 0.93 & 0.20 \\ C_2 & 0.27 & \mathbf{1} & 0.60 & 0.60 & 0.60 & 0.53 & 0.33 & 0.53 \\ C_3 & 0.27 & 0.60 & \mathbf{1} & 0.60 & 0.73 & 0.27 & 0.33 & 0.53 \\ C_4 & 0.13 & 0.60 & 0.60 & \mathbf{1} & 0.73 & 0.67 & 0.07 & 0.93 \\ C_5 & 0.13 & 0.60 & 0.73 & 0.73 & \mathbf{1} & 0.40 & 0.20 & 0.80 \\ C_6 & 0.47 & 0.53 & 0.27 & 0.67 & 0.40 & \mathbf{1} & 0.40 & 0.60 \\ C_7 & 0.93 & 0.33 & 0.33 & 0.07 & 0.20 & 0.40 & \mathbf{1} & 0.13 \\ C_8 & 0.20 & 0.53 & 0.53 & 0.93 & 0.80 & 0.60 & 0.13 & \mathbf{1} \end{array} \tag{15}$$

$$IM_4 = \begin{array}{c|cccccccc} & C_1 & C_2 & C_3 & C_4 & C_5 & C_6 & C_7 & C_8 \\ \hline C_1 & \mathbf{0} & 0.73 & 0.73 & 0.87 & 0.87 & 0.53 & 0.07 & 0.80 \\ C_2 & 0.73 & \mathbf{0} & 0.40 & 0.40 & 0.40 & 0.47 & 0.67 & 0.47 \\ C_3 & 0.73 & 0.40 & \mathbf{0} & 0.40 & 0.27 & 0.73 & 0.67 & 0.47 \\ C_4 & 0.87 & 0.40 & 0.40 & \mathbf{0} & 0.27 & 0.33 & 0.93 & 0.07 \\ C_5 & 0.87 & 0.40 & 0.27 & 0.27 & \mathbf{0} & 0.60 & 0.80 & 0.20 \\ C_6 & 0.53 & 0.47 & 0.73 & 0.33 & 0.60 & \mathbf{0} & 0.60 & 0.40 \\ C_7 & 0.07 & 0.67 & 0.67 & 0.93 & 0.80 & 0.60 & \mathbf{0} & 0.87 \\ C_8 & 0.80 & 0.47 & 0.47 & 0.07 & 0.20 & 0.40 & 0.87 & \mathbf{0} \end{array} \tag{16}$$

$IM_5 =$ (17)

	C_1	C_2	C_3	C_4	C_5	C_6	C_7	C_8
C_1	**1**	0.53	0.27	0.07	0.27	0.40	0.93	0.07
C_2	0.53	**1**	0.47	0.40	0.73	0.07	0.60	0.40
C_3	0.27	0.47	**1**	0.67	0.73	0.60	0.33	0.67
C_4	0.07	0.40	0.67	**1**	0.67	0.67	0.00	1.00
C_5	0.27	0.73	0.73	0.67	**1**	0.33	0.33	0.67
C_6	0.40	0.07	0.60	0.67	0.33	**1**	0.33	0.67
C_7	0.93	0.60	0.33	0.00	0.33	0.33	**1**	0
C_8	0.07	0.40	0.67	1	0.67	0.67	0	**1**

$IM_6 =$ (18)

	C_1	C_2	C_3	C_4	C_5	C_6	C_7	C_8
C_1	**0**	0.47	0.73	0.93	0.73	0.60	0.07	0.93
C_2	0.47	**0**	0.53	0.60	0.27	0.93	0.40	0.60
C_3	0.73	0.53	**0**	0.33	0.27	0.40	0.67	0.33
C_4	0.93	0.60	0.33	**0**	0.33	0.33	1.00	0.00
C_5	0.73	0.27	0.27	0.33	**0**	0.67	0.67	0.33
C_6	0.60	0.93	0.40	0.33	0.67	**0**	0.67	0.33
C_7	0.07	0.40	0.67	1	0.67	0.67	**0**	1
C_8	0.93	0.60	0.33	0	0.33	0.33	1	**0**

Listed above results are graphically presented in Fig. 1.

Criteria relations, arranged by μ_{C_k,C_l} values in the case of average results, are presented in Table 1. Results for the rest two cases—of the best and the worst estimates, are presented in Table 1 correspondingly to the pairs in the first column, but as obviously—not ranked. It should be noted, that for all three considered cases, there is no pairs with a degree of "uncertainty".

The definition of consonance and dissonance between each pair of criteria is done based on the scale, presented in Table 2 [16].

Looking at the average results of the examined criteria, the following pair dependencies might be outlined:

- A strong positive consonance has been observed for the pair $C_1 \leftrightarrow C_7$ (i.e. $J \leftrightarrow Y_{S/X}$). The value of the degree of "agreement" between those two criteria hits the upper boundary of the interval determining the strong positive consonance.
- There are four criteria pairs in a positive consonance—$C_4 \leftrightarrow C_5$, $C_4 \leftrightarrow C_6$, $C_4 \leftrightarrow C_8$, which show the dependencies between the specific growth rate μ_{2E} and, respectively, k_S, k_E and $Y_{E/X}$, as well as the pair $C_5 \leftrightarrow C_6$ (i.e. $k_S \leftrightarrow k_E$).
- For the pairs $C_5 \leftrightarrow C_8$ and $C_6 \leftrightarrow C_8$, which show the relations between the yield coefficient $Y_{E/X}$ and both saturation constants k_S and k_E, a weak positive consonance is identified.
- Negative consonance has been observed for eight criteria pairs that might be divided into two groups: (1) $C_1 \leftrightarrow C_4$, $C_1 \leftrightarrow C_5$, $C_1 \leftrightarrow C_6$, and $C_1 \leftrightarrow C_8$, which show the relations between the objective function value J and, respectively, the specific growth rate μ_{2E}, saturation constants k_S and k_E, and yield coefficient $Y_{E/X}$; and (2) $C_4 \leftrightarrow C_7$, $C_5 \leftrightarrow C_7$, $C_6 \leftrightarrow C_7$ and $C_7 \leftrightarrow C_8$, which show the relations

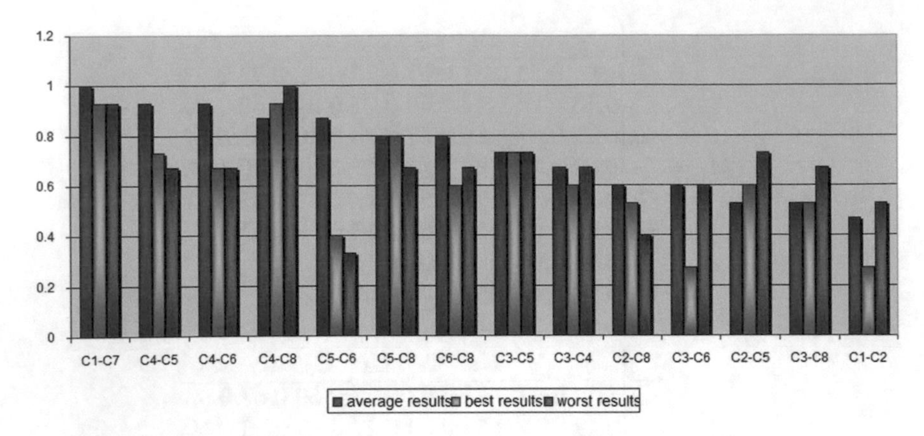

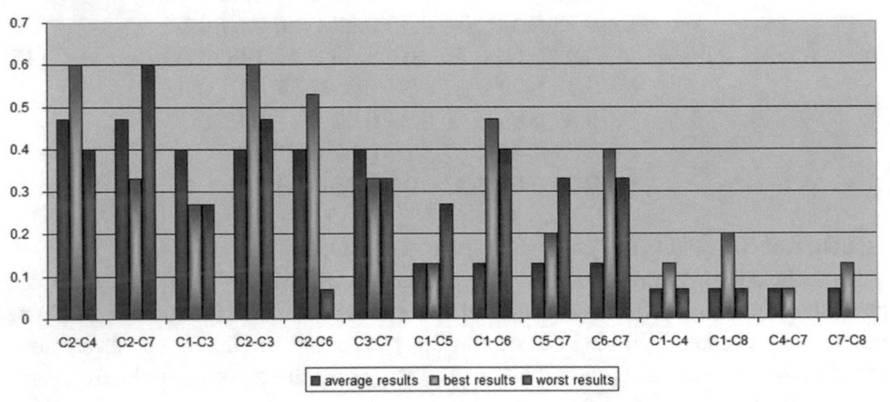

Fig. 1 Degrees of "agreement" (μ_{C_k,C_l} values) for three considered cases

between the yield coefficient $Y_{S/X}$ and, respectively, again the specific growth rate μ_{2E}, saturation constants k_S and k_E, and yield coefficient $Y_{E/X}$.

The rest criteria pairs are identified in the intervals of a weak dissonance, dissonance and strong dissonance.

Due to the stochastic nature of considered here six modifications of SGA, different criteria dependences have been observed in the cases of best and worst results. When one compares the case of best results to the case of average results, there are altogether nine coincidences: (1) for the pair $C_4 \leftrightarrow C_8$, showing a positive consonance; (2) for the pair $C_5 \leftrightarrow C_8$, showing a weak positive consonance; (3) for the pair $C_3 \leftrightarrow C_5$, showing a weak dissonance; (4) for the pair $C_3 \leftrightarrow C_4$, showing a dissonance; (5) for the pairs $C_3 \leftrightarrow C_7, C_1 \leftrightarrow C_5, C_1 \leftrightarrow C_4, C_4 \leftrightarrow C_7$ and $C_7 \leftrightarrow C_8$, showing a negative consonance. For the rest of criteria pairs weaker relations have been observed.

In the case of worst results the value of the degree of "agreement" between $C_4 \leftrightarrow C_8$ hits the upper boundary of the interval determining a strong positive consonance, in a contrast to the average results, where the only one pair with a strong positive

Table 1 Criteria relations sorted by μ_{C_k,C_l} values in the average case

Criteria relation	Obtained $\langle \mu_{C_k,C_l}, \nu_{C_k,C_l} \rangle$ values in case of:		
	Average results	Worst results	Best results
$C_1 \leftrightarrow C_7$	$\langle 1.00, 0.00 \rangle$	$\langle 0.93, 0.07 \rangle$	$\langle 0.93, 0.07 \rangle$
$C_4 \leftrightarrow C_5$	$\langle 0.93, 0.07 \rangle$	$\langle 0.73, 0.27 \rangle$	$\langle 0.67, 0.33 \rangle$
$C_4 \leftrightarrow C_6$	$\langle 0.93, 0.07 \rangle$	$\langle 0.67, 0.33 \rangle$	$\langle 0.67, 0.33 \rangle$
$C_4 \leftrightarrow C_8$	$\langle 0.87, 0.13 \rangle$	$\langle 0.93, 0.07 \rangle$	$\langle 1.00, 0.00 \rangle$
$C_5 \leftrightarrow C_6$	$\langle 0.87, 0.13 \rangle$	$\langle 0.40, 0.60 \rangle$	$\langle 0.33, 0.67 \rangle$
$C_5 \leftrightarrow C_8$	$\langle 0.80, 0.20 \rangle$	$\langle 0.80, 0.20 \rangle$	$\langle 0.67, 0.33 \rangle$
$C_6 \leftrightarrow C_8$	$\langle 0.80, 0.20 \rangle$	$\langle 0.60, 0.40 \rangle$	$\langle 0.67, 0.33 \rangle$
$C_3 \leftrightarrow C_5$	$\langle 0.73, 0.27 \rangle$	$\langle 0.73, 0.27 \rangle$	$\langle 0.73, 0.27 \rangle$
$C_3 \leftrightarrow C_4$	$\langle 0.67, 0.33 \rangle$	$\langle 0.60, 0.40 \rangle$	$\langle 0.67, 0.33 \rangle$
$C_2 \leftrightarrow C_8$	$\langle 0.60, 0.40 \rangle$	$\langle 0.53, 0.47 \rangle$	$\langle 0.40, 0.60 \rangle$
$C_3 \leftrightarrow C_6$	$\langle 0.60, 0.40 \rangle$	$\langle 0.27, 0.73 \rangle$	$\langle 0.60, 0.40 \rangle$
$C_2 \leftrightarrow C_5$	$\langle 0.53, 0.47 \rangle$	$\langle 0.60, 0.40 \rangle$	$\langle 0.73, 0.27 \rangle$
$C_3 \leftrightarrow C_8$	$\langle 0.53, 0.47 \rangle$	$\langle 0.53, 0.47 \rangle$	$\langle 0.67, 0.33 \rangle$
$C_1 \leftrightarrow C_2$	$\langle 0.47, 0.53 \rangle$	$\langle 0.27, 0.73 \rangle$	$\langle 0.53, 0.47 \rangle$
$C_2 \leftrightarrow C_4$	$\langle 0.47, 0.53 \rangle$	$\langle 0.60, 0.40 \rangle$	$\langle 0.40, 0.60 \rangle$
$C_2 \leftrightarrow C_7$	$\langle 0.47, 0.53 \rangle$	$\langle 0.33, 0.67 \rangle$	$\langle 0.60, 0.40 \rangle$
$C_1 \leftrightarrow C_3$	$\langle 0.40, 0.60 \rangle$	$\langle 0.27, 0.73 \rangle$	$\langle 0.27, 0.73 \rangle$
$C_2 \leftrightarrow C_3$	$\langle 0.40, 0.60 \rangle$	$\langle 0.60, 0.40 \rangle$	$\langle 0.47, 0.53 \rangle$
$C_2 \leftrightarrow C_6$	$\langle 0.40, 0.60 \rangle$	$\langle 0.53, 0.47 \rangle$	$\langle 0.07, 0.93 \rangle$
$C_3 \leftrightarrow C_7$	$\langle 0.40, 0.60 \rangle$	$\langle 0.33, 0.67 \rangle$	$\langle 0.33, 0.67 \rangle$
$C_1 \leftrightarrow C_5$	$\langle 0.13, 0.87 \rangle$	$\langle 0.13, 0.87 \rangle$	$\langle 0.27, 0.73 \rangle$
$C_1 \leftrightarrow C_6$	$\langle 0.13, 0.87 \rangle$	$\langle 0.47, 0.53 \rangle$	$\langle 0.40, 0.60 \rangle$
$C_5 \leftrightarrow C_7$	$\langle 0.13, 0.87 \rangle$	$\langle 0.20, 0.80 \rangle$	$\langle 0.33, 0.67 \rangle$
$C_6 \leftrightarrow C_7$	$\langle 0.13, 0.87 \rangle$	$\langle 0.40, 0.60 \rangle$	$\langle 0.33, 0.67 \rangle$
$C_1 \leftrightarrow C_4$	$\langle 0.07, 0.93 \rangle$	$\langle 0.13, 0.87 \rangle$	$\langle 0.07, 0.93 \rangle$
$C_1 \leftrightarrow C_8$	$\langle 0.07, 0.93 \rangle$	$\langle 0.20, 0.80 \rangle$	$\langle 0.07, 0.93 \rangle$
$C_4 \leftrightarrow C_7$	$\langle 0.07, 0.93 \rangle$	$\langle 0.07, 0.93 \rangle$	$\langle 0.00, 1.00 \rangle$
$C_7 \leftrightarrow C_8$	$\langle 0.07, 0.93 \rangle$	$\langle 0.13, 0.87 \rangle$	$\langle 0.00, 1.00 \rangle$

consonance was $C_1 \leftrightarrow C_7$. Many other coincidences and discrepancies might be outlined between three cases, but for the case of worst results the pairs $C_4 \leftrightarrow C_7$ and $C_7 \leftrightarrow C_8$ should be distinguished, as the only two pairs with "zero" degree of "agreement".

Going deeply in the results, presented in Table 1, altogether 4 pairs have been observed of the absolute coincidence for the three considered cases—of average, best and worst results, namely: (1) the pair $C_3 \leftrightarrow C_5$, showing a weak dissonance; (2) the pairs $C_3 \leftrightarrow C_4$ and $C_3 \leftrightarrow C_7$, showing a dissonance; (3) the pair $C_1 \leftrightarrow C_4$, showing a negative consonance.

Table 2 Scale of consonance and dissonance

Interval of μ_{C_k,C_l}	Meaning
[0.00–0.05]	Strong negative consonance
(0.05–0.15]	Negative consonance
(0.15–0.25]	Weak negative consonance
(0.25–0.33]	Weak dissonance
(0.33–0.43]	Dissonance
(0.43–0.57]	Strong dissonance
(0.57–0.67]	Dissonance
(0.67–0.75]	Weak dissonance
(0.75–0.85]	Weak positive consonance
(0.85–0.95]	Positive consonance
(0.95–1.00]	Strong positive consonance

Considering the obtained ICrA estimations in the cases of average, best and worst results, and having in mind the stochastic nature of GA, it is more reasonable to rely with a higher credibility to the results in the case of average values than to results obtained in another two cases.

6 Conclusion

Recently developed ICrA approach has been here applied to examine the performance of six SGA modifications for the purposes of fermentation process model parameter identification. Very small observed deviations (less than 1 %) of the objective function value prove all of the considered here SGA modifications as equally reliable and it is of user choice to make a compromise between model accuracy and convergence time. ICrA implementation succeeds at establishing of some relations between model parameters of a *S. cerevisiae* fed-batch fermentation process, when six SGA modifications have been applied. Three case studies have been examined—of average, best and worst results for the obtained objective function value, convergence time and model parameters. Implementing ICrA, the thorough discussion about the consonance or dissonance between criteria is presented. ICrA application leads to reveal the mutual relations between model parameters themselves, and GA outcomes, such as objective function value and convergence time. Such kind of additional knowledge about the model parameters relations might be extremely useful aiming at improvement of model accuracy in further parameter identification procedures. On the other hand, such information could be used to amend the performance of the applied optimization algorithms when, which is the case, some algorithm outcomes are considered as criteria.

Acknowledgements The work is supported by the Bulgarian National Scientific Fund under the grant DFNI-I-02-5 "InterCriteria Analysis—A New Approach to Decision Making".

References

1. Angelova, M., Melo-Pinto, P., Pencheva, T.: Modified simple genetic algorithms improving convergence time for the purposes of fermentation process parameter identification. WSEAS Trans. Syst. **11**(7), 256–267 (2012)
2. Angelova, M., Roeva, O., Pencheva, T.: InterCriteria analysis of crossover and mutation rates relations in simple genetic algorithm. Ann. Comput. Sci. Info. Syst. **5**, 419–424 (2015)
3. Angelova, M., Tzonkov, St., Pencheva, T.: Genetic Algorithms based parameter identification of yeast fed-batch cultivation. In: LNCS, vol. 6046, pp. 224–231 (2011)
4. Atanassov, K.: Generalized index matrices. C. R. Acad. Bulg. Sci. **40**(11), 15–18 (1987)
5. Atanassov, K.: On index matrices, Part 1: Standard cases. Adv. Stud. Contemp. Math. **20**(2), 291–302 (2010)
6. Atanassov, K.: On index matrices, Part 2: Intuitionistic fuzzy case. Proc. Jangjeon Math. Soc. **13**(2), 121–126 (2010)
7. Atanassov, K.: On Intuitionistic Fuzzy Sets Theory. Springer, Berlin (2012)
8. Atanassov, K., Mavrov, D., Atanassova, V.: Intercriteria Decision Making: A new approach for multicriteria decision making, based on index matrices and intuitionistic fuzzy sets. In: Issues in Intuitionistic Fuzzy Sets and Generalized Nets, vol. 11, pp. 1–8 (2014)
9. Atanassov, K., Szmidt, E., Kacprzyk, J.: On intuitionistic fuzzy pairs. Notes Int. Fuz. Sets **19**(3), 1–13 (2013)
10. Ghaheri, A., Shoar, S., Naderan, M., Hoseini, S.S.: The applications of genetic algorithms in medicine. Oman Med. J. **30**(6), 406–416 (2015)
11. Goldberg, D.E.: Genetic Algorithms in Search. Optimization and Machine Learning. Addison Wesley Longman, London (2006)
12. Ilkova, T., Petrov, M.: Intercriteria analysis for identification of *Escherichia coli* fed-batch mathematical model. J. Int. Sci. Publ.: Mater., Meth. Technol. **9**, 598–608 (2015)
13. Pencheva, T., Roeva, O., Hristozov, I.: Functional State Approach to Fermentation Processes Modelling. Prof. M. Drinov Acad. Publ. House, Sofia (2006)
14. Roeva, O. (ed.): Real-world Application of Genetic Algorithms. InTech (2012)
15. Roeva, O., Fidanova, S.: A comparison of genetic algorithms and ant colony optimization for modeling of *E. coli* cultivation process. In: Real-world Application of Genetic Algorithms, pp. 261–282. InTech (2012)
16. Roeva, O., Fidanova, S., Vassilev, P., Gepner, P.: InterCriteria analysis of a model parameters identification using genetic algorithm. Ann. Comput. Sci. Inf. Syst. **5**, 501–506 (2015)

Interval Algebraic Approach to Equilibrium Equations in Mechanics

Evgenija D. Popova

Abstract The engineering demand for more realistic and accurate models involving interval uncertainties lead to a new interval model of linear equilibrium equations in mechanics, which is based on the algebraic completion of classical interval arithmetic (called Kaucher arithmetic). Interval algebraic approach consists of three parts: representation convention, computing algebraic solution and result interpretation. The proposed approach replaces straightforward a deterministic model by an interval model in terms of proper and improper intervals, fully conforms to the equilibrium principle and provides sharper enclosure of the unknown quantities than the best known methods based on classical interval arithmetic. Numerical applications described by systems of linear interval equilibrium equations where the number of the unknowns is equal to the number of the equations are considered in details.

1 Introduction

The basic principle of static (or dynamic) equilibrium under general force systems is an essential prerequisite for many branches of engineering, such as mechanical, civil, aeronautical, bioengineering, robotics, and others that address the various consequences of forces [1].

One main challenge for the models involving interval uncertainty is the overestimation of the system response. Nowadays, the most successful approaches for overestimation reduction are those that relate the dependency of interval quantities to the physics of the problem being considered, [8]. Recently, a model of a bar subjected to multiple axial external loads, where load magnitudes are represented by intervals, is considered in [3]. Although the aim at providing interval model conforming to the principle of static equilibrium is not completely achieved by the proposed model, the problem and its challenge are presented. A similar problem in the context of robotics is discussed in the IEEE P1788 working group on standardization of inter-

E.D. Popova (✉)
Institute of Mathematics and Informatics, Bulgarian Academy of Sciences,
Acad. G. Bonchev str., block 8, 1113 Sofia, Bulgaria
e-mail: epopova@math.bas.bg; epopova@bio.bas.bg

K. Georgiev et al. (eds.), *Advanced Computing in Industrial Mathematics*,
Studies in Computational Intelligence 681, DOI 10.1007/978-3-319-49544-6_14

val arithmetic, [6]. It is shown in [3, 6] that an interval model based entirely on the classical interval arithmetic, in its set-theoretic interpretation as proposed by Moore [7], cannot provide a good estimation neither of the uncertain reaction nor of the load distribution.

The engineering demand for more accurate models involving interval uncertainties lead to an interval model of linear equilibrium equations in mechanics [14], which is based on the algebraic completion of classical interval arithmetic (called also Kaucher or generalized interval arithmetic). It is proven that the proposed interval model always yields the narrowest interval enclosure and is in full conformance with the physical meaning of static equilibrium. The work [14] is focused on justification of the proposed interval model in one dimension, comparison to the approach of [3], and applications to computing resultant forces. In this paper we further develop the interval algebraic approach to models involving linear interval equilibrium equations. Considered are models of practical applications which reduce to systems of linear interval equilibrium equations where the number of the unknowns is equal to the number of the equations. The initial interval model is expanded by considering interval algebraic solution to the system of equilibrium equations, model properties are revealed and the quality of the interval algebraic solution is compared to the best interval solution enclosure obtained by classical interval arithmetic.

The structure of the paper is as follows. In the next section some basic notions and properties of the algebraic extension [4] of classical interval arithmetic are summarized. In Sect. 3 we present the new interval model and its generalization to systems of interval equilibrium equations involving as many unknowns as is the number of the equations. Numerical applications developed in details in Sect. 4 illustrate the proposed interval algebraic approach, its conformance to the equilibrium principle, bring out its effectiveness and advantages over the approach based on classical interval arithmetic. The article ends by some conclusions.

2 The Algebraic Completion of $\mathbb{IR}$

The set of classical compact intervals $\mathbb{IR} = \{[a^-, a^+] \mid a^-, a^+ \in \mathbb{R}, a^- \leq a^+\}$, called also *proper* intervals, is extended in [4] by the set $\overline{\mathbb{IR}} := \{[a^-, a^+] \mid a^-, a^+ \in \mathbb{R}, a^- \geq a^+\}$ of *improper* intervals obtaining thus the set $\mathbb{KR} = \mathbb{IR} \bigcup \overline{\mathbb{IR}} = \{[a^-, a^+] \mid a^-, a^+ \in \mathbb{R}\}$ of all ordered couples of real numbers called *generalized* (extended or Kaucher) intervals. The inclusion order relation between classical intervals[1] $\subseteq$ is generalized for $[a], [b] \in \mathbb{KR}$ by $[a] \subseteq [b] \Longleftrightarrow b^- \leq a^-$ and $a^+ \leq b^+$. For $[a] = [a^-, a^+] \in \mathbb{KR}$ define binary variable *direction* (τ) by $\tau([a]) := \mathrm{sgn}(a^+ - a^-) = \{+ \text{ if } a^- \leq a^+, \ - \text{ if } a^- > a^+\}$. All elements of $\mathbb{KR}$ with positive direction are called proper intervals and the elements with negative direction are called improper intervals. An element-to-element symmetry between proper and improper

[1]For a better understanding we denote the classical intervals by bold face letters and the intervals from $\mathbb{KR}$ by brackets $[a]$. Of course, $\mathbf{a} \in \mathbb{IR} \subset \mathbb{KR}$, and thus $[a] = \mathbf{a} \in \mathbb{KR}$ is a correct assignment.

intervals is expressed by the "Dual" operator. For $[a] = [a^-, a^+] \in \mathbb{KR}$, $\mathrm{Dual}([a]) := [a^+, a^-]$. For $[a], [b] \in \mathbb{KR}$,

$$\mathrm{Dual}(\mathrm{Dual}([a])) = [a], \tag{1}$$

$$\mathrm{Dual}([a] \circ [b]) = \mathrm{Dual}([a]) \circ \mathrm{Dual}([b]), \quad \circ \in \{+, -, \times, /\}. \tag{2}$$

Define proper projection of a generalized interval $[a]$ onto $\mathbb{IR}$ by

$$\mathrm{pro}([a]) := \begin{cases} [a] & \text{if } \tau([a]) = +, \\ \mathrm{Dual}([a]) & \text{if } \tau([a]) = -. \end{cases} \tag{3}$$

Define binary variable "sign" (σ) by $\sigma([a]) := \begin{cases} + & \text{if } \mathrm{pro}([a])^- \geq 0, \\ - & \text{otherwise.} \end{cases}$ Denote $\mathscr{T} := \{[a] \in \mathbb{KR} \mid [a] = [0, 0] \text{ or } a^- a^+ < 0\}$. The conventional interval arithmetic and lattice operations, as well as other interval functions are isomorphically extended onto the whole set $\mathbb{KR}$, [4]. A condensed representation of the arithmetic operations is derived in [2], Thus,

$$[a] + [b] = [a^- + b^-, a^+ + b^+] \quad \text{for } [a], [b] \in \mathbb{KR},$$

$$[a] \times [b] = \begin{cases} [a^{-\sigma([b]} b^{-\sigma([a])}, a^{\sigma([b]} b^{\sigma([a])}] & \text{if } [a], [b] \in \mathbb{KR} \backslash \mathscr{T} \\ [a^{\sigma([a]\tau([b])} b^{-\sigma([a])}, a^{\sigma([a]\tau([b]} b^{\sigma([a])}] & \text{if } [a] \in \mathbb{KR} \backslash \mathscr{T}, [b] \in \mathscr{T} \\ [a^{-\sigma([b]} b^{\sigma([b])\tau([a])}, a^{\sigma([b])} b^{\sigma([b])\tau([a])}] & \text{if } [a] \in \mathscr{T}, [b] \in \mathbb{KR} \backslash \mathscr{T} \\ [\min\{a^- b^+, a^+ b^-\}, \max\{a^- b^-, a^+ b^+\}] & \text{if } [a], [b] \in \mathscr{T}, \tau([a]) = \tau([b]) \\ 0 & \text{if } [a], [b] \in \mathscr{T}, \tau([a]) \neq \tau([b]), \end{cases}$$

wherein $++ = -- = +$, $+- = -+ = -$. Interval subtraction and division can be expressed as composite operations, $[a] - [b] = [a] + (-1)[b]$ and $[a]/[b] = [a] \times (1/[b])$, where $1/[b] = [1/b^+, 1/b^-]$ if $[b] \in \mathbb{KR} \backslash \mathscr{T}$. The restrictions of the arithmetic operations to proper intervals produce the familiar operations in the conventional interval space.

The generalized interval arithmetic structure possesses group properties with respect to the operations addition and multiplication. For $[a] \in \mathbb{KR}$, $[b] \in \mathbb{KR} \backslash \mathscr{T}$,

$$[a] - \mathrm{Dual}([a]) = 0, \qquad [b]/\mathrm{Dual}([b]) = 1. \tag{4}$$

The complete set of conditionally distributive relations for multiplication and addition of generalized intervals can be found in [10, 11]. Here we present only one that will be used. For $[a], [b], [s] = ([a] + [b]) \in \mathbb{KR} \backslash \mathscr{T}$, $[c] \in \mathbb{KR}$

$$([a] + [b])[c]_{\sigma([s])} = [a]_{\sigma([a])} + [b]_{\sigma([b])}, \tag{5}$$

wherein $[a]_+ = [a]$, $[a]_- = \mathrm{Dual}([a])$. Addition operation in $\mathbb{KR}$ is commutative and associative; associativity does not hold true in (interval) floating point arithmetic.

Lattice operations are closed with respect to the inclusion relation; handling of norm and metric are very similar to norm and metric in linear spaces, [4]. Some other properties and applications of generalized interval arithmetic can be found in [2, 4, 5, 10, 11, 15, 17] and the references given therein.

For $\mathbf{a} \in \mathbb{IR} \backslash \mathscr{T}$, define $\text{Abs}(\mathbf{a}) = \{\mathbf{a} \text{ if } 0 \leq \mathbf{a}; -\mathbf{a} \text{ otherwise}\}$. Relative diameter of $\mathbf{a} \in \mathbb{IR}$ is defined as $a^+ - a^-$ if $0 \in \mathbf{a}$ and $(a^+ - a^-)/\min\{|a^-|, |a^+|\}$ otherwise.

3 Interval Model of Equilibrium Equations

In this section the algebraic approach to equilibrium equations in mechanics is derived by considering two-dimensional problems involving several forces acting on a particle. The same approach with obvious modifications is applicable to three-dimensional problems and problems whose models involve other vector physical quantities possessing magnitude and direction such as velocities, accelerations, or momenta. Such problems will be illustrated in the next section. In the text of this paper forces (and other vector quantities) are denoted by underlining the letter used to represent it. This is necessary in order to distinguish vectors from the proper intervals, which are denoted by bold-face letters, and from the real-valued scalars. The magnitude of a vector will be denoted by the corresponding italic-face letter.

In the deterministic case of two- dimensional problems involving several forces, the determination of their resultant $\underline{R}$ is best carried out by first resolving each force into rectangular components. Choosing a rectangular coordinate system (Oxy), with unit vectors $\underline{i}$, $\underline{j}$, any force vector $\underline{F}$ can be resolved into rectangular components $\underline{F}_x = F_x\underline{i}$, and $\underline{F}_y = F_y\underline{j}$, so that $\underline{F} = F_x\underline{i} + F_y\underline{j}$. The scalar component F_x is positive when the vector component $\underline{F}_x$ has the same direction as the unit vector $\underline{i}$ (i.e., the same direction as the positive x axis) and is negative when $\underline{F}_x$ has the opposite direction. A similar conclusion may be drawn regarding the sign of the scalar component F_y. Denoting by F the magnitude of the force $\underline{F}$ and by θ the angle between $\underline{F}$ and the axis x, measured counterclockwise from the positive axis, we may express the scalar components of $\underline{F}$ as follows: $F_x = F\cos(\theta)$ and $F_y = F\sin(\theta)$, cf. any textbook in statics, e.g., [1]. When more than one force act on a particle (or a rigid body), it is important to determine the resultant force, i.e., the single force $\underline{R}$ which has the same effect on the particle as the given forces. The resultant force $\underline{R}$ can be determined by:

1. choosing a rectangular coordinate system;
2. resolving the given forces into their rectangular components;
3. each scalar component R_x, R_y of the resultant $\underline{R}$ of several forces $\underline{F}_i$ acting on a particle is obtained by adding algebraically the corresponding scalar components of the given forces. That is, $R_x = \sum_i F_{x,i}$, $R_y = \sum_i F_{y,i}$, which gives $\underline{R} = R_x\underline{i} + R_y\underline{j}$.

Basing on the above, the one dimensional interval algebraic model for computing the resultant force (and reaction), developed in [14], can be applied to two- and three-dimensional problems involving vector physical quantities.

Theorem 1 ([14]) *Consider a bar subjected to a finite number of loads $\underline{p}_1, \ldots, \underline{p}_k$ that may be applied in opposite directions and have uncertain magnitude $p_1 \in \mathbf{p_1}, \ldots, p_k \in \mathbf{p_k}$, $\mathbf{p}_i \geq 0$, $i = 1, \ldots, k$. Assume that a coordinate system (Ox) is chosen. Then,*

(i) for every j, $1 \leq j \leq k$, we have $[N_j] = \sum_{i=1}^{j} [p_i]$, wherein

$$[p_i] = \begin{cases} \mathbf{p}_i & \text{if the direction of } \underline{p}_i \text{ is in the positive } x \text{ axis} \\ -\mathrm{Dual}(\mathbf{p}_i) & \text{if the direction of } \underline{p}_i \text{ is opposite to the positive } x \text{ axis,} \end{cases}$$

and $[r] = -\mathrm{Dual}([N_k]) = -\mathrm{Dual}(\sum_{i=1}^{k} [p_i])$.

(ii) The interpretation of $[N_j] \in \mathbb{KR}$, $1 \leq j \leq k$, and similarly of $[r]$, is as follows.

- *If $[N_j] \in \mathcal{T}$, then $\underline{N}_j$ may have positive or negative direction and its magnitude varies in pro($[N_j]$).*
- *If $[N_j] \in \mathbb{KR} \setminus \mathcal{T}$, the magnitude of $\underline{N}_j$ varies in Abs(pro($[N_j]$)), while the direction of $\underline{N}_j$ coincides with the sign of $[N_j]$ (if $[N_j] \geq 0$ the direction of $\underline{N}_j$ is the positive x axis, otherwise it is opposite to the positive x axis).*

Strong proof that Theorem 1 provides sharpest estimation of the resultant force and its reaction is given in [14] along with a detailed discussion and examples.

Now we consider the interval algebraic model of equilibrium equations from a more general perspective. Assume that there is a deterministic model described by some linear equilibrium equation(s) that involve uncertain parameters varying within given proper intervals. Clearly, the unknowns in this model will be also uncertain and we search for proper intervals that are the sharpest interval estimations of these unknowns and that conform to the physics of the problem (statics or dynamic equilibrium). Conformance to static (dynamic) equilibrium means that the intervals found for the unknowns when replaced in the equation(s) and all operations are performed results in true equality(ies).

Definition 1 ([16]) Interval *algebraic solution* to a (system of) interval equation(s) is an interval (interval vector) which substituted in the equation(s) and performing all interval operations in exact arithmetic[2] results in valid equality(ies).

Interval algebraic solutions do not exist in general in classical interval arithmetic [16]. Generalized interval arithmetic on proper and improper intervals $(\mathbb{KR}, +, \times, \subseteq)$ is the natural arithmetic for finding algebraic solutions to interval equations since it is obtained from the arithmetic for classical intervals $(IR, +, -, \times, /, \subseteq)$ via an algebraic completion. This is another justification of the proposed interval algebraic approach. Therefore, we embed the initial problem formulation in the interval space

[2]No round-off errors.

$(\mathbb{KR}, +\times, \subseteq)$, find an algebraic solution (if exists) and interpret the obtained generalized intervals back in the initial interval space $\mathbb{IR}$. This is a three steps procedure summarized below.

1. The **representation convention** for a model involving interval forces (and/or other physical quantities considered as vectors and possessing magnitude and direction) is:

- a scalar force component F_x (F_y, F_z) involving any kind of uncertainty is represented by proper interval $\mathbf{F}_x$ ($\mathbf{F}_y$, $\mathbf{F}_z$) if the force component $\underline{F}_x$ ($\underline{F}_y$, $\underline{F}_z$) has the same direction as the positive x (y, z) coordinate axis;
- a scalar force component F_x (F_y, F_z) involving any kind of uncertainty is represented by the improper interval Dual($\mathbf{F}_x$) (Dual($\mathbf{F}_y$), Dual($\mathbf{F}_z$)) if the force component $\underline{F}_x$ ($\underline{F}_y$, $\underline{F}_z$) has opposite direction to the corresponding positive x (y, z) coordinate axis.

2. Computing. Find the **algebraic solution** for the unknown(s) in $(\mathbb{KR}, +, \times, \subseteq)$. Conditions for existence of algebraic solution of interval linear equations are published in [10, 17]. Numerical methods finding the algebraic solution to an interval linear system are discussed in [5, 17]. For small systems, the approach based on *equivalent algebraic transformations* is transparent and will be used in this paper.

3. Interpretation of the obtained generalized intervals in the initial space $\mathbb{IR}$ is done according to the physics of the unknowns. If it is a force component, then Theorem 1 (ii) is applied. In general the interpretation projects the generalized interval solution on $\mathbb{IR}$ by (3).

Since computing a resultant $\underline{R}$ of several forces $\underline{F}_i$ can be represented as a solution of the equilibrium equation $\sum_i \underline{F}_i - \underline{R} = 0$, Theorem 1 is a special case of the above more general interval algebraic approach.

4 Numerical Applications

Here we consider models of practical applications which reduce to systems of linear interval equilibrium equations where the number of the unknowns is equal to the number of the equations. In order to avoid many technical details that will hamper

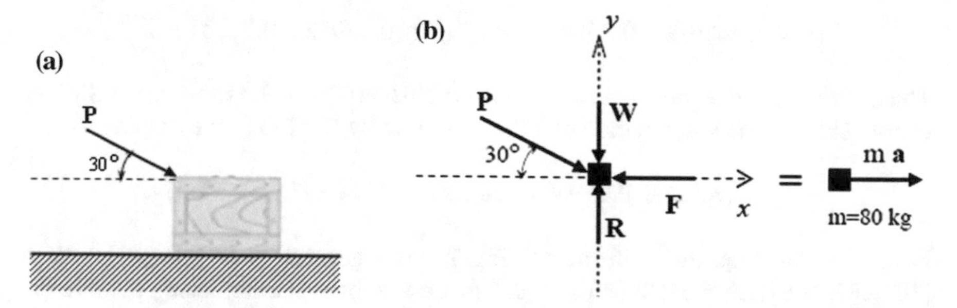

Fig. 1 **a** A force acting on a block that rests on a horizontal plane; **b** the free-body diagram

the comprehension, only two dimensional problems are considered. The numerical results presented in this section are obtained by the *Mathematica*® package `directed.m` [15]. JInterval library [9] can be used for this purpose, too.

Example 1 An 80 kg block rests on a horizontal plane, Fig. 1a. Find the magnitude of the force $\underline{P}$ required to give the block an acceleration of 2.5 m/s^2 to the right. The coefficient of kinetic friction between the block and the plane is $\mu_k = 0.25$. Assume that the mass of the block and the angle, at which the force acts on the block, are measured with 1 % uncertainty.

The chosen coordinate system is presented on the free-body diagram in Fig. 1b. Note that $\mathbf{F} = \mu_k \mathbf{R}$. The weight of the block is[3]

$$W = mg_0 \in ([79.2, 80.8]\,\text{kg})(9.80665\,\text{m/s}^2) \in [776.686, 792.378]\,\text{N}.$$

Writing Newton's second law $\sum \underline{F} = m\underline{a}$ in rectangular components and applying the representation convention, we obtain the following interval equilibrium equations

$$\mathbf{P}\cos([\theta]) - \text{Dual}(0.25\mathbf{R}) = 2.5\,\mathbf{m} \tag{6}$$

$$\mathbf{R} - \text{Dual}(\mathbf{P}\sin([\theta])) - \text{Dual}(\mathbf{W}) = 0, \tag{7}$$

where $[\theta] = [29.7^\circ, 30.3^\circ]$. We search for proper intervals **P**, **R** which satisfy these equations. Adding $\mathbf{P}\sin([\theta]) + \mathbf{W}$ to the two sides of Eq. (7) and applying property (4), we obtain

$$[R] = \mathbf{P}\sin([\theta]) + \mathbf{W}.$$

Replacing $[R]$ in the first equilibrium equation (6), we have

$$\mathbf{P}\cos([\theta]) - 0.25\text{Dual}(\mathbf{P}\sin([\theta])) - 0.25\text{Dual}(\mathbf{W}) = 2.5\,\mathbf{m}. \tag{8}$$

The distributive relation (5) holds true for the first two terms of (8) since

[3]All computed numerical intervals are outwardly rounded to the intervals presented in the paper.

$$[s] = \cos([\theta]) - 0.25\text{Dual}(\sin([\theta]) \in [0.739530, 0.7425] > 0.$$

Thus, the Eq. (8) is equivalent to $[P][s] - 0.25\text{Dual}(\mathbf{W}) = 2.5\,\mathbf{m}$. Adding $0.25\,\mathbf{W}$ to both sides of the last equation and then dividing by $\text{Dual}([s])$, we obtain

$$[P] = (2.5\,\mathbf{m} + 0.25\,\mathbf{W})/\text{Dual}([s]) \in [530.297, 538.848]\,\text{N}.$$

From the last equivalent form of Eq. (7), we get $[R] = [P]\sin([\theta]) + \mathbf{W} \in [1039.42, 1064.25]\,\text{N}$. Both $[P]$ and $[R]$ are proper intervals. Replacing them in the initial Eqs. (6) and (7) we obtain $[-1.71 \times 10^{-13}, 1.14 \times 10^{-13}]$, $[-4.55 \times 10^{-13}, 4.55 \times 10^{-13}]$, respectively. These intervals are almost but not exactly zero due to the round-off errors and show that the equilibrium equations are completely satisfied.

Now, we compare the solution $\mathbf{P}, \mathbf{R}$, obtained by the discussed algebraic approach, to the solution obtained by classical interval arithmetic. In classical interval arithmetic the goal is to find the smallest interval vector enclosing the so-called united solution set[4] of the interval system

$$\begin{pmatrix} \cos([\theta]) & -0.25 \\ -\sin([\theta]) & 1 \end{pmatrix} \begin{pmatrix} P \\ R \end{pmatrix} = \begin{pmatrix} 2.5\text{ m} \\ 9.80665\text{ m} \end{pmatrix}, \qquad \begin{matrix} \theta \in [29.7^\circ, 30.3^\circ], \\ m \in ([79.2, 80.8]. \end{matrix}$$

The smallest interval vector that encloses the united solution set of this system is $(\tilde{\mathbf{P}}, \tilde{\mathbf{R}})^\top = ([526.56, 542.68], [1037.58, 1066.18])^\top$. The percentage by which $(\tilde{\mathbf{P}}, \tilde{\mathbf{R}})^\top$ overestimates $(\mathbf{P}, \mathbf{R})^\top$ is $(46.9, 13.2)^\top\%$.

If we consider the same problem with 2 % relative uncertainty in the angle and 1 % relative uncertainty in the mass of the block, then the percentage by which $(\tilde{\mathbf{P}}, \tilde{\mathbf{R}})^\top$ overestimates $(\mathbf{P}, \mathbf{R})^\top$ is $(70.2, 20.9)^\top\%$.

Since we are looking for proper algebraic solutions of the interval equilibrium system, this restriction may not be always satisfied. The latter case is illustrated by the next example.

Example 2 A $[100 \pm 1]$ kg crate is suspended from a pulley that can roll freely on the support cable ACB and is pulled at a constant speed by cable CD, Fig. 2. If $\alpha = 30^\circ$, $\beta = 10^\circ$ and the angles are measured with 1 % uncertainty, determine the tension (a) in the support cable ACB, (b) in the traction cable CD.

The chosen coordinate system is presented on the free-body diagram in Fig. 2. The deterministic equilibrium equations of force x and y components are

$$F_{\text{ACB}}\cos(10^\circ) - F_{\text{ACB}}\cos(30^\circ) - F_{\text{CD}}\cos(30^\circ) = 0, \tag{9}$$

$$F_{\text{ACB}}\sin(10^\circ) + F_{\text{ACB}}\sin(30^\circ) + F_{\text{CD}}\sin(30^\circ) - 100 \times 9.80665 = 0. \tag{10}$$

The representation convention gives the interval equilibrium equations

[4]For $A(p)x = b(p)$, $p \in \mathbf{p}$, the united solution set is $\Sigma = \{x \in \mathbb{R}^n \mid (\exists p \in \mathbf{p})(A(p)x = b(p))\}$.

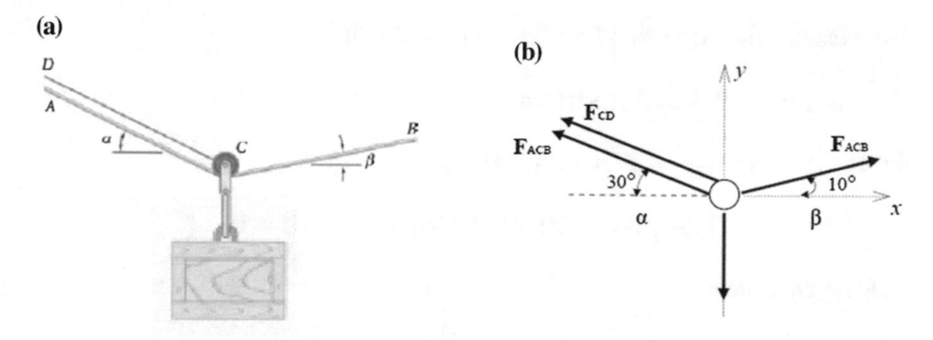

Fig. 2 **a** A crate suspended from a pulley can roll freely on the support cable ACB and is pulled at a constant speed by cable CD; **b** Free-body diagram

$$[F_{\text{ACB}}]\cos([\beta]) - \text{Dual}(\mathbf{F}_{\text{ACB}}\cos([\alpha])) - \text{Dual}(\mathbf{F}_{\text{CD}}\cos([\alpha])) = 0, \quad (11)$$

$$[F_{\text{ACB}}]\sin([\beta]) + [F_{\text{ACB}}]\sin([\alpha]) + [F_{\text{CD}}]\sin([\alpha]) - \text{Dual}([99, 100] \times 9.80665) = 0, \quad (12)$$

wherein $[\alpha] = [29, 31]°$, $[\beta] = [9, 11]°$. We search for proper intervals $\mathbf{F}_{\text{ACB}}$, $\mathbf{F}_{\text{CD}}$, that satisfy (11) and (12). First, we check the validity of the distributive relations for the first two additive terms in Eqs. (11), (12). Since

$$[s_1] = \cos([\beta]) - \text{Dual}(\cos([\alpha])) \in [0.121107, 0.116479] > 0,$$
$$[s_2] = \sin([\beta]) + \sin([\alpha]) \in [0.667387, 0.679895] > 0,$$

by (5), the system (11) and (12) is equivalent to the system

$$[F_{\text{ACB}}][s_1] - \text{Dual}(\mathbf{F}_{\text{CD}}\cos([\alpha])) = 0,$$
$$[F_{\text{ACB}}][s_2] + [F_{\text{CD}}]\sin([\alpha]) - \text{Dual}([99, 100] \times 9.80665) = 0.$$

Remark 1 It is important that we check the distributive relations for every expression where we want to take a common interval variable out of brackets. For example, due to (5), and because $\cos([\alpha]) - \text{Dual}(\cos([\beta])) < 0$, the expression $[F_{\text{ACB}}]\cos([\alpha]) - \text{Dual}([F_{\text{ACB}}]\cos([\beta]))$ is equivalent to

$$\text{Dual}([F_{\text{ACB}}])\,(\cos([\alpha]) - \text{Dual}(\cos([\beta]))).$$

We add $[F_{\text{CD}}]\cos([\alpha])$ to the two sides of Eq. (11) and by (4) obtain the equivalent equation

$$[F_{\text{ACB}}][s_1] = [F_{\text{CD}}]\cos([\alpha]).$$

Dividing both sides of the last equation by $\text{Dual}(\cos([\alpha]))$, and due to (4), we obtain

$$[F_{\text{CD}}] = [F_{\text{ACB}}][s_1]/\text{Dual}(\cos([\alpha])). \quad (13)$$

We substitute the expression for $[F_{\text{CD}}]$ in Eq. (12). Since

$$[s_3] = [s_2] + \sin([\alpha])[s_1]/\text{Dual}(\cos([\alpha])) \in [5.25337, 5.57483] > 0,$$

due to the distributive relation, Eq. (12) is equivalent to

$$[F_{\text{ACB}}][s_3] - \text{Dual}([99, 100] \times 9.80665) = 0,$$

which is equivalent to

$$[F_{\text{ACB}}] = [99, 100] \times 9.80665/\text{Dual}([s_3]) \in [1317.51, 1324.97]. \quad (14)$$

Substituting (14) in (13), we obtain the second component of the algebraic solution to interval system (11) and (12)

$$[F_{\text{CD}}] \in [184.806, 177.669].$$

Substituting $[F_{\text{ACB}}]$ and $[F_{\text{CD}}]$ into left sides of the Eqs. (11) and (12), we obtain respectively $[-2.27 \times 10^{-13}, 2.27 \times 10^{-13}]$ and $[-5.68 \times 10^{-13}, 4.54 \times 10^{-13}]$. These intervals are almost but not exactly zero due to the round-off errors. We have to interpret $[F_{\text{ACB}}]$ and $[F_{\text{CD}}]$ in $\mathbb{IR}$ as the corresponding proper intervals, namely,

$$\mathbf{F}_{\text{ACB}} = \text{Abs}(\text{pro}([F_{\text{ACB}}])) \in [1317.51, 1324.97] \text{ N},$$
$$\mathbf{F}_{\text{CD}} = \text{Abs}(\text{pro}([F_{\text{CD}}])) \in [177.669, 184.806] \text{ N}.$$

However, $[F_{\text{CD}}]$ is an improper interval. Therefore, substituting $\mathbf{F}_{\text{ACB}}$ and $\mathbf{F}_{\text{CD}}$ into left sides of the Eqs. (11) and (12), we obtain much wider intervals involving zero, namely, $[6.16307, -6.20045]$ and $[-3.53667, 3.60141]$, respectively. The relative diameters of $\mathbf{F}_{\text{ACB}}$ and $\mathbf{F}_{\text{CD}}$ are 0.00565 and 0.0402, respectively.

Remark 2 Proper algebraic solution to the system (11) and (12) can be obtained if, for example, we squeeze the interval $[\alpha]$ to the interval $[30 - 0.1, 30 + 0.1]$.

Now, we compare the solution $\mathbf{F}_{\text{ACB}}$, $\mathbf{F}_{\text{CD}}$, obtained by the discussed algebraic approach, to the solution obtained by classical interval arithmetic. The Eqs. (9) and (10) are rearranged to

$$F_{\text{ACB}}\,(\cos(10°) - \cos(30°)) - F_{\text{CD}}\cos(30°) = 0,$$
$$F_{\text{ACB}}\,(\sin(10°) + \sin(30°)) + F_{\text{CD}}\sin(30°) = 100 \times 9.80665$$

and the corresponding interval linear system that has to be solved is

$$\begin{pmatrix} \cos([\beta]) - \cos([\alpha]), & \cos([\alpha]) \\ \sin([\beta]) + \sin([\alpha]), & \sin([\alpha]) \end{pmatrix} \begin{pmatrix} F_{\text{ACB}} \\ F_{\text{CD}} \end{pmatrix} = \begin{pmatrix} 0 \\ [99, 100] \times 9.80665 \end{pmatrix}.$$

Since some interval parameters, e.g., $[\alpha]$, $[\beta]$, appear in more than one element of the matrix and/or the right-hand side vector, this is a parametric interval linear system. In classical interval arithmetic we search for a minimal outer interval estimation of the so-called united parametric solution set to the system. It can be proven, by method discussed in [12], that the united parametric solution set of the above system depends linearly on the interval parameters involved there. Therefore, one can find the minimal interval vector containing the united parametric solution set by finding the interval hull of the set of solutions to the point linear systems of equations obtained for the parameters taking values at all combinations of the corresponding interval end-points, the so-called combinatorial approach. Applying this approach, we found $\tilde{\mathbf{F}}_{\mathrm{ACB}} = [1293.33, 1349.74]$, $\tilde{\mathbf{F}}_{\mathrm{CD}} = [175.743, 186.773]$, whose relative diameters are respectively 0.04361 and 0.06276. Replacing $\tilde{\mathbf{F}}_{\mathrm{ACB}}$, $\tilde{\mathbf{F}}_{\mathrm{CD}}$ in the left-hand sides of the generalized interval equilibrium equations (11)–(12), we obtain much wider intervals involving zero $[4.89652, -5.02244]$, $[-20.6303, 21.4392]$. There is no inclusion relation between $\mathbf{F}_{\mathrm{ACB}}$, $\mathbf{F}_{\mathrm{CD}}$ and $\tilde{\mathbf{F}}_{\mathrm{ACB}}$, $\tilde{\mathbf{F}}_{\mathrm{CD}}$. Nevertheless, judging from the value of the relative diameters and the extent to which the interval equilibrium equations are satisfied, we conclude that the interval algebraic approach applied to the equilibrium equations provides sharper interval estimations than the traditional approach based on classical interval arithmetic.

In some deterministic models, e.g., when determine the forces in the members of a truss, in order to write the equilibrium equations one has to choose the direction of each of the unknown forces, cf. [1, Chap. 6]. It cannot be determined until the solution is completed whether the guess was correct. To do that, the value found for each of the unknowns is considered: a positive sign means that the selected direction was correct; a negative sign means that the direction is opposite to the assumed direction. This convention is transparently applicable to the corresponding interval algebraic model which delivers the correct sign together with the interval magnitude.

5 Conclusion

The engineering demand for more accurate models involving interval uncertainties, that conform to the physics of the modeled problem, lead to a new interval algebraic model of equilibrium equations in mechanics. The latter is based on the algebraic completion $(\mathbb{KR}, +, \times, \subseteq)$ of classical interval arithmetic. By a simple representation convention one can easily transform a deterministic formulation into a unique interval arithmetic formulation in the interval space $(\mathbb{KR}, +, \times, \subseteq)$. Then in the same rich algebraic space one finds a sharp algebraic solution for the unknown quantities and interpret them in the original physical setting of the problem. If the algebraic solution is a proper interval (vector), it is assured that the equilibrium equations are completely satisfied and the obtained interval enclosures are the sharpest ones. It is demonstrated at the end of Example 2 that if (part of) the algebraic solution is not proper interval vector, its proper projection (3) provides narrower interval estimation for the unknowns than the best solution enclosure (the exact interval hull of the united parametric solution set) in classical interval arithmetic. Contrary to classical interval approach, the algebraic one provides satisfaction of the linear equilibrium equations even for very large parameter uncertainties. Therefore, for large uncertainties the algebraic approach is essential in obtaining sharp interval estimates.

If the deterministic model involves more unknowns that the number of equilibrium equations, other relations are obtained from the information contained in the statement of the problem. In this case a hybrid approach is necessary which will be considered in a forthcoming paper [13].

The most attractive in the interval algebraic approach to linear equilibrium equations in mechanics is its transparent application and full conformance to the deterministic model. Along with guaranteed quantification of all sources of uncertainties, the new algebraic approach provides also sharper enclosure of the unknown quantities than the best known methods based on classical interval arithmetic.

References

1. Beer, F.P., Johnston, E.R., Mazurek, D.F., Cornwell, P.J., Eisenberg, E.R.: Vector Mechanics for Engineers: Statics and Dynamics, 9th edn. McGraw-Hill (2010)
2. Dimitrova, N., Markov, S.M., Popova, E.D.: Extended interval arithmetics: new results and applications. In: Atanassova, L., Herzberger, J. (eds.) Computer Arithmetic and Enclosure Methods, pp. 225–232. Elsevier Science Publishers B. V. (1992)
3. Elishakoff, I., Gabriele, S., Wang, Y.: Generalized Galileo Galilei problem in interval setting for functionally related loads. Arch. Appl. Mech. **86**(7), 1203–1217 (2015)
4. Kaucher, E.: Interval analysis in the extended interval space IR. Comput. Suppl. **2**, 33–49 (1980)
5. Markov, S.M., Popova, E.D., Ullrich, C.P.: On the solution of linear algebraic equations involving interval coefficients. In: Margenov, S., Vassilevski, P. (eds.) Iterative Methods in Linear Algebra, II, IMACS Series in Computational and Applied Mathematics, vol. 3, pp. 216–225 (1996)
6. Mazandarani, M.: (2015) IEEE standard 1788-2015 versus multidimensional RDM interval arithmetic, posting to IEEE P1788 working group. http://grouper.ieee.org/groups/1788/email/msg08439.html (2016). Accessed 25 Jan 2016
7. Moore, R.E.: Interval Analysis. Prentice-Hall, Englewood Cliffs, N.J. (1966)
8. Muhanna, R.L., Rama Rao, M.V., Mullen, R.L.: Advances in interval finite element modelling of structures. Life Cycle Reliab. Saf. Eng. **2**(3), 15–22 (2013)
9. Nadezhin, D.Y., Zhilin, S.I.: JInterval library: principles, development, and perspectives. Reliable Comput. **19**, 229–247 (2014)
10. Popova, E.D.: Algebraic solutions to a class of interval equations. J. Univers. Comput. Sci. **4**(1), 48–67 (1998)
11. Popova, E.D.: Multiplication distributivity of proper and improper intervals. Reliable Comput. **7**(2), 129–140 (2001). doi:10.1023/A:1011470131086
12. Popova, E.D.: Computer-assisted proofs in solving linear parametric problems. In: the Proceedings of SCAN'06, p. 35. IEEE Computer Society Press (2006)
13. Popova, E.D.: Interval model of equilibrium equations in mechanics. In: S. Freitag, R.L. Muhanna, R.L. Mullen (eds.) Proceedings of REC'2016, Ruhr University Bochum, pp. 241–255 (2016). http://rec2016.rub.de/downloads/rec2016_proceedings.pdf
14. Popova, E.D. Improved solution to the generalized Galilei's problem with interval loads. Arch. Appl. Mech. online (2016). doi:10.1007/s00419-016-1180-2

15. Popova, E.D., Ullrich. C.P.: Directed interval arithmetic in Mathematica. Implementation and applications. Technical report 96-3, Universität Basel, Switzerland. http://www.math.bas.bg/~epopova/papers/tr96-3.pdf (1996). Accessed 25 Jan 2016
16. Ratschek, H., Sauer, W.: Linear interval equations. Computing **26**, 105–115 (1982)
17. Shary, S.P.: A new technique in systems analysis under interval uncertainty and ambiguity. Reliable Comput. **8**(5), 321–418 (2002)

InterCriteria Analysis of Relations Between Model Parameters Estimates and ACO Performance

Olympia Roeva and Stefka Fidanova

Abstract In this paper we apply the approach InterCriteria Analysis (ICrA) to establish the existing relations and dependencies of defined parameters in non-linear model of an *E. coli* fed-batch fermentation process. Moreover, based on results of series of Ant Colony Optimization (ACO) identification procedures we observe the mutual relations between model parameters and ACO outcomes (execution time and objective function value). We perform a series of model identification procedures applying ACO. To estimate the model parameters we apply consistently 11 differently tuned ACO algorithms. We use various population sizes—from 5 to 100 ants in the population. In terms of ICrA we define five criteria, namely model parameters (maximum specific growth rate, μ_{max}; saturation constant, k_S and yield coefficient, $Y_{S/X}$) and ACO outcomes (execution time, T and objective function value, J). Based on ICrA we examine the obtained parameters estimates and discuss the conclusions about existing relations and dependencies between defined criteria. The obtained here results we compare with the ICrA results achieved using Genetic Algorithms (GA) as optimization techniques. Thus, based on the results of ACO and GA (the worst, best and average estimates) we define more precisely in which group (negative consonance, dissonance or positive consonance) fall the given ICrA criteria pairs.

1 Introduction

The InterCriteria Analysis (ICrA) is developed with the aim to gain additional insight into the nature of the criteria involved and discover on this basis existing relations between the criteria themselves [6]. It is based on the apparatus of the Index

O. Roeva
Institute of Biophysics and Biomedical Engineering, Bulgarian Academy of Sciences, Acad. G. Bonchev Str., bl. 105, 1113 Sofia, Bulgaria
e-mail: olympia@biomed.bas.bg

S. Fidanova (✉)
Institute of Information and Communication Technologies, Bulgarian Academy of Sciences, Acad. G. Bonchev Str., bl. 25A, 1113 Sofia, Bulgaria
e-mail: stefka@parallel.bas.bg

K. Georgiev et al. (eds.), *Advanced Computing in Industrial Mathematics*, Studies in Computational Intelligence 681, DOI 10.1007/978-3-319-49544-6_15

Matrices (IM) [2, 3], and the Intuitionistic Fuzzy Sets [4] and can be applied for decision making in different areas of science and practice. The approach has been discussed in a several papers considering parameter estimation problems. In [12] ICrA has been applied for the first time in the field of parameter identification of fermentation processes (FP) models. The ICrA implementation allowed to establish relations and dependencies between two of the main genetic algorithms (GA) parameters—numbers of individuals and number of generations, on the one hand, and convergence time, model accuracy and model parameters on the other hand. In [11] ICrA is applied to establish fundamental correlation between the kinetic variables of fed-batch processes for *E. coli* fermentation. Further, ICrA is applied to explore the existing relations and dependencies of defined model parameters and GA outcomes—execution time and objective function value—in case of *S. cerevisiae* FP [1] and *E. coli* FP [16]. Moreover, ICrA is applied for establishing the relations between GAs parameter generation gap, convergence time, model accuracy and model parameters [13]. Finally ICrA is applied to establish the dependencies of considered parameters based on different criteria referred to various metaheuristic algorithms, namely hybrid schemes using GA and Ant Colony Optimization (ACO) [15].

Results of these applications of the ICrA proved that in the case of modelling of FP ICrA approach could be very useful. FP are characterized with complex, non-linear dynamic and their modelling is a hard combinatorial optimization problem. On the one hand, the parameter identification is of key importance for modelling process and additional knowledge about the model parameters relations will be extremely useful to improve the model accuracy. On the other hand, the information may be used to improve the performance of the used optimization algorithms if, for instance, some algorithm outcomes are added to the considered criteria. Thus, the relations between model parameters and optimization algorithm performance will be established.

In this paper we applied the ICrA to establish the basic relations between the parameters in the model of an *E. coli* fed-batch FP. The existing relations are identified based on results of a series of parameters identification procedures. The use of metaheuristic techniques such as ACO has received more and more attention [9, 10]. This method offer good solutions, even global optima, within reasonable computing time, so we choose to use ACO for estimation of the *E. coli* fed-batch FP model parameters.

The paper is organized as follows. The background of InterCriteria Analysis is given in Sect. 2. The problem formulation is described in Sect. 3. The numerical results and a discussion are presented in Sect. 4. Conclusion remarks are done in Sect. 5.

2 InterCriteria Analysis

Here we expand on the idea proposed in [6]. Following [4, 6] we will obtain an Intuitionistic Fuzzy Pair (IFP) [7] as the degrees of "agreement" and "disagreement" between two criteria applied on different objects. We remind briefly that an IFP is an ordered pair of real non-negative numbers $\langle a, b\rangle$ such that:

$$a + b \leq 1.$$

For clarity, let us be given an Index Matrix (IM) (see [2]) whose index sets consist of the names of the criteria (for rows) and objects (for columns). The elements of this IM are further supposed to be real numbers (in the general case, this is not required). We will obtain an IM with index sets consisting of the names of the criteria (for rows and for columns) with elements IFPs corresponding to the "agreement" and "disagreement" of the respective criteria.

Two things are further supposed (which are not always guaranteed in practice and, when not fulfilled, present an interesting direction for new research in themselves):

1. All criteria provide an evaluation for all objects (i.e. there are no inapplicable criteria for a given object) and all these evaluations are available (no missing evaluations).
2. All the evaluations of a given criteria can be compared amongst themselves.

Further by O we denote the set of all objects $O_1, O_2, \ldots, O_n$ being evaluated, and by $C(O)$ the set of values assigned by a given criteria C to the objects, i.e.

$$O \stackrel{\text{def}}{=} \{O_1, O_2, \ldots, O_n\},$$
$$C(O) \stackrel{\text{def}}{=} \{C(O_1), C(O_2), \ldots, C(O_n)\}.$$

Let $x_i = C(O_i)$. Then the following set can be defined:

$$C^*(O) \stackrel{\text{def}}{=} \{\langle x_i, x_j\rangle | i \neq j \,\&\, \langle x_i, x_j\rangle \in C(O) \times C(O)\}.$$

Further, if $x = C(O_i)$ and $y = C(O_j)$, $x \prec y$ will be written iff $i < j$.

In order to compare two criteria we must construct the vector of all internal comparisons of each criteria, which fulfill exactly one of three relations R, $\overline{R}$ and $\tilde{R}$. In other words, we require that for a fixed criterion C and any ordered pair $\langle x, y\rangle \in C^*(O)$ it is true:

$$\langle x, y\rangle \in R \Leftrightarrow \langle y, x\rangle \in \overline{R}, \quad (1)$$
$$\langle x, y\rangle \in \tilde{R} \Leftrightarrow \langle x, y\rangle \notin (R \cup \overline{R}), \quad (2)$$
$$R \cup \overline{R} \cup \tilde{R} = C^*(O). \quad (3)$$

From the above it is seen that we need only consider a subset of $C(O) \times C(O)$ for the effective calculation of the vector of internal comparisons (denoted further by $V(C)$) since from (1), (2) and (3) it follows that if we know what is the relation between x and y we also know what is the relation between y and x. Thus we will only consider lexicographically ordered pairs $\langle x, y\rangle$. Let, for brevity:

$$C_{i,j} = \langle C(O_i), C(O_j) \rangle.$$

Then for a fixed criterion C we construct the vector:

$$V(C) = \{C_{1,2}, C_{1,3}, \ldots, C_{1,n}, C_{2,3}, C_{2,4}, \ldots, \\ C_{2,n}, C_{3,4}, \ldots, C_{3,n}, \ldots, C_{n-1,n}\}.$$

It can be easily seen that it has exactly $\frac{n(n-1)}{2}$ elements. Further, to simplify our considerations, we replace the vector $V(C)$ with $\hat{V}(C)$, where for each $1 \leq k \leq \frac{n(n-1)}{2}$ for the k-th component it is true:

$$\hat{V}_k(C) = \begin{cases} 1 \text{ iff } V_k(C) \in R, \\ -1 \text{ iff } V_k(C) \in \overline{R}, \\ 0 \text{ otherwise.} \end{cases}$$

Then when comparing two criteria we determine the "degree of agreement" between the two as the number of matching components (divided by the length of the vector for normalization purposes). This can be done in several ways, e.g. by counting the matches or by taking the complement of the Hamming distance. The "degree of disagreement" is the number of components of opposing signs in the two vectors (again normalized by the length). This also may be done in various ways. A pseudocode of the Algorithm 1 [16] used in this study for calculating the degrees of agreement and disagreement between two criteria C and C' is presented below.

It is obvious (from the way of calculation) that for $\mu_{C,C'}$, $\nu_{C,C'}$, we have:

$$\mu_{C,C'} = \mu_{C',C}, \nu_{C,C'} = \nu_{C',C}.$$

Also, $\langle \mu_{C,C'}, \nu_{C,C'} \rangle$ is an IFP.

In the most of the obtained pairs $\langle \mu_{C,C'}, \nu_{C,C'} \rangle$, the sum $\mu_{C,C'} + \nu_{C,C'}$ is equal to 1. However, there may be some pairs, for which this sum is less than 1. The difference

$$\pi_{C,C'} = 1 - \mu_{C,C'} - \nu_{C,C'} \tag{4}$$

is considered as a degree of "uncertainty".

3 Problem Formulation

Let us use the following non-linear differential equation system to describe the *E. coli* fed-batch FP [15]:

$$\frac{dX}{dt} = \mu X - \frac{F_{in}}{V} X, \tag{5}$$

Algorithm 1 Calculating "agreement" and "disagreement" between two criteria

Require: Vectors $\hat{V}(C)$ and $\hat{V}(C')$

1: **function** DEGREE OF AGREEMENT($\hat{V}(C)$, $\hat{V}(C')$)
2: $V \leftarrow \hat{V}(C) - \hat{V}(C')$
3: $\mu_{C,C'} \leftarrow 0$
4: **for** $i \leftarrow 1$ to $\frac{n(n-1)}{2}$ **do**
5: **if** $V_i = 0$ **then**
6: $\mu_{C,C'} \leftarrow \mu_{C,C'} + 1$
7: **end if**
8: **end for**
9: $\mu_{C,C'} \leftarrow \frac{2}{n(n-1)} \mu_{C,C'}$
10: **return** $\mu_{C,C'}$
11: **end function**

12: **function** DEGREE OF DISAGREEMENT($\hat{V}(C)$, $\hat{V}(C')$)
13: $V \leftarrow \hat{V}(C) - \hat{V}(C')$
14: $\nu_{C,C'} \leftarrow 0$
15: **for** $i \leftarrow 1$ to $\frac{n(n-1)}{2}$ **do**
16: **if** abs$(V_i) = 2$ **then** ▷ abs: absolute value
17: $\nu_{C,C'} \leftarrow \nu_{C,C'} + 1$
18: **end if**
19: **end for**
20: $\nu_{C,C'} \leftarrow \frac{2}{n(n-1)} \nu_{C,C'}$
21: **return** $\nu_{C,C'}$
22: **end function**

$$\frac{dS}{dt} = -q_S X + \frac{F_{in}}{V}(S_{in} - S), \tag{6}$$

$$\frac{dV}{dt} = F_{in}, \tag{7}$$

where

$$\mu = \mu_{max} \frac{S}{k_S + S}, \quad q_S = \frac{1}{Y_{S/X}} \mu \tag{8}$$

and X is the biomass concentration, [g/l]; S is the substrate concentration, [g/l]; F_{in} is the feeding rate, [l/h]; V is the bioreactor volume, [l]; S_{in} is the substrate concentration in the feeding solution, [g/l]; μ and q_S are the specific rate functions, [1/h]; μ_{max} is the maximum value of the μ, [1/h]; k_S is the saturation constant, [g/l]; $Y_{S/X}$ is the yield coefficient, [-].

For the model (Eqs. 5–8) the parameters that will be identified are μ_{max}, k_S and $Y_{S/X}$.

Let $Z_{\text{mod}} \stackrel{\text{def}}{=} [X_{\text{mod}}\ S_{\text{mod}}]$ (model predictions for biomass and substrate) and $Z_{\text{exp}} \stackrel{\text{def}}{=} [X_{\text{exp}}\ S_{\text{exp}}]$ (known experimental data for biomass and substrate). Then putting $Z = Z_{\text{mod}} - Z_{\text{exp}}$, we define the objective function as:

Table 1 Parameters of ACO algorithm

Parameter	Value
Number of ants (nind)	5–100
Initial pheromone	0.5
Evaporation	0.1
Maximum generations (maxgen)	100

$$J = \|Z\|^2 \rightarrow \min, \tag{9}$$

where $\|\|$ denotes the ℓ^2-vector norm [16].

For the model parameters identification we use experimental data for biomass and glucose concentration of an *E. coli* MC4110 fed-batch fermentation process.

To estimate the model parameters we applied consistently 11 differently tuned ACO algorithms. We use various ant numbers—from 5 to 100 ants, namely ACO_5, ACO_{10}, ACO_{20},..., ACO_{90}, ACO_{100}. The number of generations is fixed to 100. The main ACO parameters are summarized in Table 1.

Due to the stochastic nature of the applied algorithm we perform series of 30 runs for each differently tuned ACO algorithm. Thus, we obtain the average, best and worst estimate of the parameters, as well as of the algorithm execution time and value of objective function. The detailed description of identification procedure is given in [15].

To perform ICrA three IMs are constructed—the IM A_1 (Eq. 10) with the obtained average results, the IM A_2 (Eq. 11) with the best obtained results and IM A_3 (Eq. 12) with the worst obtained results.

$$A_1 = \begin{array}{r|rrrrr} & C_1 & C_2 & C_3 & C_4 & C_5 \\ \hline ACO_5 & 0.5700 & 0.0285 & 2.0260 & 6.2523 & 16.9105 \\ ACO_{10} & 0.5436 & 0.0237 & 2.0180 & 6.0527 & 29.4062 \\ ACO_{20} & 0.4893 & 0.0148 & 2.0261 & 5.4330 & 56.0980 \\ ACO_{30} & 0.4968 & 0.0116 & 2.0320 & 5.2849 & 90.6210 \\ ACO_{40} & 0.5112 & 0.0180 & 2.0262 & 5.2853 & 109.6219 \\ ACO_{50} & 0.5148 & 0.0156 & 2.0263 & 5.2206 & 131.3684 \\ ACO_{60} & 0.4962 & 0.0127 & 2.0220 & 5.2184 & 151.6018 \\ ACO_{70} & 0.5049 & 0.0171 & 2.0180 & 5.1350 & 173.7539 \\ ACO_{80} & 0.4719 & 0.0105 & 2.0280 & 5.1324 & 197.6533 \\ ACO_{90} & 0.5082 & 0.0152 & 2.0221 & 5.1415 & 237.5271 \\ ACO_{100} & 0.4737 & 0.0108 & 2.0240 & 5.0885 & 260.1005 \end{array} \tag{10}$$

$A_2 =$ (11)

	C_1	C_2	C_3	C_4	C_5
ACO_{5}	0.4911	0.0139	2.0200	5.0652	16.7857
ACO_{10}	0.4986	0.0130	2.0220	4.8083	29.7494
ACO_{20}	0.4806	0.0116	2.0260	4.9293	55.5208
ACO_{30}	0.4956	0.0146	2.0221	4.7408	90.7302
ACO_{40}	0.4794	0.0101	2.0262	4.8004	109.4347
ACO_{50}	0.4959	0.0145	2.0201	4.6598	131.4308
ACO_{60}	0.4854	0.0110	2.0263	4.8983	152.1166
ACO_{70}	0.4977	0.0114	2.0222	4.7739	173.7383
ACO_{80}	0.4827	0.0109	2.0240	4.8078	196.7641
ACO_{90}	0.4800	0.0100	2.0241	4.7856	234.3915
ACO_{100}	0.4884	0.0126	2.0261	4.8382	260.1473

$A_3 =$ (12)

	C_1	C_2	C_3	C_4	C_5
ACO_{5}	0.5532	0.0183	2.0120	7.9011	16.7857
ACO_{10}	0.5115	0.0146	2.0040	8.0956	29.6246
ACO_{20}	0.5733	0.0278	2.0140	6.5924	55.5052
ACO_{30}	0.4641	0.0104	2.0240	6.2202	90.7146
ACO_{40}	0.5292	0.0148	2.0221	6.0784	112.8667
ACO_{50}	0.5043	0.0169	2.0160	5.7156	131.9924
ACO_{60}	0.5376	0.0198	2.0161	5.7759	151.1026
ACO_{70}	0.5232	0.0161	2.0200	5.6652	175.1891
ACO_{80}	0.4746	0.0125	2.0220	5.7891	199.2601
ACO_{90}	0.5187	0.0211	2.0180	5.6120	236.0919
ACO_{100}	0.4881	0.0122	2.0340	5.4866	258.4781

In addition to the presented in [14] results here the average, worst and best estimates for the three model parameters in all 11 cases are given too. Thus, five criteria are considered—C_1 is the parameter μ_{max}, C_2 is the parameter k_S, C_3 is the parameter $Y_{S/X}$, C_4 is the objective function value J and C_5 is the resulting execution time T.

4 Numerical Results and Discussion

Computer specifications to run all identification procedures are Intel Core i5-2329 3.0 GHz, 8 GB Memory, Windows 7 (64bit) operating system.

Based on the presented Algorithm 1 the ICrA is implemented in the Matlab 7.5 environment. We obtain IMs that determine the degrees of "agreement" ($\mu_{C,C'}$) and "disagreement" ($\nu_{C,C'}$) between criteria for the average, worst and best ACO results.

Average results

Resulting degrees of "agreement" ($\mu_{C,C'}$) (IM_1) and degrees of "disagreement" ($\nu_{C,C'}$) (IM_2) are as follows:

$\mathrm{IM}_1 =$

	C_1	C_2	C_3	C_4	C_5
C_1	**1**	0.89	0.42	0.76	0.29
C_2	0.89	**1**	0.35	0.76	0.25
C_3	0.42	0.35	**1**	0.47	0.49
C_4	0.76	0.76	0.47	**1**	0.05
C_5	0.29	0.25	0.49	0.05	**1**

, $\mathrm{IM}_2 =$

	C_1	C_2	C_3	C_4	C_5
C_1	**0**	0.11	0.56	0.24	0.71
C_2	0.11	**0**	0.64	0.24	0.75
C_3	0.56	0.64	**0**	0.51	0.49
C_4	0.24	0.24	0.51	**0**	0.95
C_5	0.71	0.75	0.49	0.95	**0**

ICrA analysis of the average results shows some degrees of "uncertainty" ($\pi_{C,C'}$, Eq. (4)) as follows:

$\mathrm{IM}_3 =$

	C_1	C_2	C_3	C_4	C_5
C_1	**0**	0	0.02	0	0
C_2	0	**0**	0.02	0	0
C_3	0.02	0.02	**0**	0.02	0.02
C_4	0	0	0.02	**0**	0
C_5	0	0	0.02	0	**0**

Worst results

Resulting degrees of "agreement" ($\mu_{C,C'}$) (IM_4) and degrees of "disagreement" ($\nu_{C,C'}$) (IM_5) are as follows:

$\mathrm{IM}_4 =$

	C_1	C_2	C_3	C_4	C_5
C_1	**1**	0.82	0.31	0.60	0.35
C_2	0.82	**1**	0.27	0.49	0.45
C_3	0.31	0.27	**1**	0.31	0.75
C_4	0.60	0.49	0.31	**1**	0.09
C_5	0.35	0.45	0.75	0.09	**1**

, $\mathrm{IM}_5 =$

	C_1	C_2	C_3	C_4	C_5
C_1	**0**	0.18	0.69	0.40	0.65
C_2	0.18	**0**	0.73	0.51	0.55
C_3	0.69	0.73	**0**	0.69	0.25
C_4	0.40	0.51	0.69	**0**	0.91
C_5	0.65	0.55	0.25	0.91	**0**

Best results

Resulting degrees of "agreement" ($\mu_{C,C'}$) (IM_6) and degrees of "disagreement" ($\nu_{C,C'}$) (IM_7) are as follows:

$\mathrm{IM}_6 =$

	C_1	C_2	C_3	C_4	C_5
C_1	**1**	0.78	0.27	0.44	0.42
C_2	0.78	**1**	0.27	0.51	0.31
C_3	0.27	0.27	**1**	0.62	0.71
C_4	0.44	0.51	0.62	**1**	0.40
C_5	0.42	0.31	0.71	0.40	**1**

, $\mathrm{IM}_7 =$

	C_1	C_2	C_3	C_4	C_5
C_1	**0**	0.22	0.73	0.56	0.58
C_2	0.22	**0**	0.73	0.49	0.69
C_3	0.73	0.73	**0**	0.38	0.29
C_4	0.56	0.49	0.38	**0**	0.60
C_5	0.58	0.69	0.29	0.60	**0**

ICrA analysis of the best and the worst results do not shows degrees of "uncertainty", i.e. $\pi_{C,C'} = 0$ for all criteria pairs.

The obtained degrees of "agreement" ($\mu_{C,C'}$ values) between considered criteria for average, best and worst results are presented in Fig. 1.

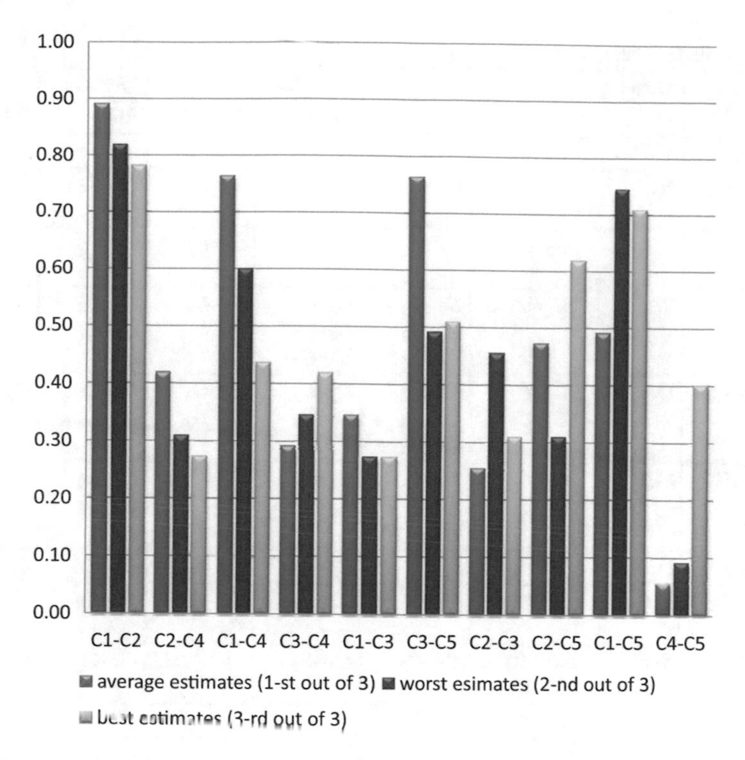

Fig. 1 Degrees of "agreement" ($\mu_{C,C'}$ values) for all cases

Let us consider the following scheme for defining the consonance and dissonance between each pair of criteria (see Table 2), where SNC is strong negative consonance, NC is negative consonance, WNC is weak negative consonance, WD is weak dissonance, D is dissonance, SD is strong dissonance, WPC is weak positive consonance, PC is positive consonance and SPC is strong positive consonance [5].

In Table 2 the results obtained in this research are compared to the results presented in [16]. In [16] for parameters identification of an *E. coli* MC4110 FP model consistently 14 differently tuned GA are applied. Various population sizes—from 5 to 200 chromosomes in the population are used. The number of generations is fixed to 200. The same five criteria are defined: C_1 is the parameter μ_{max}, C_2 is the parameter k_S, C_3 is the parameter $Y_{S/X}$, C_4 is the objective function value J and C_5 is the resulting execution time T.

Analysis of the both average results (here presented and the average results presented in [16]) shows the following criteria pair relations:

- For the pair C_5–C_4 (i.e., $T \leftrightarrow J$) a negative consonance is identified for average and worst results. Such dependence is logical—for a large number of algorithm iterations (i.e., greater execution time T) it is more likely to find a more accurate solution, i.e. a small value of J.

Table 2 Criteria relations sorted by $\mu_{C,C'}$ values

$\mu_{C,C'}$	Meaning	Average results		Best results		Worst results	
		ACO	GA [16]	ACO	GA [16]	ACO	GA [16]
[0–0.05]	SNC	C_5–C_4					
(0.05–0.15]	NC		C_5–C_4			C_5–C_4	C_5–C_3 C_5–C_4
(0.15-0.25]	WNC	C_2–C_3			C_5–C_4		C_2–C_3 C_4–C_3
(0.25–0.33]	WD	C_5–C_3	C_2–C_3 C_5–C_3	C_2–C_3 C_4–C_3 C_2–C_4	C_5–C_2	C_3–C_4 C_3–C_1 C_2–C_3	C_5–C_1
(0.33–0.43]	D	C_2–C_4 C_4–C_3	C_5–C_2 C_2–C_4 C_4–C_3	C_5–C_3 C_5–C_4	C_5–C_3 C_4–C_3	C_5–C_1	C_2–C_4
(0.43–0.57]	SD	C_5–C_2 C_5–C_1	C_5–C_1	C_1–C_3 C_1–C_4	C_1–C_3 C_2–C_3 C_2–C_4	C_2–C_4 C_2–C_5	
(0.57–0.67]	D			C_5–C_1	C_1–C_4	C_1–C_4	C_5–C_2
(0.67–0.75]	WD		C_1–C_4	C_5–C_2	C_1–C_2 C_5–C_1	C_5–C_3	
(0.75–0.85]	WPC	C_1–C_4 C_1–C_3	C_1–C_3	C_1–C_2		C_1–C_2	C_1–C_2 C_1–C_3
(0.85–0.95]	PC	C_1–C_2	C_1–C_2				C_1–C_4
(0.95–1]	SPC						

- For most results we established strong correlation between criteria C_1–C_2 (i.e., $\mu_{max} \leftrightarrow k_S$) and C_1–C_3 (i.e., $\mu_{max} \leftrightarrow Y_{X/S}$). There are two exceptions—GA best results and ACO worst results, where the criteria pair C_1–C_3 is respectively in SD and WD. Considering the physical meaning of the model parameters [8] it is clear that there is dependence between these criteria. A strong correlation is expected between criteria C_1–C_2 [8]. Considering relation C_1–C_3 we found some exceptions—GA best and ACO best and worst results.

Due to stochastic nature of considered here meta-heuristic techniques (GA and ACO) we observed some different criteria dependences based on the ICrA of the worst and best results:

- Based on the worst ACO and GA results we found weaker relation between criteria pairs C_1–C_5, C_2–C_4 and C_2–C_5. There are some small discrepancies for the pairs C_3–C_4 and C_2–C_3, and some larger discrepancies—for the pairs C_1–C_3, C_1–C_4, and C_5–C_3. For the pairs C_1–C_4, C_1–C_2 and C_1–C_3 we observed higher value of $\mu_{C,C'}$, i.e. PC or WPC.
- Compared to the ICrA of average results there are some discrepancies too. For example, average results show that the pair C_4–C_3 is in dissonance, while worst results—in WD (ACO results) or WNC (GA results).

- In the ICrA of the best results we identify the same results—some discrepancies are observed. There are some small discrepancies for the pairs C_3–C_4, C_1–C_4, C_2–C_1, C_2–C_4, C_5–C_4 and C_5–C_1, and some larger discrepancies—for the pairs C_2–C_3 and C_5–C_2. Here we observed a higher value of $\mu_{C,C'}$ only for the pair C_1–C_2, i.e. WPC (ACO results).

Taking into account the nature of the GA and ACO we consider that the ICrA of the average results has the highest significance.

5 Conclusion

In this paper, based on the apparatus of the Index Matrices and the Intuitionistic Fuzzy Sets, InterCriteria Analysis of a model parameters identification using Ant Colony Optimization is performed. A non-linear model of an *E. coli* fed-batch fermentation process is considered. Series of model identification procedures using Ant Colony Optimization are done. The InterCriteria Analysis is applied to explore the existing relations and dependencies of defined model parameters and Ant Colony Optimization outcomes—execution time and objective function value. Three case studies are examined considering average, worst and best results for the obtained model parameters, execution time and objective function value. The obtained results are compared to the results from a model parameters identification using Genetic Algorithms. Applying the InterCriteria Analysis and analyzing the results we establish relations and dependencies between the defined criteria. Based on the used scale for defining the consonance and dissonance between each pair of criteria, we discuss which criteria are in consonance and dissonance, as well as the degree of their dependence.

Acknowledgements Work presented here is partially supported by the Bulgarian National Scientific Fund under Grants DFNI-I02/5 InterCriteria Analysis. A New Approach to Decision Making and DFNI I02/20 Efficient Parallel Algorithms for Large Scale Computational Problems.

References

1. Angelova, M., Roeva, O., Pencheva, T.: InterCriteria analysis of crossover and mutation rates relations in simple genetic algorithm. Ann. Comput. Sci. Inf. Syst. **5**, 419–424 (2015)
2. Atanassov, K.: On index matrices, Part 1: Standard cases. Adv. Stud. Contemp. Math. **20**(2), pp. 291–302 (2010)
3. Atanassov, K.: On index matrices, Part 2: Intuitionistic fuzzy case. In: Proceedings of the Jangjeon Mathematical Society, vol. 13, no. 2, pp. 121–126 (2010)
4. Atanassov, K.: On Intuitionistic Fuzzy Sets Theory. Springer, Berlin (2012)
5. Atanassov, K., Atanassova, V., Gluhchev, G.: InterCriteria analysis: ideas and problems. In: Notes on Intuitionistic Fuzzy Sets, vol. 21, no. 2, pp. 81–88 (2015)

6. Atanassov, K., Mavrov, D., Atanassova, V.: Intercriteria decision making: a new approach for multicriteria decision making, based on index matrices and intuitionistic fuzzy sets. Issues in IFSs and GNs **11**, 1–8 (2014)
7. Atanassov, K., Szmidt, E., Kacprzyk, J.: On intuitionistic fuzzy pairs. In: Notes on Intuitionistic Fuzzy Sets, Vol. 19, no. 3, pp. 1–13 (2013)
8. Bastin, G., Dochain, D.: On-line estimation and adaptive control of bioreactors. Els. Sc. Publ. (1991)
9. Boussaid, I., Lepagnot, J., Siarry, P.: A survey on optimization metaheuristics. Inf. Sci. **237**, 82–117 (2013)
10. Dorigo, M., Stutzle, T.: Ant Colony Optimization. MIT Press (2004)
11. Ilkova, T., Petrov, M.: Intercriteria analysis for identification of Escherichia coli fed-batch mathematical model. J. Int. Sci. Publications: Mater. Methods Technol. **9**, 598–608 (2015)
12. Pencheva, T., Angelova, M., Atanassova, V., Roeva, O.: InterCriteria analysis of genetic algorithm parameters in parameter identification. In: Notes on Intuitionistic Fuzzy Sets, vol. 21, no. 2, pp. 99–110 (2015)
13. Pencheva, T., Angelova, M., Vassilev, P., Roeva, O.: InterCriteria analysis approach to parameter identification of a fermentation process model. Adv. Intell. Syst. Comput. **401**, 385–397 (2016)
14. Roeva, O., Fidanova, S., Paprzycki, M.: Influence of the population size on the genetic algorithm performance in case of cultivation process modelling. In: Proceedings of the Federated Conference on Computer Science and Information Systems (FedCSIS), WCO 2013, Poland, pp. 371–376 (2013)
15. Roeva, O., Fidanova, S., Paprzycki, M.: InterCriteria analysis of ACO and GA hybrid algorithms. Stud. Comput. Intell. **610**, 107–126 (2016)
16. Roeva, O., Fidanova, S., Vassilev, P., Gepner, P.: InterCriteria analysis of a model parameters identification using genetic algorithm. Ann. Comput. Sci. Inf. Syst. **5**, 501–506 (2015)

Newtonian and Non-Newtonian Pulsatile Blood Flow in Arteries with Model Aneurysms

S. Tabakova, P. Raynov, N. Nikolov and St. Radev

Abstract The cardiovascular diseases depend directly on the blood flow dynamics. The mathematical modeling and numerical simulations are expected to play an important role to predict the genesis of the atherosclerosis and the formation and rupture of the aneurysms. In the present work the numerical solutions for the oscillatory flow velocity due to the Newtonian and the non-Newtonian (Carreau) model are constructed for a straight long tube and for a tube (artery) with a model aneurysm. The numerical solutions are obtained by the finite-difference method (FDM) for the straight tube and by the software ANSYS/FLUENT for both geometries. The numerical results obtained by the ANSYS/FLUENT for a straight long tube are validated by the analytical and numerical solutions using the FDM for the Newtonian and Carreau models for different Womersley numbers, correspondent to different tube radii. The obtained peak wall shear stresses from the oscillatory flow in the straight long tube are lower than those in the tube with the model aneurysm, which can be used as an indicator for further clinical examinations.

S. Tabakova (✉) · N. Nikolov · St. Radev
Institute of Mechanics, BAS, Acad. G. Bontchev str., bl. 4, 1113 Sofia, Bulgaria
e-mail: stabakova@gmail.com

N. Nikolov
e-mail: n.nikolov@imbm.bas.bg

St. Radev
e-mail: stradev@imbm.bas.bg

S. Tabakova
Technical University - Sofia, Branch Plovdiv, 25 Tzanko Djustabanov str., 4000 Plovdiv, Bulgaria

P. Raynov
University of Food Technologies, 26 Maritsa Blvd., 4000 Plovdiv, Bulgaria
e-mail: plamsky@mail.bg

K. Georgiev et al. (eds.), *Advanced Computing in Industrial Mathematics*, Studies in Computational Intelligence 681, DOI 10.1007/978-3-319-49544-6_16

1 Introduction

In most cases the in vivo measurement techniques are unable to prevent the cardio-vascular diseases evolution, which depends directly on the blood flow dynamics. One of the most dangerous diseases is that of the formation and rupture of different artery aneurysms. The study of the blood flow in tubes can be treated as a flow model in different types of arteries.

The blood is a suspension of particles and plasma, which has a non-Newtonian character as a fluid. It is a typical representative of the shear thinning fluids with an apparent viscosity dependent on the shear rate, i.e. its viscosity continuously decreases or increases with the shear rate increase or decrease reaching two different upper and lower plateaus independent of the further change of the shear rate. Several non-Newtonian models are used to express the blood rheology: the Carreau model [1–5], the Carreau-Yasuda model [1, 6–9], the Casson model [1, 3, 8], the Power law model [1, 3, 4] and others. Some of these models, such as that of Carreau, give a non-linear dependence of the shear stress on the shear rate. Since the shear rate changes significantly in the arteries with non-constant cross section, the viscosity could not be taken as a constant. This means that there are no analytical solutions for the blood flow in arteries. The proper knowledge of the viscosity leads to a proper knowledge of the Wall Shear Stresses (WSS), which are of a major importance for the prediction of an aneurysm rupture. The problem becomes more complicated if the pulsatile character of the blood flow is considered. It occurs that the blood flow can be approximated with the Newtonian fluid flow in the larger arteries, e.g. in the aorta, while in the narrow arteries the non-Newtonian character of the blood flow is essential. The analysis of non-Newtonian flows in infinitely long tubes is very important when studying the blood flow in different types of arteries. The well known analytical solution proposed by Womersley [10] is often applied to approximate the pulsative velocity of blood flow in arteries, when the fluid is regarded as Newtonian. However, if non-Newtonian models are applied for the blood viscosity, the flow solution can be obtained only numerically. For example, the Lattice Boltzmann Method is applied in [8] for the 2D oscillatory Newtonian and non-Newtonian flows in straight and curved tubes. The viscosity is given by the models of Casson and Carreau-Yasuda. The authors show that the difference between the velocity and the Wall Shear Stresses (WSS) calculated by the Newtonian viscosity model and by the non-Newtonian models increases with the decrease of the Womersley number, which expresses the relation between the oscillatory inertia and viscous forces.

In this paper we investigate numerically the non-Newtonian oscillatory flow of blood in a long straight tube and in a tube with a model aneurysm, using the Carreau viscosity model. The numerical simulations are performed by the software ANASYS/FLUENT for different Womersley numbers correspondent to different tube radii. The obtained solution for the velocity in the straight tube will be compared with the analytical solution for the Newtonian fluid model and with the numerically obtained solution (by the FDM) for the Carreau viscosity model given in [11].

2 Problem Statement

The blood is assumed incompressible with constant density ρ and apparent viscosity μ_{app} (constant for the Newtonian blood model and defined by a non-linear function of the shear rate for the non-Newtonian blood model).

Two different axisymmetrical geometries are considered in cylindrical coordinates (x, r, φ), where x is the axial coordinate: a straight circular tube with radius R and a circular tube with entry radius R and a model aneurysm given by the Gaussian shape function [12], as shown in Fig. 1:

$$r(x) = R + H\exp(-\frac{x^2}{2W^2}), \tag{1}$$

where H and W are the aneurysm height and width.

The equations of motion and continuity in a vector form are:

$$\rho(\frac{\partial \mathbf{v}}{\partial t} + \mathbf{v}\cdot\nabla\mathbf{v}) = -\nabla p + \nabla\cdot\mathbf{T}, \tag{2}$$

$$\nabla\cdot\mathbf{v} = 0, \tag{3}$$

where $\mathbf{v} = (u, v, w)$ is the velocity vector, p is the pressure, $\mathbf{T} = f(\dot{S})$ is the viscous stress tensor, with $\dot{S}$– the shear rate tensor.

For a very long (infinite) tube, we obtain $v = w = 0$ and $u = u(r, t)$ from the Eq. (3). In this case the shear stress tensor has only one non-zero term $\tau = \mu_{app}(\dot{\gamma})\dot{\gamma}$, where $\dot{\gamma} = \frac{\partial u}{\partial r}$. The system (2) transforms into a single equation for the axial velocity u:

$$\rho\frac{\partial u}{\partial t} = -\frac{\partial p}{\partial x} + \frac{1}{r}\frac{\partial}{\partial r}(\mu_{app} r\frac{\partial u}{\partial r}) \tag{4}$$

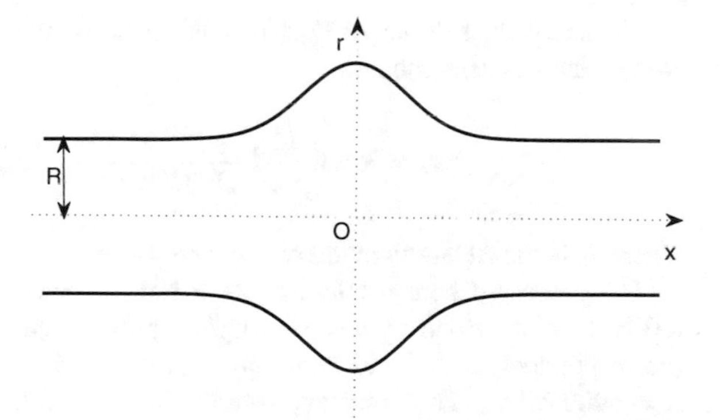

Fig. 1 The Gaussian model of an aneurysm at $R = 1$, $H = R$ and $W = R$

The boundary conditions for the velocity u are the no-slip condition $u = 0$ at $r = R$ and the symmetry condition $\frac{\partial u}{\partial r} = 0$ at $r = 0$. For the pressure gradient we consider the case of an oscillatory function in time $-\frac{\partial p}{\partial x} = A\cos(nt)$, where A is the pulse amplitude and n is the angular frequency.

The Carreau model of blood when treated as a non-Newtonian fluid is chosen with apparent viscosity μ_{app} that is usually given [2] by the following expression - further denoted by μ_c:

$$\mu_c = \mu_\infty + (\mu_0 - \mu_\infty)[1 + \lambda^2\dot{\gamma}^2]^{(n_c-1)/2}, \tag{5}$$

where λ and n_c are empirically determined. For human blood [2]: $\mu_0 = 0.056\,\text{Pa s}$, $\mu_\infty = 0.00345\,\text{Pa s}$, $\lambda = 3.313\,\text{s}$ and $n_c = 0.3568$.

3 Analysis of the Results

3.1 Long Straight Tube

The Eq. (4) is dimensionlized using the following characteristic scales: R as a characteristic length ($r = RY$), $1/n$ as a characteristic time ($t = T/n$), μ_∞ as a characteristic viscosity ($\mu_c = \bar{\mu}_c\mu_\infty$):

$$\frac{1}{Y}\frac{\partial}{\partial Y}(\bar{\mu}_c Y\frac{\partial u}{\partial Y}) - \alpha^2\frac{\partial u}{\partial T} + \frac{R^2 A}{\mu_\infty}\cos(T) = 0, \tag{6}$$

where $0 \le Y \le 1$, $\frac{\partial u(0,T)}{\partial Y} = 0$, $u(1,T) = 0$ and $\alpha = R\sqrt{\frac{\rho n}{\mu_\infty}}$ is the Womersley number.

The analytical solution of Eq. (6) is the so called Womersley solution [10] for the Newtonian viscosity model:

$$u_n = Real\left[\frac{iA}{n\rho}\left(\frac{J_0(i^{3/2}\alpha Y)}{J_0(i^{3/2}\alpha)} - 1\right)exp(iT)\right], \tag{7}$$

where J_0 is the Bessel function of 'zero-th' order.

The presented here results are for a human carotid artery with a radius $R = 0.0031$ m at blood density $\rho = 1000\,\text{kg/m}^3$, pulse frequency of oscillations $n = 2.4\pi$ (correspondent to 72 heart beats per minute) and a pressure gradient amplitude $A = 6000$ Pa/m (45 mm mercury column per meter). In this case $\alpha = 4.58$ and the maximum Reynolds number achieved during the blood oscillatory flow is around 530. The cross-section mean velocity based on the solution (7) is:

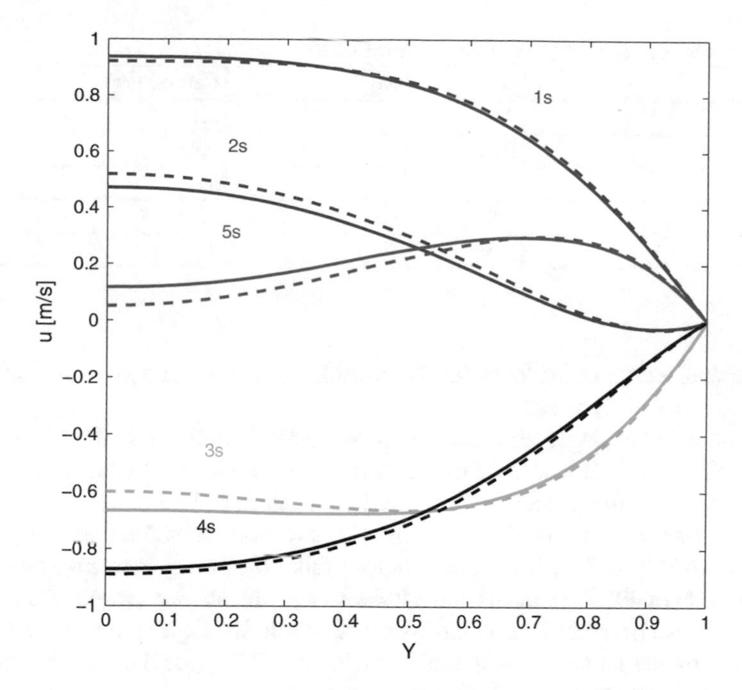

Fig. 2 Comparison between the axial velocities u of the Newtonian model (*dotted lines*) and of the Carreau model (*solid lines*) at different instants of time t: 1, 2, 3, 4, 5 s. (the different colors correspond to different times)

$$\bar{u}_n = 0.5868 \sin(T + 0.3605) \tag{8}$$

For the Carreau viscosity model Eq. (6) is solved numerically by the FDM of Crank-Nicholson with time and space steps $O(10^{-3})$ giving relative error $O(10^{-6})$. The numerical solution has been approximated to obtain its mean velocity:

$$\bar{u}_c = 0.5799 \sin(T + 0.3867) \tag{9}$$

Newtonian and Carreau velocities are presented in Fig. 2 for different times t. It is well seen the fast change of the velocity profile in time. However, the Carreau velocities are quite similar to the Newtonian ones, except in the symmetry axis region, i.e., near to $Y = 0$, which is due to the big difference in viscosities at the small velocity gradient there. As a whole, the difference between the two solutions increases with the decrease of the Womersley number, i.e., with the decrease of the tube radius (artery), found in our previous paper [11]. This has been observed also in the 2D case [8, 13, 14].

The WSS can be obtained from the velocity solution by the following formula:

$$WSS = \mu_{app} \frac{\partial u}{\partial r} |_{r=R} \tag{10}$$

Table 1 Absolute values of the Wall Shear Stress (WSS)

	Newtonian (Pa)	Carreau (Pa)
1 s	3.25	3.5
2 s	0.88	0.95
3 s	3.75	4
4 s	1.5	1.7
5 s	2.85	3

The absolute values of the WSS for the considered example are given in Table 1 for the times $t = 1, 2, 3, 4, 5$ s.

The obtained peaks of WSS for the human carotid artery are in the experimental limits [15]: 2.5–4.3 Pa. The WSS of the Carreau model are slightly higher from those of the Newtonian model and are in a small phase shift, as found in [11].

The full system of equations (2) and (3) have been solved numerically by the software ANSYS/FLUENT for a straight long tube (1000 times longer than its radius) using a mesh of 40000 elements and 84042 nodes. The obtained results for the axial velocity u have been verified by the Womersley solution Eq. (7) (for the Newtonian fluid) and by the numerical solution found by the FDM (for the Carreau viscosity model). The relative error for both cases is less than 4 %.

3.2 *Artery with Model Aneurysm*

The numerical calculations of the Eqs. (2) and (3) in the case of the model aneurysm given by the shape formula (1) has been performed at $H = W = R = 0.0031$ m. The aneurysm is situated in the middle of the tube, which is long enough to achieve the straight tube flow (discussed in the previous subsection) in the regions before and after the aneurysm. Here the length is taken to be 0.62 m, such that the axial coordinate is $-0.31 \text{ m} \leq x \leq 0.31$ m. The used mesh for the calculations with the ANSYS/FLUENT in this domain consists by 156000 cells and 160040 nodes. The boundary condition at the inlet of the tube for the velocity is to be equal to the mean axial velocity u_n from Eq. (8) for the Newtonian model and by u_c from Eq. (9) for the Carreau model. The other boundary conditions are a constant pressure at the outlet and the usual no-slip condition on the wall. The results show that besides the axial velocity u, the velocity vector has also a radial non-zero component v in the aneurysm region. The appearance of v is connected to the toroidal vortices in the aneurysm part. This can be seen from Fig. 3, where the velocity vectors are colored according to the axial velocity magnitude at time $t = 5$ s in the Newtonian viscosity case.

It is interesting to show the axial velocity distribution along the symmetry axis $r = 0$ for different times. In the aneurysm region its character is uneven depending

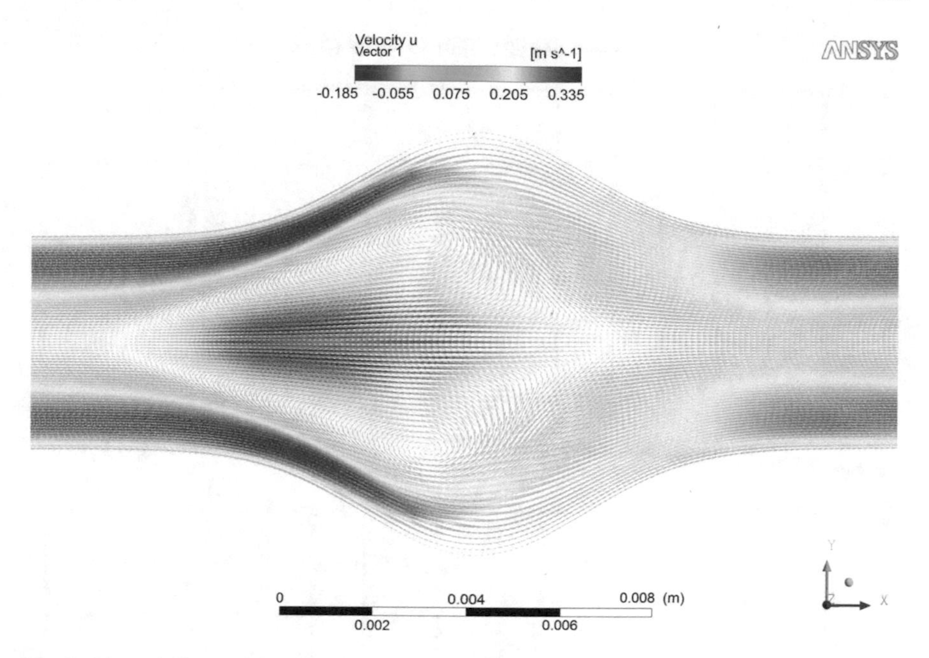

Fig. 3 The velocity vectors of the Newtonian model at time $t = 5$ s colored by the magnitude of the axial velocity u

on the time instant, which is shown in Fig. 4. As it is seen the Newtonian and Carreau velocity profiles are similar in the straight tube region (before and after the aneurysm), which is in a good comparison with the corresponding plots in Fig. 2.

The WSS in the aneurysm region are calculated by the full shear rate, as the radial velocity as well as its gradient on the wall are non-zeros:

$$WSS = \mu_{app}(\frac{\partial v}{\partial x} + \frac{\partial u}{\partial r}) |_{r=R} \quad (11)$$

The obtained absolute values of the WSS ($|WSS|$) in the tube with the aneurysm are quite different from those in the straight tube region, which is plotted in Fig. 5 for different times. The comparison between the results in Fig. 5 and those in Table 1 shows that some of the peak $|WSS|$ for the aneurysm are more than two times higher than those of the straight tube. In general the Carreau $|WSS|$ are slightly higher than the Newtonian ones, but in the aneurysm region they are almost equal in some places along the aneurysm width, while in others are quite different. This fact can be used as an indicator for further clinical studies.

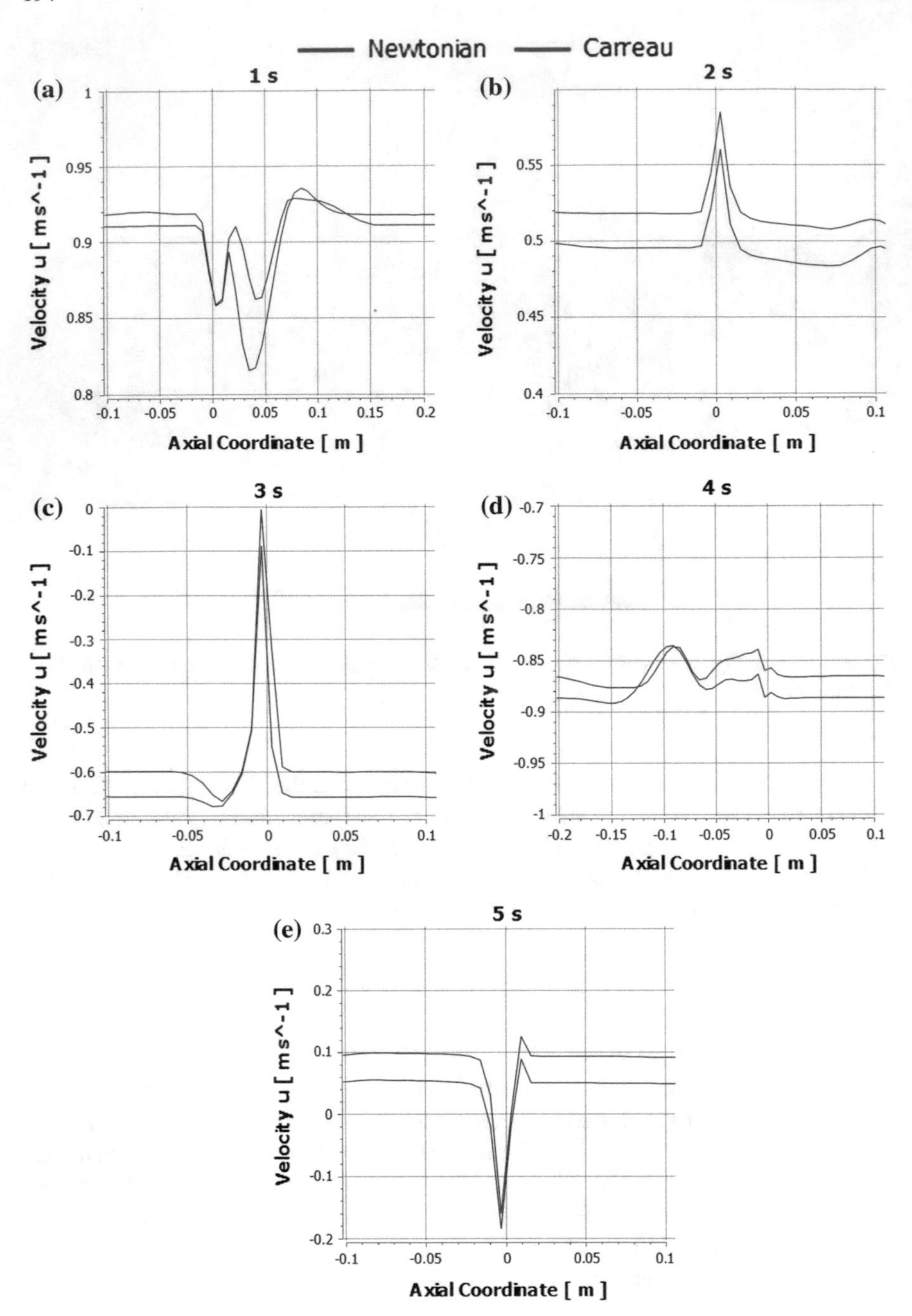

Fig. 4 Comparison between the axial velocities $u(x, 0, 0)$ on the centerline of the Newtonian model (*red lines*) and of the Carreau model (*blue lines*) at different instants of time t: (a) 1 s, (b) 2 s, (c) 3 s, (d) 4 s and (e) 5 s

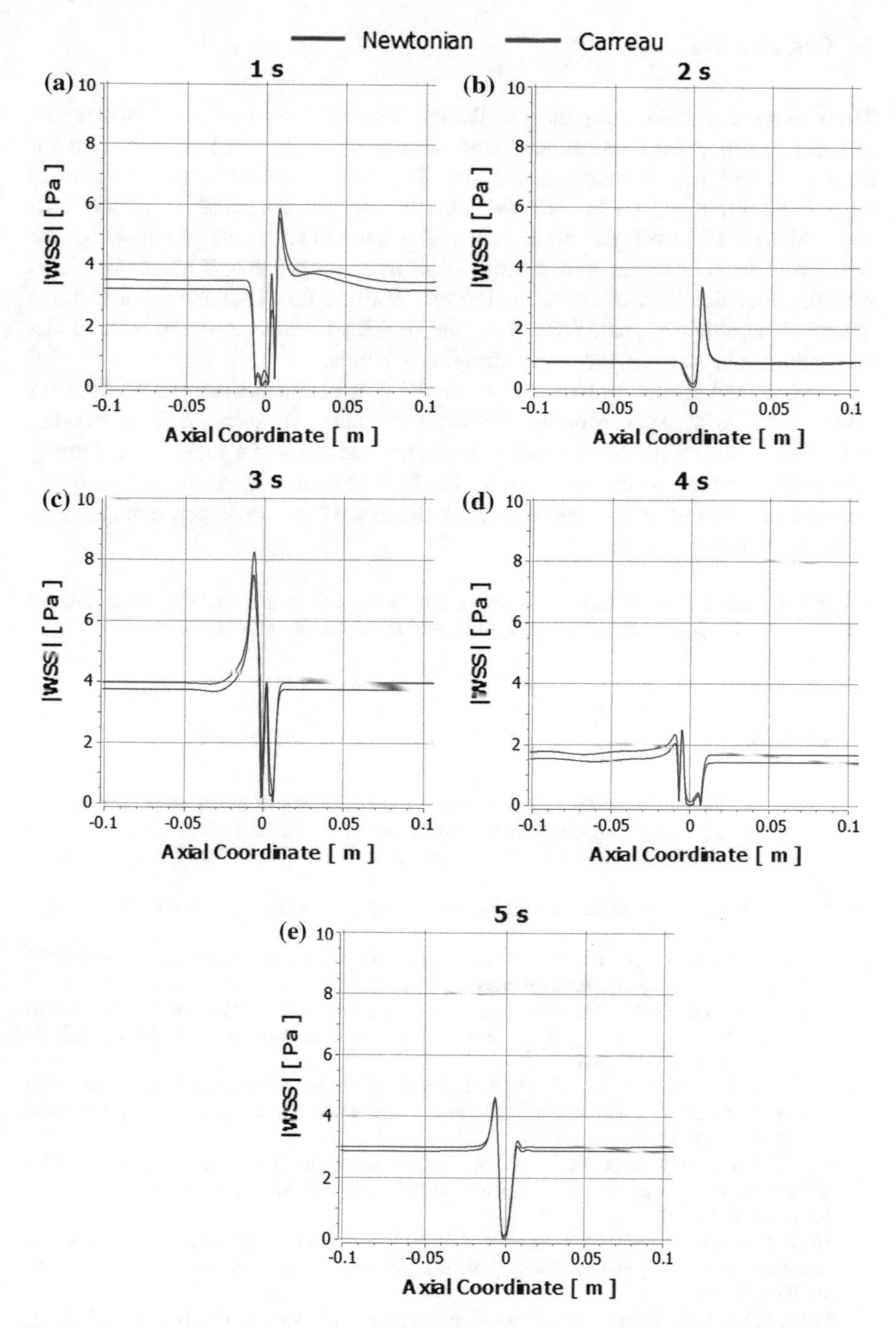

Fig. 5 Comparison between the absolute values of the WSS of the Newtonian model (*red lines*) and of the Carreau model (*blue lines*) at different instants of time t: (a) 1 s, (b) 2 s, (c) 3 s, (d) 4 s and (e) 5 s

4 Conclusion

The numerical solutions for the oscillatory flow velocity due to the Newtonian and Carreau model are constructed numerically for a straight long tube and for a tube (artery) with a model aneurysm. The numerical solutions are obtained by a finite-difference method (FDM) for the straight tube and using the software ANSYS/FLUENT for both cases. The numerical results obtained by the ANSYS/FLUENT for the velocity and WSS in a straight long tube are validated with the analytical and numerical solutions by FDM for Newtonian and Carreau models. The obtained peak WSS from the oscillatory flow in a tube with model aneurysm are higher than those in a straight long tube.

The obtained results for the Carreau model flow characteristics can be used for future studies with experimentally registered oscillatory pressure gradient. In order to predict the real WSS it is necessary to use the geometry of a patient based artery with aneurysm and to take into account the fluid structure interaction of the blood flow in a deformable artery, whose characteristics are based on experimental results of the wall artery structure.

Acknowledgements The authors have been partially supported for this research by the National Science Fund of Bulgarian Ministry of Education and Research: Grant DFNI-I02/3.

References

1. Razavi, A., Shirani, E., Sadeghi, M.R.: Numerical simulation of blood pulsatile flow in a stenosed carotid artery using different rheological models. J. Biomech. **44**, 2021–2030 (2011)
2. Myers, T.G.: Application of non-Newtonian models to thin film flow. Phys. Rev. E **72**, 066302 (2005)
3. Shibeshi, S.S., Collins, W.E.: The rheology of blood flow in a branched arterial system. Appl. Rheol. **15**, 398–405 (2005)
4. Liu, B., Tangbemph, D.: Influence of non-Newtonian properties of blood on the wall shear stress in human atherosclerotic right coronary arteries. Mol. Cell Biomech. **8**, 73–90 (2011)
5. Valencia, A., Ledermann, D., Rivera, R., Bravo, E., Galvez, M.: Blood flow dynamics and fluid-structure interaction in patient-specific bifurcating cerebral aneurysms. Num. Methods Fluids. **58**, 1081–1100 (2008)
6. Gijsen, F.J.H., Allanic, E., van de Vosse, F.N., Janssen, J.D.: The influence of the non-Newtonian properties of blood on the flow in large arteries: unsteady flow in a 90° curved tube. J. Biomech. **32**, 705–713 (1999)
7. Gijsen, F.J.H., van de Vosse, F.N., Janssen, J.D.: The influence of the non-Newtonian properties of blood on the flow in large arteries: steady flow in a carotid bifurcation model. J. Biomech. **32**, 601–608 (1999)
8. Boyd, J., Buick, J.M., Green, S.: Analysis of the Casson and Carreau-Yasuda non-Newtonian blood models in steady and oscillatory flows using the lattice Boltzmann method. Phys. Fluids. **19**, 093103 (2007)
9. Chen, J., Lu, X.-Y.: Effect of non-Newtonian and pulsatile blood flow on mass transport in the human aorta. J. Biomech. **39**, 818–832 (2006)
10. Womersley, J.R.: Method for the calculation of velocity, rate of flow and viscous drag in arteries when the pressure gradient is known. J. Physiol. **127**, 553–563 (1955)

11. Tabakova, S., Nikolova, E., Radev, St.: Carreau model for oscillatory blood flow in a tube. AIP Conf. Proc. **1629**, 336–343 (2014)
12. Gopalakrishnan, S.S., Pier, B., Biesheuvel, A.: Dynamics of pulsatile flow through model abdominal aortic aneurysms. J. Fluid Mech. **758**, 150–179 (2014)
13. Kutev, N., Tabakova, S., Radev, St.: Approximation of the oscillatory blood flow using the Carreau viscosity model. In: Tikhonov, A.A. (ed.) Proceedings, IEEE, IEEE Catalog Number CFP15A24-ART (2015)
14. Tabakova, S., Kutev, N., Radev, St.: Application of the Carreau viscosity model to the oscillatory flow in blood vessels. AIP Conf. Proc. **1690**, 040019 (2015)
15. Reneman, R.S., Hoeks, A.P.G.: Wall shear stress as measured in vivo: consequences for the design of the arterial system. Med. Biol. Eng. Comput. **46**, 499–507 (2008)

Reduced Rule-Base Fuzzy-Neural Networks

Margarita Terziyska and Yancho Todorov

Abstract In this paper two different fuzzy-neural systems with reduced fuzzy rules bases, namely Distributed Adaptive Neuro Fuzzy Architecture (DANFA) and Semi Fuzzy Neural Network (SFNN), are presented. Both structures are realized with Takagi-Sugeno fuzzy inference mechanism and they posses reduced number of parameters for update during the learning procedure. Thus, the computational time for algorithm execution is additionally reduced, which make the modeling structures a promising solution for real time applications. As a learning approach for the designed structures a simplified two-step gradient descent approach is implemented. To demonstrate the potentials of both models, simulation experiments with two benchmark chaotic time systems—Mackey-Glass and Rossler are studied. The obtained results show accurate models performance with minimal prediction error.

1 Introduction

Neural networks and fuzzy logic are proven as universal approximators, which can estimate any nonlinear function to a prescribed accuracy. For identification of complex nonlinear processes different kinds of fizzy-neural architectures are also used. These structures have an advantage over traditional statistical estimation and adaptive control approaches. They estimate a function without the need of a detailed mathematical description on the functional dependency between inputs and outputs. Combining neural networks and fuzzy systems in one unified framework has become popular in the last decades. The fusion of both combine the learning and

M. Terziyska
Department of Informatics and Statistics, University of Food Technologies,
26 Maritza blvd., 4000 Plovdiv, Bulgaria
e-mail: m.terziyska@uft-plovdiv.bg

Y. Todorov (✉)
Institute of Information and Communication Technologies,
Bulgarian Academy of Sciences, Acad. G. Bontchev st., bl.2,
1113 Sofia, Bulgaria
e-mail: yancho.todorov@ieee.org

K. Georgiev et al. (eds.), *Advanced Computing in Industrial Mathematics*,
Studies in Computational Intelligence 681, DOI 10.1007/978-3-319-49544-6_17

computational ability of neural networks with the human like IF-THEN thinking and reasoning of a fuzzy system. This could be compared with the human brain [1] neural network concentrate on the structure of human brain, i.e., on the hardware whereas fuzzy logic system concentrate on software. A lot of architectures have been proposed in the literature that combines fuzzy logic and neural network. Some of the most popular are ANFIS [2] and DENFIS [3]. They are all composed of a set of if-then rules. In principle, the number of fuzzy rules depends exponentially on the number of inputs and membership functions. If n is the number of inputs of a fuzzy-neural system and m is the number of the membership functions, then the number of generated fuzzy rules is m^n. Thus, the huge number of generated rules requires the determination of a large number of parameters during the learning procedure. For instance, for a fuzzy inference system with 10 inputs, each with two membership functions, the grid partitioning leads to $1024(=2^{10})$ rules, which is extremely large number of rules for any practical application.

In order to reduce the number of fuzzy rules without loss of accuracy, different fuzzy clustering approaches as fuzzy C-means [4, 5] and K-means [6] can be used. As well, subtractive clustering and hyperplane clustering are proposed in [7, 8]. Evolving fuzzy systems [3, 9], such as DENFIS, use evolving clustering and dynamically form bases of fuzzy rules generated during the past instance of the learning process. The new AnYa fuzzy-neural structure also belongs to the evolving fuzzy systems. This architecture works with the so-called clouds instead of fuzzy sets. This removes the need for training of membership functions parameters. However, a priori data is needed to form the clouds [10]. Another possibility to reduce the number of fuzzy rules gives the self-constructing and self-organizing fuzzy-neural network structures [11, 12]. In this type of structures, during the training procedure, inactive rules are being removed, which consequently leads to reduction in the number of trained parameters. In order to deal with the rule-explosion problem, hierarchical fuzzy neural networks could be used but they employ a very complex learning method. A method that compresses a fuzzy system with an arbitrarily large number of rules into a smaller fuzzy system by removing the redundancy in the fuzzy rule base is presented in [13]. As a result of this compression, the number of on-line operations during the fuzzy inference process is significantly reduced without compromising the solution. Review of the most of existing rule base reduction methods for fuzzy systems, summary of their attributes and introduction of advanced techniques for formal presentation of fuzzy systems based on Boolean matrices and binary relations, which facilitate the overall management of complexity, is made in [14].

In this paper two different fuzzy-neural structures with reduced number of the fuzzy rules are proposed, namely Distributed Adaptive Neuro-Fuzzy Architecture (DANFA) and Semi Fuzzy Neural Network (SFNN). As a learning procedure for the designed structures, a simplified two-step gradient algorithm based on minimization of an instant quadratic error measurement function is applied. Thus, in each sampling period of a model's operation, two groups of parameters are being scheduled: the premise—the parameters of the fuzzy membership functions and the consequent—the parameters of the liner modeling functions. Applying the proposed principles lead to reduction of the both associated fuzzy rule's premise and consequent parameters. This facilitates the on-line learning procedure and reduces the computational effort without a great loss of modeling accuracy. To demonstrate the potentials of the proposed networks, simulation experiments with two benchmark chaotic time systems—Mackey-Glass and Rossler are studied. They produce fast changing by amplitude and frequency signals which often are hard to be modeled in real time. The obtained results show accurate models performance with minimal modeling error. Thus, their application areas may be extended to modeling of fast changing plant processes and process control in the framework of model based control.

2 Classical Fuzzy-Neural Network

In this section the Classical Fuzzy Neural Network (CFNN) using the Takagi-Sugeno approach is discussed. Its structure is shown on Fig. 1.

Layer 1: This layer accepts the input variables and then nodes in this layer only transmit the input values to the next layer directly.

Layer 2: Each node in this layer implements the fuzzification procedure via Gaussian membership function, where X_p is the input value, $c_{Xp,m}$ and $\sigma_{Xp,m}$ are the center and the standard deviation of the function:

$$\mu_{Xp,m}^{(n)} = \exp \frac{-(x_p - c_{Xp,m})^2}{2\sigma_{Xp,m}^2} \tag{1}$$

Layer 3: This layer is a kind of rules generator as it forms the fuzzy logic rules. Their number depends on the number of inputs p and the number of their fuzzy sets m, and it is calculated according to the expression $N = m^p$. In this layer, each node represents a fuzzy rule in following form:

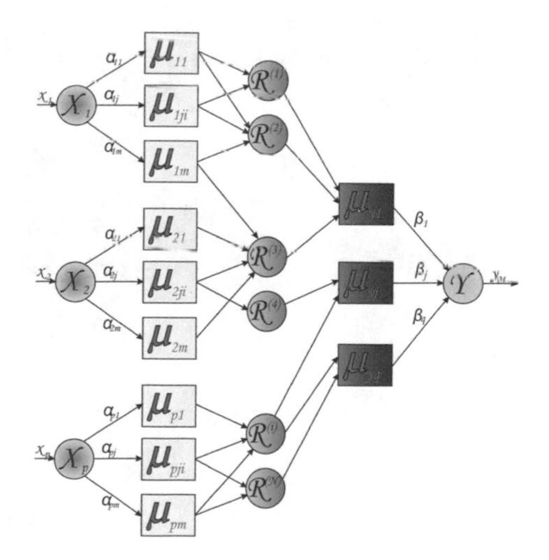

Fig. 1 Structure of Classical Fuzzy Neural Network

$$R^{(i)} : if\ x_1\ is\ \tilde{A}_1^{(i)}\ and\ x_p\ is\ \tilde{A}_p^{(i)}\ then\ f_y{}^{(i)}(k) \quad (2)$$

$$\begin{aligned} f_y^{(i)}(k) &= a_1^{(i)}y(k-1) + a_2^{(i)}y(k-2) + \cdots + a_{ny}^{(i)}y(k-n_y) + \\ &+ b_1^{(i)}u(k) + b_2^{(i)}u(k-1) + \cdots + b_{nu}^{(i)}u(k-n_u) + c_0^{(i)} \end{aligned} \quad (3)$$

Layer 4: In the fourth layer the operation of fuzzy implication is realized as:

$$\mu_{yq}^{(n)}(k+j) = \mu_{x_1,m}^{(n)}(k+j) * \mu_{x_2,m}^{(n)}(k+j) * \cdots * \mu_{x_p,m}^{(n)}(k+j) \quad (4)$$

Layer 5: In the fifth, last layer the final decision which consists in determining the value of the model output is taken. It can be described by the following expression:

$$\hat{y}(k+j) = \frac{\sum_{i=1}^{q} f_y^{(i)}(k+j)\mu_y^{(i)}(k+j)}{\sum_{i=1}^{q} \mu_y^{(i)}(k+j)} \quad (5)$$

3 Distributed Adaptive Neuro-Fuzzy Architecture

The structure of the proposed DANFA model with a second order Takagi-Sugeno inference mechanism is shown on Fig. 2. The model represents a modification of the CFNN with a second order Takagi-Sugeno inference mechanism, already described in the previous section. Actually, the structure is a network from q CFNNs distributed in six-layered architecture. The main idea behind is to distribute the input signals to separate fuzzy neural structures in order to decrease the number of the fuzzy rules.

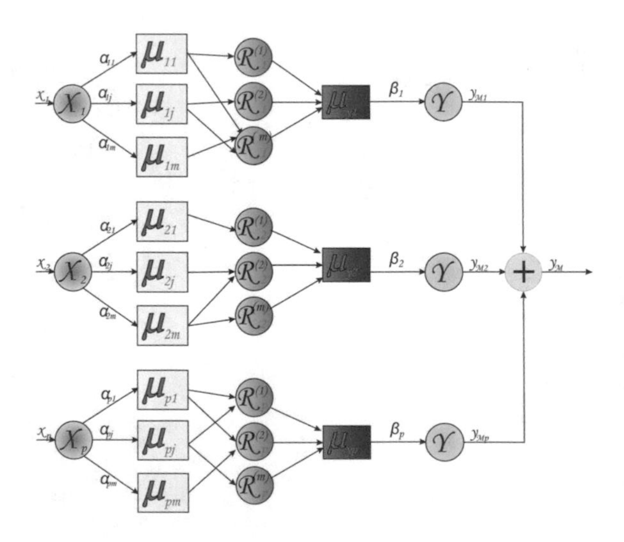

Fig. 2 Block scheme of the proposed DANFA structure

Each of these sub-structures act as a separate sub-model within the DANFA structure. The output signal of the DANFA model is computed as a sum of the output signals of all CFNN's and it is obtained by implementing Eqs. (1)–(5). Thus, the output of the model is expressed as:

$$\hat{y}_M(k+j) = \hat{y}_{M1}(k+j) + \hat{y}_{M2}(k+j) + \cdots + \hat{y}_{Mq}(k+j) \tag{6}$$

where $\hat{y}_{Mr}$ for r = 1:q is obtained as follow:

$$\hat{y}_{Mr}(k+j) = \frac{\sum_{i=1}^{q} f_r^{(i)}(k+j)\mu_r^{(i)}(k+j)}{\sum_{i=1}^{q} \mu_r^{(i)}(k+j)} \tag{7}$$

For simplicity, a DANFA structure with $q = 2$, as it shown on Fig. 3 is studied. Its output expressed as:

$$\hat{y}_M = \hat{y}_{M1} + \hat{y}_{M2} = \frac{\sum_{i=1}^{q} f_u{}^{(i)} \mu_u{}^{(i)}}{\sum_{i=1}^{q} \mu_u{}^{(i)}} + \frac{\sum_{i=1}^{q} f_y{}^{(i)} \mu_y{}^{(i)}}{\sum_{i=1}^{q} \mu_y{}^{(i)}} \tag{8}$$

where $\hat{y}_{M1}$ and $\hat{y}_{M2}$ are the output parameters respectively to CFNN1 and CFNN2 (see Fig. 3); f_u and f_y are the Sugeno output functions; μ_u and μ_y are the corresponding membership degrees of the quantization levels, all defined as:

$$f_u{}^{(i)} = b_{1u}{}^{(i)} u(k) + b_{2u}{}^{(i)}(k-1) + \cdots + b_{n_u u}{}^{(i)} u(k-n_u) + b_{ou}{}^{(i)} \tag{9}$$

$$\mu_u^{(i)} = \mu_{u1j}{}^{(i)} * \mu_{u2j}{}^{(i)} * \ldots\ldots \mu_{upj}{}^{(i)} \tag{10}$$

$$f_y{}^{(i)} = a_{1y}{}^{(i)} y(k-1) + a_{2y}{}^{(i)} y(k-2) + \cdots + a_{n_y y}(k-n_y) + a_{oy}{}^{(i)} \tag{11}$$

$$\mu_y{}^{(i)} = \mu_{y1j}{}^{(i)} * \mu_{y2j}{}^{(i)} * \ldots\ldots \mu_{ypj}{}^{(i)} \tag{12}$$

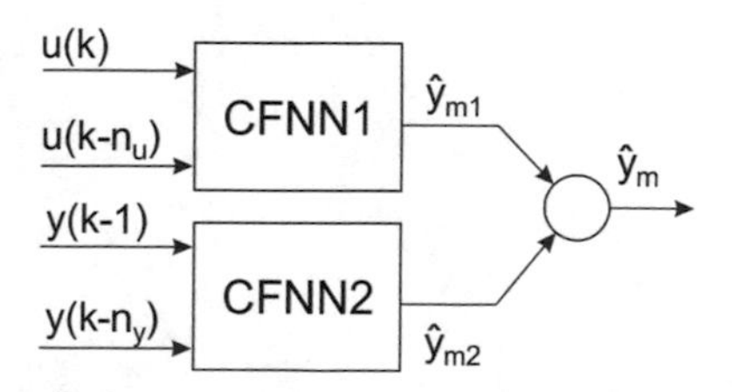

Fig. 3 Block scheme of the inference distribution within a DANFA structure

From Fig. 3 it is clear that CFNN1 is a fuzzy-neural network which cares for vector regressor u and CFNN2 is a fuzzy-neural network which cares for vector regressor y. For the considered case each of these networks have two input variables with three fuzzy sets, respectively $N = 3^2 = 9$ rules.

The main advantage of the proposed DANFA structure is its ability to operate with smaller number of parameters compared to the case of pure CFNN model. On the other hand, the reduced number of fuzzy rules induce the reduction of the model parameters (coefficients, membership parameters), as well. The number of the generated fuzzy rules in a DANFA structure is $N = q * m^p$, where q is the number of used CFNNs, p is the number of input variables and m is the number of their fuzzy sets. Obviously, if one wants to realize a model with 4 inputs with 3 fuzzy sets each, then the CFNN model will generate 81 fuzzy rules, while the DANFA model will generate only 18 fuzzy rules. Also, during the learning procedure in CFNN model are computed the values of 1053 parameters, while in the DANFA model are computed the values of only 156 parameters.

4 Semi-fuzzy Neural Network

The structure of the proposed SFNN model is a modification of the CFNN model described in Sect. 2 and its structure is shown on Fig. 4. It represents five-layered architecture with the Takagi-Sugeno inference mechanism. However, in SFNN model a part of input signals are not fuzzified and they come with their crisp values, weighted by an appropriate coefficient, into the third layer (fuzzy rules layer), i.e. directly into the consequent (then) part of the fuzzy rules. This approach enables the possibility

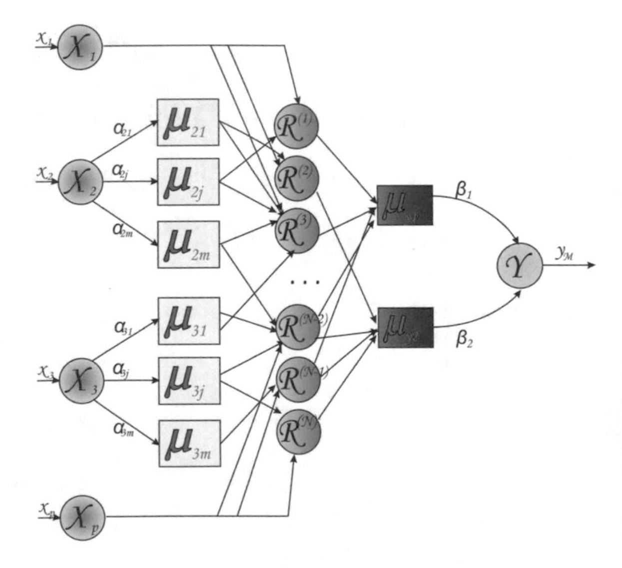

Fig. 4 Block scheme of the proposed SFNN structure

to reduce flexibly the number of the fuzzy rules and their associated parameters, determined by the learning procedure.

The considered SFNN model is realized with NARX representation of the consequents. Therefore, three types of SFNN structures may be generated: the first one is a case where the values of the vector-regressor $y[y(k-1), y(k-2)]$ are fuzzified inputs and the values of the vector-regressor $u[u(k), u(k-1)]$ are nonfuzzified inputs—SFNN Type I; the second one is an opposite the values of the vector-regressor $y[y(k-1), y(k-2)]$ are nonfuzzified inputs and the values of the vector-regressor $u[u(k), u(k-1)]$ are fuzzified inputs—SFNN Type II; in the third one the values of $u(k)$ and $y(k-1)$ are fuzzified inputs, while the $u(k-1)$ and $y(k-2)$ are nonfuzzified inputs—SFNN Type III.

5 Learning Approach

As a learning procedure for DANFA model, a simplified two-step gradient algorithm is applied. It is based on minimization of an instant error measurement function between the real and modeled outputs defined as:

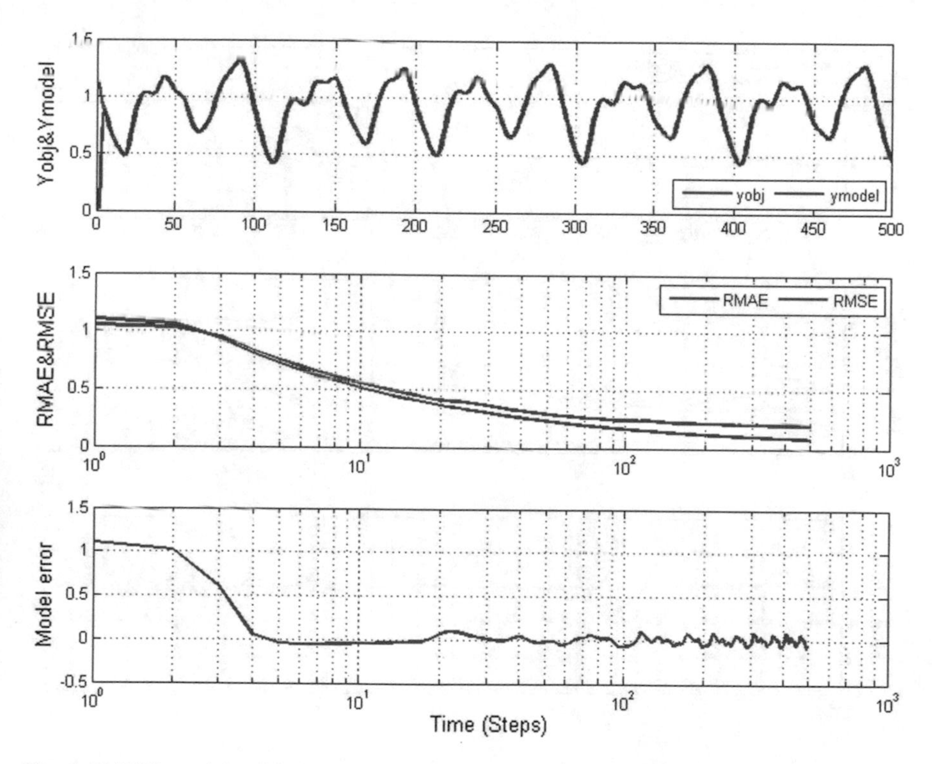

Fig. 5 DANFA model validation by using Mackey-Glass chaotic time series

$$E(k) = \frac{(y(k) - \hat{y}(k))^2}{2} = \frac{(y(k) - (\hat{y}_{M1} + \hat{y}_{M2}))^2}{2} \tag{13}$$

where $y(k)$ denotes the measured real output and $\hat{y}(k)$ is sum calculated by the two fuzzy-neural networks output parameters $\hat{y}_{M1}$ and $\hat{y}_{M2}$. The algorithm performs two steps gradient learning procedure. Assuming, that β_{uij} is an adjustable i-th coefficient for the Sugeno function f_u into the j-th activated rule for the CFNN1 then the general parameter learning rule for the consequent parameters is:

$$\beta_{uij}(k+1) = \beta_{uij}(k) + \eta\left(-\frac{\partial E}{\partial \beta_{uij}}\right) \tag{14}$$

After calculating the partial derivatives, the final recurrent predictions for each adjustable coefficient β_{uij} and the free coefficient are obtained by the following equations:

$$\beta_{uij}(k+1) = \beta_{uij}(k) + \eta\varepsilon(k)\bar{\mu}_u^{(j)}(k)x_{1i}(k) \tag{15}$$

$$\beta_{0uj}(k+1) = \beta_{0uj}(k) + \eta\varepsilon(k)\bar{\mu}_u^{(j)}(k) \tag{16}$$

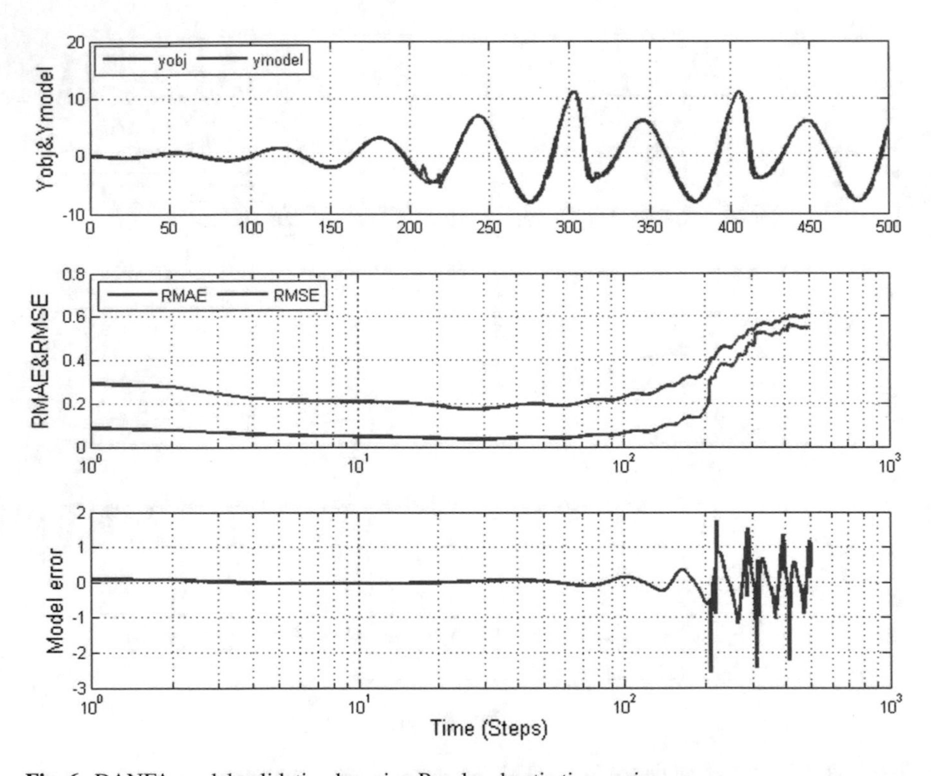

Fig. 6 DANFA model validation by using Rossler chaotic time series

The output error E can be used back directly to the input layer, where there are the premise adjustable parameters (center—c_{ij} and the deviation—σ_{ij} of a Gaussian fuzzy set). The error E is propagated through the links composed by the corresponded membership degrees, where the link weights are unit. Hence, the learning rule for the second group adjustable parameters in the input layer can be done by the same learning rule:

$$c_{uij}(k+1) = c_{uij}(k) + \eta\varepsilon(k)\bar{\mu}_u^{(i)}(k)[f_u^{(i)} - \hat{y}_M(k)]\frac{[x_{1i}(k) - c_{uij}(k)]}{c_{uij}^2(k)} \tag{17}$$

$$\sigma_{uij}(k+1) = \sigma_{uij}(k) + \eta\varepsilon(k)\bar{\mu}_u^{(i)}(k)[f_u^{(i)} - \hat{y}_M(k)]\frac{[x_{1i}(k) - \sigma_{uij}(k)]^2}{\sigma_{uij}^3(k)} \tag{18}$$

The parameters for CFNN2 using the same approach can be expressed as:

$$\beta_{ylj}(k+1) = \beta_{yij}(k) + \eta\varepsilon(k)\bar{\mu}_y^{(j)}(k)x_{2i}(k) \tag{19}$$

$$\beta_{0yj}(k+1) = \beta_{0yj}(k) + \eta\varepsilon(k)\bar{\mu}_y^{(j)}(k) \tag{20}$$

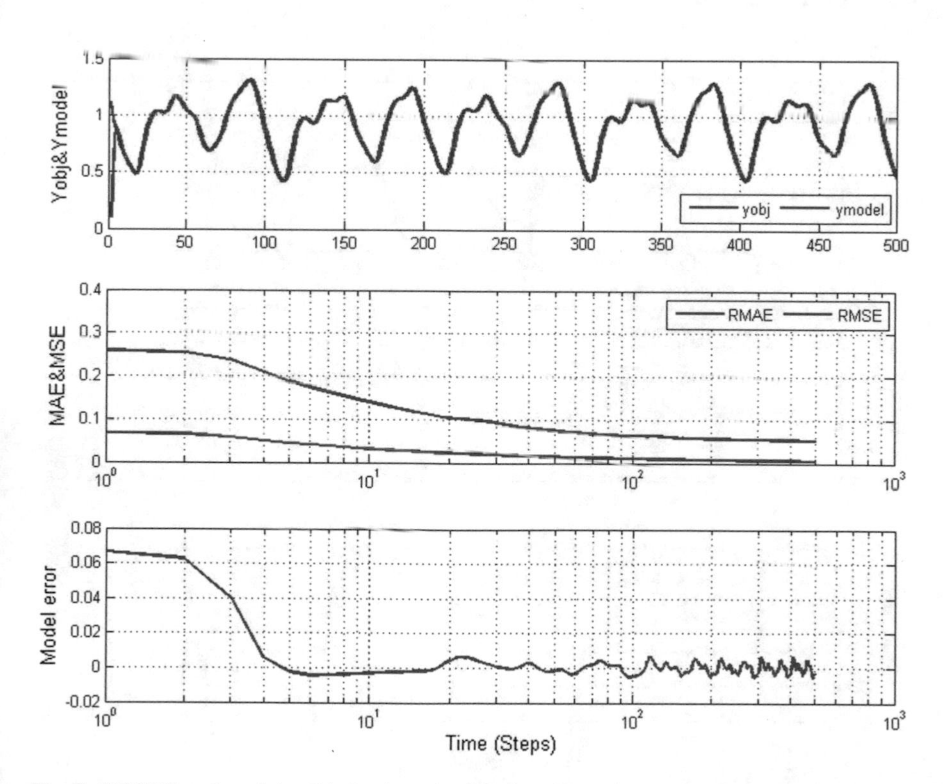

Fig. 7 SFNN Type I model validation by using Mackey-Glass chaotic time series

$$c_{yij}(k+1) = c_{yij}(k) + \eta\varepsilon(k)\bar{\mu}_y^{(i)}(k)[f_y^{(i)} - \hat{y}_M(k)]\frac{[x_{2i}(k) - c_{yij}(k)]}{c_{yij}^2(k)} \quad (21)$$

$$\sigma_{yij}(k+1) = \sigma_{yij}(k) + \eta\varepsilon(k)\bar{\mu}_y^{(i)}(k)[f_y^{(i)} - \hat{y}_M(k)]\frac{[x_{2i}(k) - \sigma_{yij}(k)]^2}{\sigma_{yij}^3(k)} \quad (22)$$

For the proposed SFNN structure, the training algorithm is quite similar to this for DANFA model. An instant error measurement function is defined again by (13) and then the values of the linear and the non-linear parameters are determined according to expressions (14)–(18).

6 Experimental Results

To investigate the modeling potentials of the proposed DANFA and SFNN models, Mackey-Glass (MG) and Rossler chaotic system series benchmark models, have been used. The used time series will not converge or diverge, and the trajectory is highly sensitive to initial conditions. The MG time series is described by time-delay differential equation:

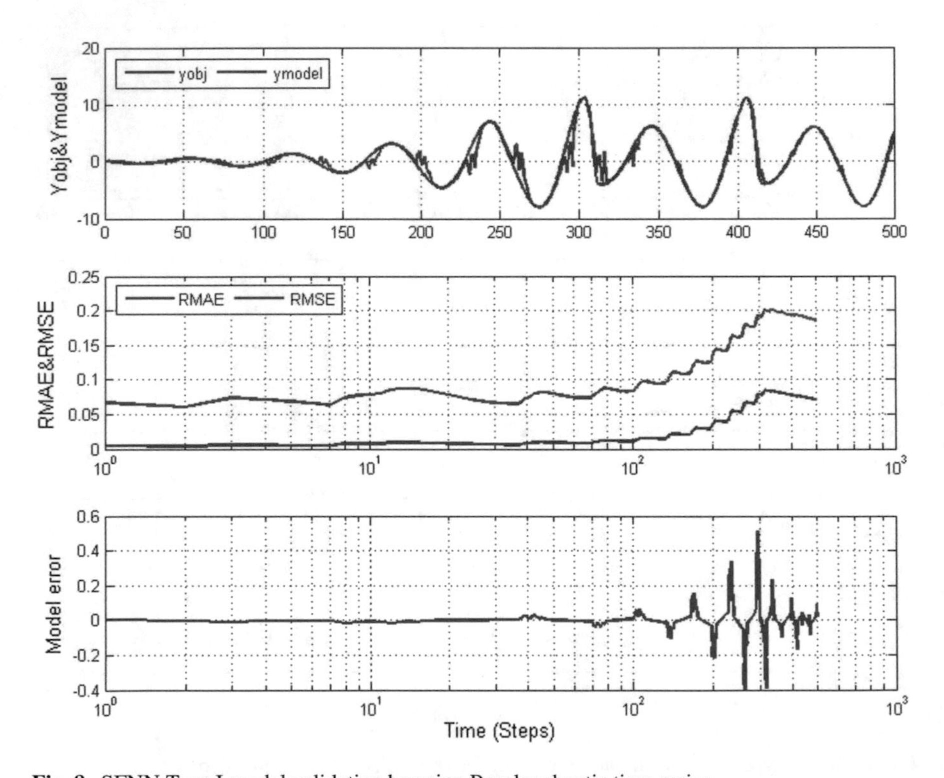

Fig. 8 SFNN Type I model validation by using Rossler chaotic time series

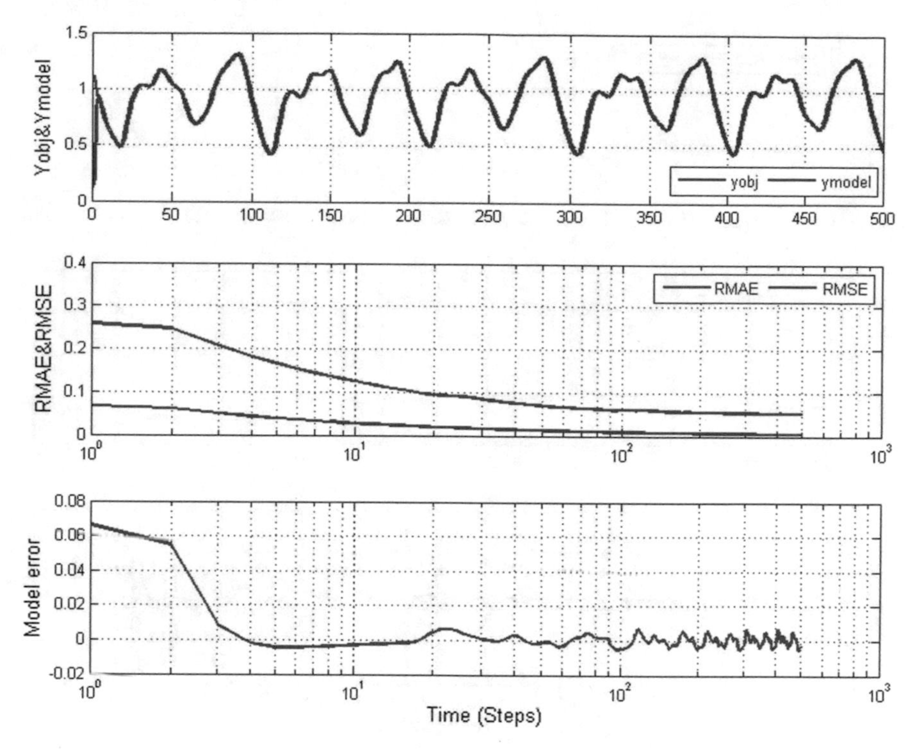

Fig. 9 SFNN Type II model validation by using Mackey-Glass chaotic time series

$$x(i+1) = \frac{x(i) + ax(i-s)}{(1 + x^c(i-s)) - bx(i)} \tag{23}$$

where $a = 0.2$; $b = 0.1$; $C = 10$; initial conditions $x_0 = 0.1$ and $s = 17\,\text{s}$. Results with DANFA model and the three types of SFNN models validation by using Mackey-Glass chaotic time series are shown on Figs. 5, 7, 9 and 11, respectively. As it can be seen, the proposed model structures predict accurately the generated time series, with minimum modeling error and fast transient responses of the RMSE (Root Mean Squared Error) and RMAE (Root Mean Absolute Error),(the last two presented in logarithmic scale) reaching very small values.

Additional experiments with the proposed DANFA and SFNN models for modeling of Rossler chaotic time series are made. These series is described by three coupled first-order differential equations:

$$\frac{dx}{dt} = -y - z\frac{dy}{dt} = x + ay\frac{dz}{dt} = b + z(x - c) \tag{24}$$

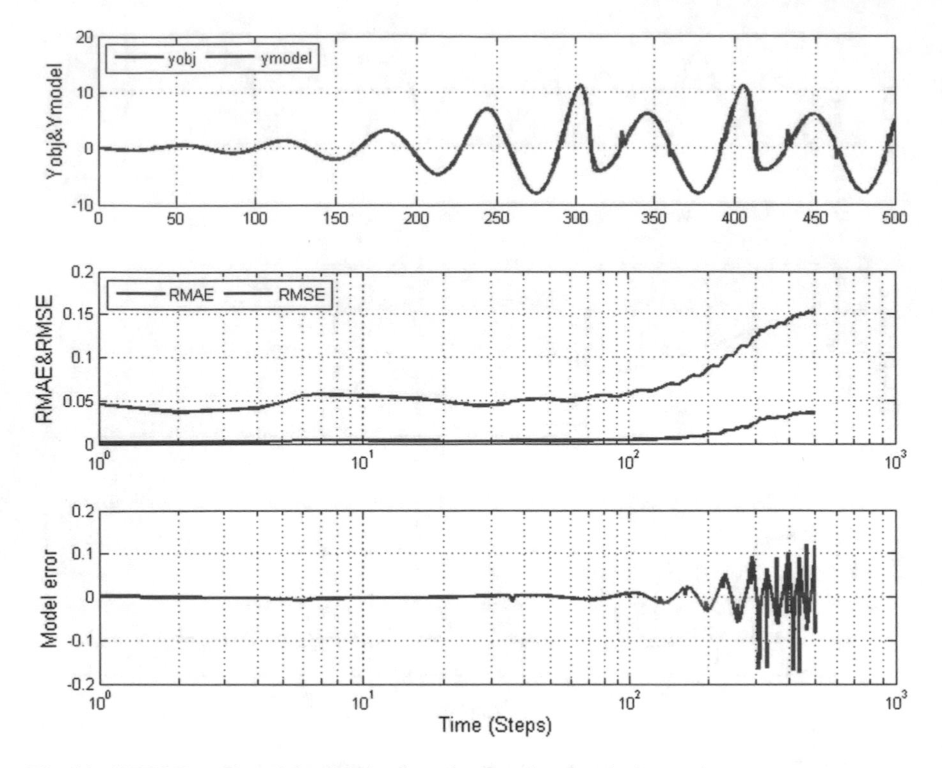

Fig. 10 SFNN Type II model validation by using Rossler chaotic time series

where $a = 0.2$; $b = 0.4$; $c = 5.7$; initial conditions $x_0 = 0.1$; $y_0 = 0.1$; $z_0 = 0.1$. The results are given on Figs. 6, 8, 10 and 12, respectively. As can be seen again, the achieved values of the studied error terms are very small and closer to zero (Figs. 7, 8, 9, 10, 11 and 12).

A comparison between the CFNN and the DANFA models is made and it is shown in Table 1. The results in Table 1 are for Mackey-Glass chaotic time series prediction. As it can be seen the absolute value of current prediction error (Model_err in Table 1) obtained with the DANFA model is smaller than this obtained with the CFNN model. The greater accuracy of a DANFA model can be easily explained. In the expression (3) the free parameter c_0 plays the role of a disturbance filter. This can easily be proved if the expression (3) is transformed into the following form:

$$\begin{aligned} f_y^{(i)}(k) &= a_1^{(i)} y(k-1) + a_2^{(i)} y(k-2) + \cdots + a_{ny}^{(i)} y(k-n_y) + \cdots \\ &+ b_1^{(i)} u(k) + b_2^{(i)} u(k-1) + \cdots + b_{nu}^{(i)} u(k-n_u) + d^{(i)}(k) \end{aligned} \tag{25}$$

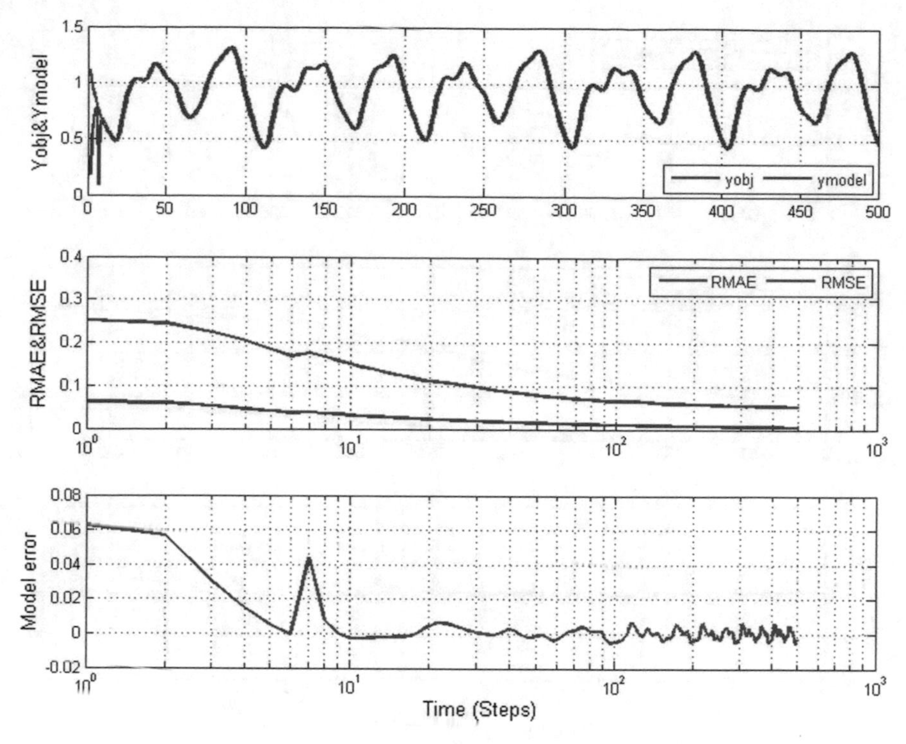

Fig. 11 SFNN Type III model validation by using Mackey-Glass chaotic time series

This conversion is valid in $T(q^{-1}) = 1$. In (25) $d(k)$ is unknown disturbance that is defined by:

$$d(k) = \frac{T(q^{-1})}{\Delta(q^{-1})}\upsilon(k) \tag{26}$$

Thus, the CFNN model works with one disturbance filter while DANFA model has two filters (see expressions (9) and (11)) and this is the reason the latter is more accurate.

The values of RMAE and RMSE in interval of 500 steps for the three types SFNN are summarized in Table 2. From these data it can be concluded that the SFNN Type III model is more accurate than the SFNN Type I and type II models, i.e. the case that u(k), y(k − 1) are fuzzified inputs.

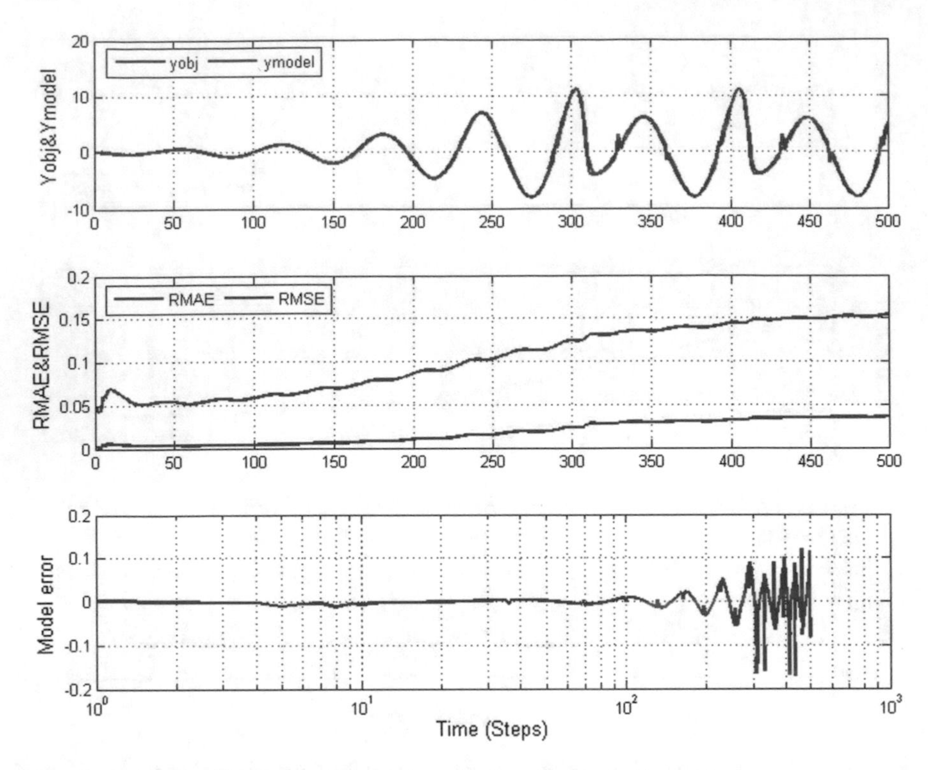

Fig. 12 SFNN Type III model validation by using Rossler chaotic time series

Table 1 Comparison between CFNN and DANFA model

Steps	DANFA model			Classical FN Network		
	Model_err	RMAE	RMSE	Model_err	RMAE	RMSE
50	−0.0196	0.2963	0.2314	−0.0470	0.2883	0.1193
100	−0.0672	0.2450	0.1659	−0.1512	0.2780	0.1019
150	−0.0129	0.2250	0.1375	−0.0513	0.2746	0.0991
200	−0.0588	0.2172	0.1209	−0.1274	0.2755	0.0962
250	−0.0240	0.2010	0.1097	−0.0512	0.2753	0.0947
300	−0.0484	0.2090	0.1015	−0.1254	0.2782	0.0945
350	−0.0397	0.2043	0.0953	−0.1017	0.0770	0.0930
400	−0.0452	0.2030	0.0904	−0.1186	0.0783	0.0932
450	−0.0344	0.1999	0.0862	−0.0951	0.0765	0.0919
500	−0.0407	0.1992	0.0828	−0.1122	0.2747	0.0922

Table 2 SFNN Type I, SFNN Type II and SFNN Type III comparison

Steps	SFNN Type I model		SFNN Type II		SFNN Type III model	
	RMAE	RMSE	RMAE	RMSE	RMAE	RMSE
50	0.0625	0.0145	0.0541	0.0156	0.0522	0.0129
100	0.0432	0.0104	0.0362	0.0111	0.0347	0.0093
150	0.0370	0.0087	0.0323	0.0092	0.0313	0.0078
200	0.0304	0.0077	0.0262	0.0081	0.0252	0.0069
250	0.0272	0.0070	0.0233	0.0074	0.0225	0.0063
300	0.0225	0.0065	0.0185	0.0068	0.0176	0.0059
350	0.0225	0.0061	0.0191	0.0064	0.0184	0.0056
400	0.0194	0.0058	0.0159	0.0061	0.0152	0.0053
450	0.0187	0.0056	0.0153	0.0058	0.0141	0.0051
500	0.0172	0.0054	0.0141	0.0056	0.0134	0.0049

7 Conclusions

In this paper are presented two fuzzy-neural architectures with reduced fuzzy rules bases and Takagi-Sugeno inference mechanism. Both models are realized as a Distributed Adaptive Neuro-Fuzzy Architecture (DANFA), where the input space is distributed along a set of fuzzy inferences and a Semi-Fuzzy Neural Network (SFNN), with selective fuzzification of the input space. Two benchmark chaotic systems (Mackey-Glass and Rossler chaotic time series) are chosen to demonstrate the modeling properties of the models. The obtained results show that the proposed DANFA and SFNN models predict accurately the generated time series, with minimum prediction error and fast transient response of the RMSE, reaching values closer to zero. Their main advantage is that they operate with small number of rules and respectively, they have a smaller number of parameters for learning. Thus, it can be facilitated the modeling of nonlinear systems with considerably less calculations in comparison to the CFNN model. Furthermore, DANFA and SFNN models have other advantages they are more accurate than the CFNN model and they not require a priori data and they are not bound by additional procedures, such as clustering. This makes them more suitable for real-time applications such as modeling and control of fast plant processes.

References

1. Chennakesava, R.A.: Fuzzy Logic and Neural Networks: Basic Concepts and Applications. New Age International Pvt Ltd Publishers (2008)
2. Jang, R.: ANFIS: adaptive-network-based fuzzy inference system. IEEE Trans. Syst. Man Cybern. **23**(5), 665–685 (1993)

3. Kasabov, N., Qun, S.: DENFIS: dynamic evolving neural-fuzzy inference system and its application for time-series prediction. IEEE Trans. Fuzzy Syst. **10**(2) (2002)
4. Chuang, K.-S., et al.: Fuzzy c-means clustering with spatial information for image segmentation. Comput. Med. Imaging Graph. **30**, 9–15 (2006)
5. Zalik, K.: Fuzzy C-Means Clustering and Facility Location Problems. Artificial Intelligence and Soft Computing (2006)
6. Gallucc, L., et al.: Graph based k-means clustering. Sign. Proces. **92**(9), 1970–1984 (2012)
7. Mehrabian, A.R., et al.: Neuro-fuzzy modeling of super-heating system of a steam power plant. Artif. Intell. Appl. 13–16 (2006)
8. Panella, M.: A hierarchical procedure for the synthesis of ANFIS networks. Adv. Fuzzy Syst. (2012)
9. Kasabov, N., Filev, D.: Evolving intelligent systems: methods, learning and applications. In: International Symposium of Evolving Fuzzy Systems, Sept 2006 (2006)
10. Angelov, P.: Autonomous Learning Systems. Wiley (2013)
11. Allende-Cid, H., et al.: Self-organizing neuro-fuzzy inference system. In: Iberoamerican Congress on Pattern Recognition CIARP, pp. 429–436 (2008)
12. Ferreyra, A., Rubio, J.: A new on-line self-constructing neural fuzzy network. In: Proceedings of the 45th IEEE Conference on Decision and Control, San Diego, CA, USA (2006)
13. Gegov, A.: Complexity Management in Fuzzy Systems, Studies in Fuzziness and Soft Computing, vol. 211 (2007)
14. Gegov, A., Gobalakrishnan, N.: Advanced inference in fuzzy systems by rule base compression. Mathware Soft Comput. **14**, 201–216 (2007)

Accuracy of Linear Craniometric Measurements Obtained from Laser Scanning Created 3D Models of Dry Skulls

Diana Toneva, Silviya Nikolova, Ivan Georgiev and Assen Tchorbadjieff

Abstract The aim of this study was to establish the reliability of directly taken linear measurements on dry skulls and corresponding measurements taken on the 3D digital models created by laser scanning as well as to assess the agreement between both measuring methods. Four skulls were measured in two competitive methods—a direct measuring, based on the conventional craniometric method, and a digital measuring, accomplished on 3D models created by laser scanning. Thirteen cranial measurements were taken on both dry skulls and 3D models. The intra- and inter-examiner reliability was estimated using intraclass correlation coefficient. The agreement between both measuring methods was assessed applying the Bland-Altman method for replicated measurements. A Bland-Altman plot was constructed for each of the 13 parameters. The 3D model and directly taken measurements were assessed as highly reliable and reproducible, excepting the orbital height. Our results showed that 96 % of all digital measurements differ from the directly taken ones with less than 2 mm and respectively 67.6 % differ with less than 1 mm. Based on the results of the Bland-Altman plots, most of the measurements obtained by both measuring methods could be accepted as comparable, since the majority of differences were within the constructed limits of agreement. However, there were digital measurements, particularly these with landmarks situated on bone margins, which systematically overestimated the directly taken ones.

D. Toneva (✉) · S. Nikolova
Institute of Experimental Morphology, Pathology and Anthropology with Museum,
Bulgarian Academy of Sciences, 1113 Sofia, Bulgaria
e-mail: ditoneva@abv.bg

S. Nikolova
e-mail: sil_nikolova@abv.bg

I. Georgiev
Institute of Information and Communication Technologies,
Bulgarian Academy of Sciences, 1113 Sofia, Bulgaria
e-mail: ivan.georgiev@parallel.bas.bg

I. Georgiev · A. Tchorbadjieff
Institute of Mathematics and Informatics,
Bulgarian Academy of Sciences, 1113 Sofia, Bulgaria
e-mail: atchorbadjieff@math.bas.bg

K. Georgiev et al. (eds.), *Advanced Computing in Industrial Mathematics*,
Studies in Computational Intelligence 681, DOI 10.1007/978-3-319-49544-6_18

Keywords Laser scanning · Craniometry · 3D models · Accuracy · Reliability

1 Introduction

The pursuit for a higher precision and quality in scientific research has been growing very fast in the past decades. This requires designing and implementation of new types of research equipment and development of new methods for research data analysis. The results from the new developed research methods are expected to be more precise and less disputable. However, as a first step every new method or equipment must be evaluated and compared to the theory and existing methods and data. This first and basic procedure is the inter-comparison between equivalent data obtained by new and old methods.

Because of the various applications and increasing accessibility of the imaging technologies in the medical and scientific fields, the accuracy of the digitally obtained metrical characteristics has been discussed in many studies. The most reported technology in the literature with regard to the reliability of the digital measurements is the computed tomography (CT) and especially cone beam computed tomography (CBCT), because of its growing application in the dental and orthodontic practice. The use of dry human skulls has been a traditional approach to validate new craniofacial imaging modalities [3]. The accuracy of linear measurements has been estimated on 2D tomographic slices and 2D cephalograms [11, 14, 16, 17, 19], but the results show some drawbacks, such as perspective limitations and positioning errors. Other studies have been dedicated to the assessment of the accuracy of cranial measurements made on 3D volumetric representations from CT and CBCT scans [2, 3, 10, 14, 15, 18, 23]. It has been established that the craniometric CBCT measurements are accurate and reliable, although most of the authors have noticed that the CT measuring data tend to be slightly lower than the conventional ones. Other authors have compared the measurements of the skulls with their replica models produced by rapid prototyping [9, 21], concluding that the models are extremely accurate, although slightly bigger than the original objects.

The CT technology is a suitable imaging method for diagnostic and therapeutic purposes in the medical practice and gives opportunity for various additional metrical analyses as well as for creating a large virtual database from patients' data. However, it should be applied more cautiously in the work with bone remains from archaeological excavations or forensic contexts. The exposure of a bone to clinical levels of radiation has been presumed to reduce the amount of amplifiable DNA [12] and thus, the extraction of ancient DNA from bone remains revealed in archaeological excavations could be obstructed in case of a foregoing CT-scanning. Therefore, if only surface data are needed for the purposes of an investigation, the laser scanning appears to be more suitable method for capturing data, than the hazardous exposure of the bones to X-rays.

The hand-held laser scanners appear to be very useful for digitizing bone samples because of their portability, easy manipulating with digital models and reduced risk of damages of the real objects [4]. The three-dimensional (3D) digital models created by laser scanning have an increasing application in the field of physical anthropology in recent years. Since the skull is the most investigated part of the skeleton, being an important source of information for a variety of anthropological studies, the agreement between the conventional and digitally taken standard cranial measurements is a crucial point for the future craniometric investigations. However, despite the widening usage of hand-held laser scanners in paleoanthropology and forensic anthropology and lots of applications of the created 3D bone models, there have not been many studies, concerning the precision of the cranial measurements obtained on laser scanning created 3D models [22, 23, 26]. The use of the cranial measurements in sex determination, race investigations, personal identification, etc., indicates the necessity of more studies comparing the accuracy between conventional and 3D digital measuring methods.

In this study, we aimed to establish the reliability of the directly taken linear measurements on dry skulls and corresponding measurements taken on the 3D digital models created by laser scanning as well as to assess the agreement between both measuring methods. There have been used different statistical methods for assessment of the methods agreement—by absolute difference, different correlation coefficients, linear regression, Bland-Altman plot, etc. We chose to apply the Bland-Altman method for measuring agreement using replicated measurements, which has been rarely performed because of its more complicated computation, but appearing to be most suitable for our purpose. Moreover, the traditional Bland-Altman plot, being a very illustrative method, has been widely used in clinical research.

2 Material and Methodology

2.1 *Material*

The subject of this inter-comparison study were four skulls from the osteological collection at the Institute of Experimental Morphology, Pathology and Anthropology with Museum, Bulgarian Academy of Sciences. The skulls belonged to adult individuals from both sexes [6].

2.2 *Methodology*

2.2.1 Measurements

The skulls were measured in two competitive methods—a direct measuring, based on the conventional craniometric method, and a digital measuring, accomplished on

3D models created by laser scanning. Thirteen cranial measurements were measured on both dry skulls and 3D models. The measurements represent direct distances between definite craniometric landmarks described according to Martin and Saller [20] (Table 1). The conventional measurements were taken with standard sliding and spreading calipers.

The 3D models for the digital measuring were created using hand-held laser scanner Creaform VIUscan^{TM}. The skulls were scanned without mandibles (Fig. 1). The scanning was set at a resolution of 0.7 mm and a texture resolution of 150 DPI. The accuracy of the laser scanner was to 0.050 mm. The surface image data collected by the laser scanning were post-processed in the scanner software platform VXelements^{TM}. The measurements on the 3D models (.stl) were taken using the free software Geomagic Verify Viewer (3D Systems, Inc).

All of the dry skulls and 3D models were measured three times by two examiners to test the intra- and inter-examiner reliability. Each set of 13 measurements were taken on a separate day in a random order. The replicated measurements were taken independently of each other.

2.2.2 Data

A total of 624 measurements were performed on the dry skulls and 3D models by both examiners. The data acquired from the direct and digital measuring method consisted of two series of 312 measurements. Each method series was separated by two equal parts corresponding to the measuring data of both examiners. For the purpose of comparing both methods, the data for each of the 13 measurements were grouped, as the measurements of each examiner obtained by the one method were paired with these received by the other method in accordance with the succession of the readings (1-st, 2-nd, 3-rd measuring).

Means of the measurements performed by both examiners on the dry skulls and 3D digital models were calculated over all 13 measurements (Table 2). It should be noticed that the data from the digital measurements were averaged from data with a higher precision, and the measured distances were with values to one hundredth of a millimeter, due to the advantages provided by the contemporary digital technologies. Conversely, the precision of the data acquired from the direct measurements was constrained up to the scale bar of the calipers, which is to a millimeter.

2.2.3 Statistics

The first step in the analysis was to evaluate consistency in the results between involved examiners. The intra- and inter-examiner reliability was assessed using intraclass correlation coefficient (ICC). Since in our study each of a random sample of n targets was rated independently by k judges and each target was rated by each of the same k judges, the "two-way mixed" model of ICC [27] was used:

Table 1 Craniometric measurements

Measurements	Definition	Landmarks—definition/Type[a]
Cranial length (CL)	Direct distance from glabella to opisthocranion	Glabella (g)—the most prominent point in the midsagittal plane between the superciliary arches (Type III)
Cranial base length (CBL)	Direct distance from nasion to basion	Opisthocranion (op)—the most posterior point of the skull in the midsagittal plane (Type III)
Length of *foramen magnum* (LFM)	Direct distance from basion to opisthion	Nasion (n)—the point of intersection between the frontonasal suture and the midsagittal plane (Type I)
Occipital breadth (OCB)	Direct distance between both asterion	Basion (ba)—the midline point on the anterior margin of the *foramen magnum* (Type II)
Breadth of *foramen magnum* (BFM)	Greatest distance between the lateral margins of *foramen magnum*[b]	Opisthion (o)—the midline point at the posterior margin of the *foramen magnum* (Type II)
Cranial height (CH)	Direct distance from bregma to basion	Asterion (ast)—the point of intersection of the lambdoid, occipitomastoid and parietomastoid sutures (Type I)
Facial length (FL)	Direct distance from basion to prosthion	Bregma (b)—the point of intersection between the coronal and sagittal sutures (Type I)
Upper facial breadth (UFB)	Direct distance between both frontomalare temporale	Prosthion (pr)—the most anterior point in the midline on the alveolar processes of the maxillae (Type II)
Upper facial height (UFH)	Direct distance from nasion to prosthion	Frontomalare temporale (fmt)—the most laterally positioned point on the fronto-zygomatic suture (Type II)
Middle facial breadth (MFB)	Direct distance between both zygomaxillare	Zygomaxillare (zm)—the most lower point on the zygomaticomaxillary suture (Type II)
Length of the nasal bones (LNB)	Direct distance from nasion to rhinion	Rhinion (rhi)—the point of intersection between the internasal suture and the upper end of the piriform aperture (Type II)
Orbital breadth (OBB)	Direct distance from maxillofrontale to ektoconchion	Maxillofrontale (mf)—the point of intersection between the maxillofrontal suture and the medial rim of the orbit (Type II)
Orbital height (OBH)	Direct distance between the superior and inferior orbital margins, measured perpendicular to the orbital breadth[b]	Ektoconchion (ek)—the point of intersection between the lateral rim of the orbit and the line beginning from mf crossing the orbit parallel to the upper orbital margin (Type III)

[a]The type of the landmarks is given with Roman numerals (Type I, II, and III), according to the categorical classification of the landmarks [7]
[b]BFM and OBH are measured between Type III landmarks

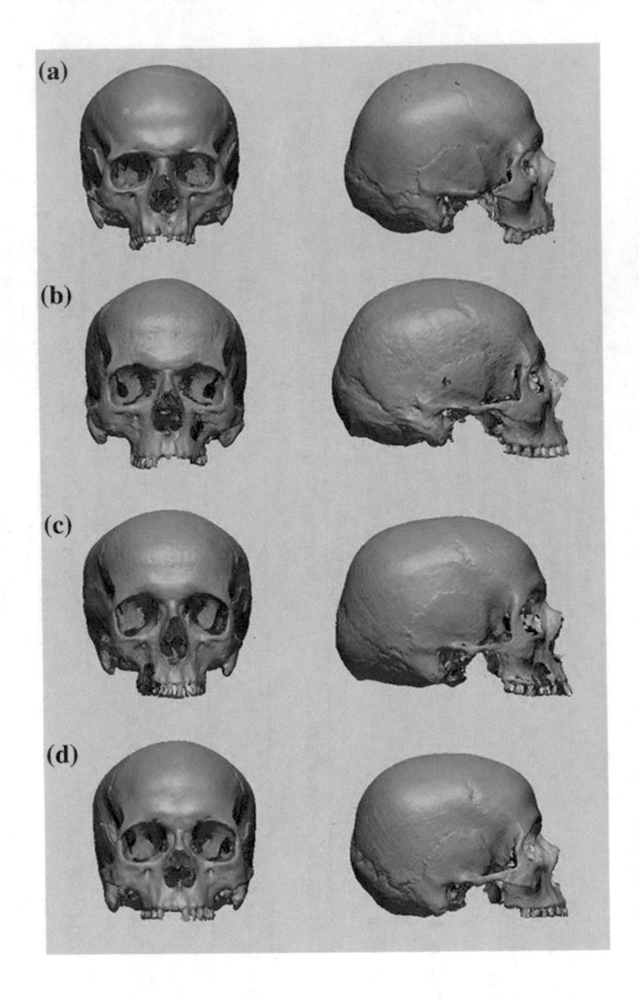

Fig. 1 Frontal and lateral views of the 3D models of the four skulls (**a**–**d**)

$$ICC\,(3,1) = \frac{BMS - EMS}{BMS + (k-1)\,EMS}$$

where with *BMS* is denoted between-targets mean k square with $n - 1$ degree of freedom, yielded from two-way ANOVA. The notation *EMS* is used for within-target residual sum of squares, with $(n - 1)(k - 1)$ degree of freedom.

The intra-examiner reliability was calculated for each measurement on the base of the triple measuring of the four skulls and 3D models, respectively ($k = 3$, $n = 4$). The inter-examiner reliability was calculated for each measurement separately for the digital and direct measuring methods based on the juxtaposition of the individual values of the triple measurements of both examiners for the four samples ($k = 2$, $n = 12$) (Table 3).

In the inter-methods comparison, the accuracy was assessed by an examination of the differences between digital and direct measurements performed by both examin-

Table 2 Means of the digital and direct measurements performed by both examiners. Means of the measurements for each measuring method based on the combined measurements of both examiners and corresponding mean differences for each parameter

Measurements	Digital measurements /mm/			Direct measurements /mm/			Mean difference /mm/
	Examiner I	Examiner II	Mean	Examiner I	Examiner II	Mean	
CL	180.48	180.46	180.47	180.92	180.83	180.88	−0.41
CBL	99.05	99.18	99.12	99.75	99.58	99.67	−0.55
LFM	35.98	35.77	35.88	35.21	34.63	34.92	0.96
OCB	110.65	110.51	110.58	110.38	110.42	110.40	0.18
BFM	30.84	30.61	30.73	29.58	29.21	29.40	1.33
CH	131.53	131.60	131.56	132.46	131.50	131.98	−0.42
FL	70.44	70.41	70.43	69.79	70.50	70.15	0.28
UFB	95.00	95.03	95.01	95.67	95.13	95.40	−0.38
UFH	101.59	101.86	101.73	102.17	102.25	102.21	−0.48
MFB	91.51	92.04	91.77	92.33	92.42	92.38	−0.60
LNB	20.90	21.12	21.01	21.21	21.58	21.40	−0.38
OBB	38.78	38.62	38.70	39.92	38.79	39.35	−0.66
OBH	33.89	34.00	33.95	33.25	32.67	32.96	0.99

Table 3 Intra- and inter-examiner reliability (ICCs)

Measurements	Intra-examiner reliability				Inter-examiner reliability	
	Digital measurements		Direct measurements		Digital measurements Examiner I/ Examiner II	Direct measurements Examiner I/ Examiner II
	Examiner I	Examiner II	Examiner I	Examiner II		
CL	0.998765	0.999417	0.999104	0.998946	0.998350	0.998027
CBL	0.998730	0.995827	0.999319	0.999095	0.997115	0.998389
LFM	0.959601	0.973110	0.988976	0.984081	0.963756	0.970985
OCB	0.992870	0.985779	0.993300	0.990424	0.982697	0.989921
BFM	0.994893	0.984771	0.972303	0.982387	0.985435	0.965900
CH	0.999877	0.999809	0.999381	0.997948	0.999635	0.996953
FL	0.958926	0.971117	0.958993	1	0.836640	0.979043
UFB	0.997212	0.995597	0.992610	0.991939	0.996457	0.986715
UFH	0.963741	0.843542	0.981203	0.980186	0.833535	0.971626
MFB	0.992704	0.997732	0.998818	0.995975	0.987033	0.987061
LNB	0.994014	0.987694	0.990733	0.997278	0.990700	0.989085
OBB	0.919665	0.956470	0.961538	0.848708	0.804901	0.834675
OBH	0.696814	0.595882	0.500000	–	0.323467	0.406780

–ICC value could not be calculated for OBH from the direct measurements of the second examiner. It was due to the lack of between-subject variance. However, the model was correct and all values on the OBH for the triple measurements of all four skulls were with a difference of no more than 1 mm

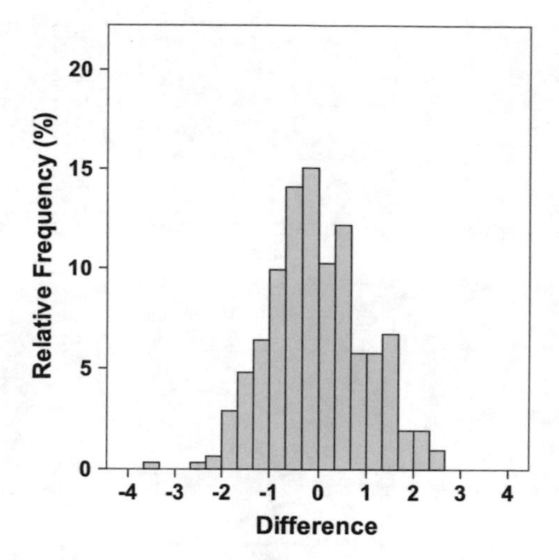

Fig. 2 Histogram of the differences of all paired measurements (n = 312)

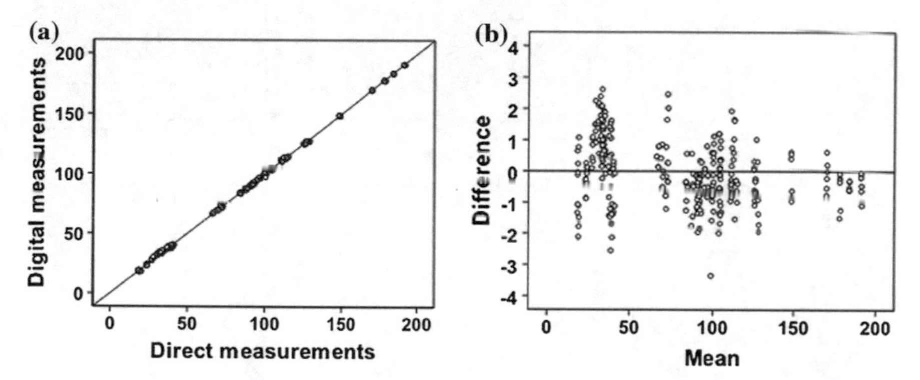

Fig. 3 Scatterplots of **a** the direct versus the digital cranial measurements with the line of equality; **b** the means versus the differences between digital and direct measurements

ers. The differences between the two methods were measured as absolute differences in millimeters /mm/. Firstly, the inter-comparison was assessed based on the mean differences between both measuring methods for all 13 parameters (Table 2). As a second step, the comparison of the methods was founded on the individual differences from the paired triple measurements of the two examiners on all four samples. The frequency of the differences between all paired measurements was illustrated by a percentage histogram (Fig. 2). A scatterplot was used to graph the correlation between the data of direct and digital measurements (Fig. 3a). A graph plotting the means against the differences between both measuring methods was used to illustrate the systematic diversion of the differences in accordance to the magnitude of the measurements [5] (Fig. 3b). The assumption of normality for the digital-direct differences of each parameter (n = 24) was tested by the Shapiro-Wilk test.

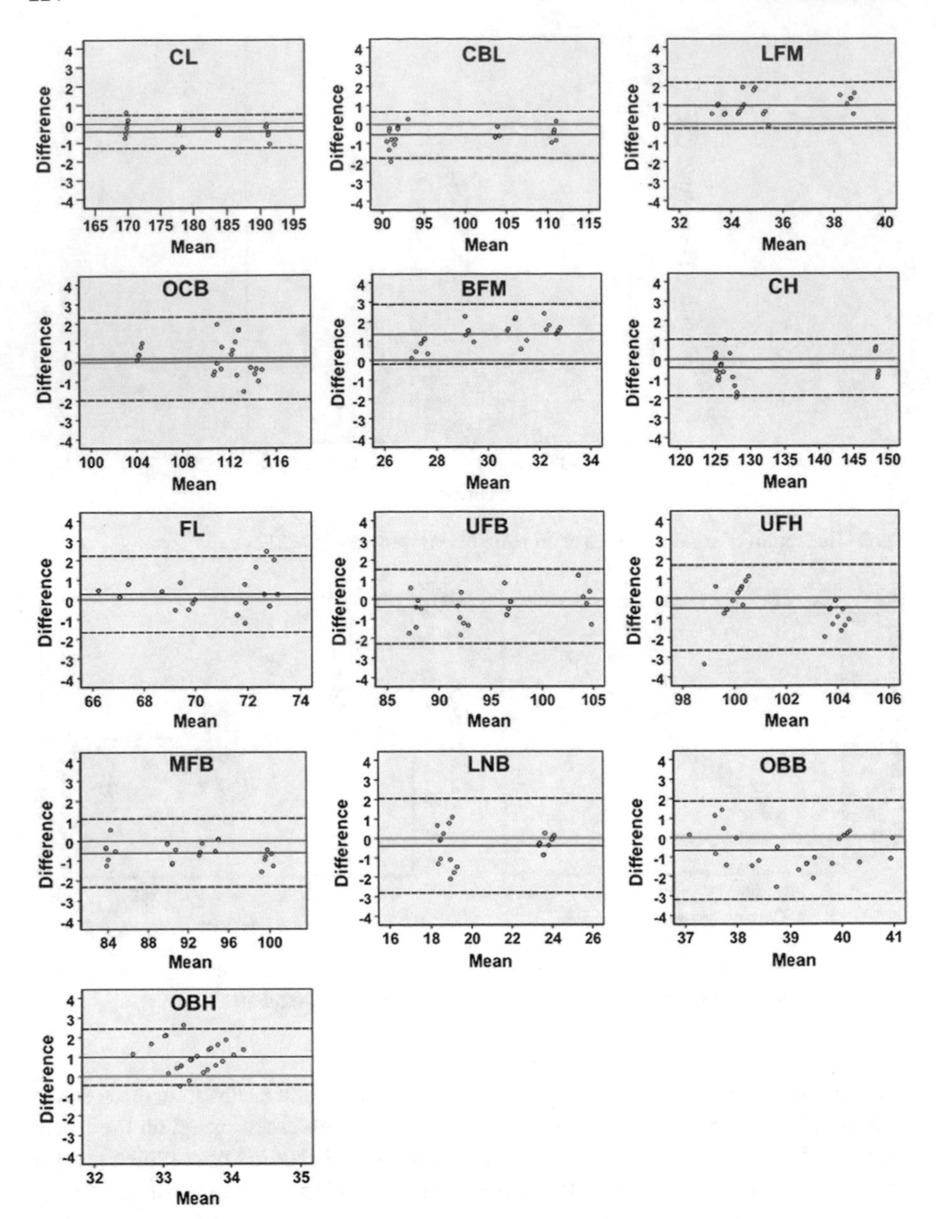

Fig. 4 Bland-Altman plots representing the bias (*continuous black line*) and the 95 % limits of agreement (*dotted lines*), based on the replicated measurements of each cranial measurement

The digital and direct measuring methods were compared applying the Bland–Altman method [1], which plots the means of the results of both methods (x-axis) and the differences between the two methods (y-axis). The methods agreement was quantified by constructing 95 % limits of agreement (LoA) for each of the 13 measurements (Fig. 4). Because of the repeated measurements, the LoA were built in the assumption of compound variance of the observed variance of the differences between the within-subject means σ_d and the within-subject variances σ_{Xw} and σ_{Yw} from measurements by the used methods (X and Y) with equal number of replicates m. The variance of the differences between means was yielded by the following formula [5]:

$$\sigma_{X-Y}^2 = \sigma_d^2 + \left(1 - \frac{1}{m}\right)\sigma_{Xw}^2 + \left(1 - \frac{1}{m}\right)\sigma_{Yw}^2$$

The values for the standard deviation were computed using the Linear Mixed-Effects Models library in R [8]. The used data set consisted of 4 columns of parameters—method, item, replication number and distance measurement in mm.

3 Results

The ICCs calculated for each examiner and between examiners showed almost perfect intra- and inter-examiner reliability (Table 3). The only exception was observed for the OBH with ICC values indicating moderate intra-examiner reliability and fair inter-examiner agreement for both measuring methods.

The means of the digitally and directly taken measurements and the corresponding mean differences for all 13 parameters are given in Table 2. Eight of the craniometric measurements on the 3D digital models were slightly smaller than the directly measured ones, as the mean differences ranged from −0.38 to −0.66 mm. The remaining five measurements, including the LFM, BFM, OCB, FL and OBH, showed higher readings on the 3D digital models than these on the dry skulls. Unlike the OCB and FL, which had the lowest absolute mean differences, the mean differences of the other three features had the highest values of all calculated ones. The bigger values of the LFM, BFM, and OBH, obtained with the software probably were due to the location of the landmarks for these measurements on bone margins, which made very difficult their placement exactly on the margin in the 3D model. Thus, these measurements required the placement of the landmarks slightly above/below the margin within the bone area so as to be caught by the program, and respectively the obtained values got higher than the directly measured ones. The digital measurements of the BFM and OBH exceeded the direct ones by an average of 1.33 mm and 0.99 mm, respectively.

The percentage histogram of the differences obtained from the paired repeated measurements of all parameters showed that they were normally distributed as the most common were differences within 1 mm (Fig. 2). Overall, our results showed that 96 % of all digital measurements differed from the directly taken ones with less than 2 mm and respectively 67.6 % differed with less than 1 mm. The scatterplot between the grouped data of both methods indicated the lack of a systematic bias, as almost all points lied on the line of equality (Fig. 3a). According to the graph plotting the means against the differences between both measuring methods (Fig. 3b), it could be observed that the divergence of the differences decreased with the magnitude of measurements. The most reasonable explanation for this trend was the observed difficulties in the precise measuring of the dimensions with small sizes. The hypothesis that the digital-direct differences of each parameter were normally distributed was confirmed by the Shapiro-Wilk test's p-values, which were higher than 0.05 ($p > 0.05$).

The Bland-Altman plot as a method for comparing different measuring methods was constructed for each of the 13 measurements. The biggest bias was observed for LFM, BFM and OBH (Fig. 4). The differences in seven of the craniometric measurements were entirely distributed in the LoA. Four of the measurements (CBL, CH, FL, and UFH) had 23/24 or 95.8 % of the differences falling within the limits. The CL and OBH showed 22/24 or 91.7 % of the differences distributed in the LoA. The analysis of the data that left outside the intervals of agreement showed that the outliers were produced completely random and there was not any relation to the selected methods of measuring or the personality of examiners. However, it should be notices that this method of analysis is not very sensitive to one or two large outlying differences and there is no need to be removed from the analysis [5].

The smallest width of the 95 % LoA was established for the CL (1.78 mm) and the biggest one was observed for the OBB (5.04 mm). The very wide LoA observed for the LNB and OBB were due respectively to the bigger between-subject variance and the bigger within-subject variance of the direct method. Most of the Bland-Altman plots evidenced for an agreement between both measuring methods, except for the LFM, BFM and OBH, because of the very big bias. However, only five measurements had a width of the LoA less than 3 mm (CL, CBL, LFM, CH and OBH), which indicated quite wide LoA for the most measurements. A reason for the wide LoA intervals in our study could be the small sample size. Besides, it has been noticed that when the method is applied using replicated measurements, the LoA intervals are wider compared to the variant with the means of each measuring method [5].

4 Discussion

The reliability and accuracy of the digital measurements obtained on laser scanning created 3D models are key points at the time of an increasing usage of hand-held laser scanners in paleoanthropology and forensic anthropology. Concerning the reliability of the 3D laser scanning method, there have been established excellent results for the linear craniometric measurements [22] as well as precise ones for the surface area and volume measurements [26]. Our study is not an exception with almost perfect intra- and inter-examiner reliability for nearly all digitally and directly taken measurements, except for the OBH indicating moderate intra-examiner reliability and fair inter-examiner agreement. The results obtained for this parameter could due to the type of landmarks (Type III) defining this measurement as well as to the small sample size. However, namely Type III landmarks have been reported to yield the most precise coordinate data on 3D laser scanner models [25], as the measurements between them have been found to be very consistent [23]. On the other hand, Type

III landmarks have been established to be lowly reproducible according to a study based on the coordinate landmarks data obtained by 3D digitizing [24]. Although, in our study, exactly a measurement between these landmarks showed poor results on the reliability, the other two measurements defined by the same type landmarks had excellent ones. So it should be considered that the separate Type III landmarks and the measurements defined by them could show varying reproducibility, depending not only on the choice of the 3D technique but also on the choice of the landmarks, and thus to lead to different results and conclusions.

According to the used inter-comparison technology, the accuracy of the cranial measurements was found to differ to a varying degree. Because of the manual measuring and the easier landmark identification directly on the real objects, the dry skulls and their produced prototypes have shown least differences in their metrical characteristics [23]. Comparing the digital technologies, the laser scans have provided more accurate measurements compared to the CT ones, due to the interpolation of the data between CT scan slices at the 3D rendering [23]. The 3D laser scanning method has been reported to give a slightly lower reading compared to the conventional measuring [22]. However, such a tendency has not been observed in other studies [23, present study]. Our results showed that there were even a few digital measurements such as LFM, BFM, and OBH, showing a systematic overestimation of the direct measurements.

It worth noting that the providing of an accurate digital metrical analysis requires a very good quality of the 3D models with well captured surface in the places of all investigated landmarks. The suture-based landmarks (or Type I) have been reported to be very problematic for identification on 3D models, as the measurements between such landmarks have shown a greater variation in the measurement error [23]. In our study, the measurements with one or two Type I landmarks did not indicate consistent big differences between both measuring methods, but some of them were also highly variable. As a whole, Type I landmarks have been reported to be more precisely identified when they are collected with a digitizer than on 3D models [24, 25].

The landmark location has been suggested to be a major source of variability in the measurements depending mostly on the human judgement [13]. This could be a reason to some extent for the inaccuracy in the measuring of the BFM and OBH in our study, which were described indefinitely as a largest diameter and a perpendicular to some another measurement (i.e. include only Type III landmarks), and thus, the personal assessment of the examiner is supposed to be a substantial factor. The lower intra- and inter-examiner reliability observed for the OBH could cause to a certain degree the inconsistency between both measuring methods, but this cannot be an explanation in the case of the BFM. However, as we previously noticed there were software caused difficulties in the landmark location for these measurements, so the measuring differences could not be specified as only human dependent. It should be also taken into account the big difference in the precision of the used measuring techniques, although our results showed that all digital-direct differences with few exceptions were within 2 mm, which have been considered as an acceptable amount of error in forensic anthropology [28].

5 Conclusion

The ICCs indicated that digital and directly taken measurements were highly reliable and reproducible, except for the OBH. Overall, almost all digital measurements differed from the directly taken ones with less than 2 mm and respectively 2/3 of them differed with less than 1 mm.

Based on the results of the Bland-Altman plots, most of the measurements obtained by both measuring methods could be accepted as comparable, since the majority of differences were within the LoA intervals. However, there were digital measurements, particularly these with landmarks situated on bone margins, which systematically overestimated the directly taken ones and should be considered with more attention when it concerns to a 3D digital measuring method.

Acknowledgements This study was supported by the "Program for Career Development of Young Scientists, Bulgarian Academy of Sciences", research grant DFNP—75/27.04.2016.

References

1. Altman, D.G., Bland, J.M.: Measurement in medicine: the analysis of method comparison studies. Statistician **32**, 307–317 (1983)
2. Baumgaertel, S., Palomo, J.M., Palomo, L., Hans, M.G.: Reliability and accuracy of cone-beam computed tomography dental measurements. Am. J. Orthod. Dentofac. Orthop. **136**(1), 19–25 (2009)
3. Berco, M., Rigali Jr., P.H., Miner, M.R., DeLuca, S., Anderson, N.K., Will, L.A.: Accuracy and reliability of linear cephalometric measurements from cone-beam computed tomography scans of a dry human skull. Am. J. Orthod. Dentofac. Orthop. **136**, 17.e1–17.e9 (2009)
4. Bibliowicz, J., Khan, A., Agur, A., Singh, K.: High-precision surface reconstruction of human bones from point-sampled data. In: ISHS 2011 Conference Proceedings: International Summit on Human Simulation, pp. 1–10 (2011)
5. Bland, J.M., Altman, D.G.: Measuring agreement in method comparison studies. Stat. Methods Med. Res. **8**, 135–160 (1999)
6. Buikstra, J.E., Ubelaker, D.H.: Standards for data collection from human skeletal remains. In: Arkansas Archeological Survey Research Series No. 44, Fayetteville (1994)
7. Bookstein, F.L.: Morphometric Tools for Landmark Data: Geometry and Biology. Cambridge University Press, Cambridge (1991)
8. Cartensen, B., Simpson, J., Gurrin, L.C.: Statistical models for assessing agreement in method comparison studies with replicate measurements. Int. J. Biostat. **4**(1), 1–26 (2008)
9. Choi, J.Y., Choi, J.H., Kim, N.K., Kim, Y., Lee, J.K., Kim, M.K., Lee, J.H., Kim, M.J.: Analysis of errors in medical rapid prototyping models. Int. J. Oral Maxillofac. Surg. **31**(1), 23–32 (2002)
10. Damstra, J., Fourie, Z., Huddleston Slater, J.J., Ren, Y.: Accuracy of linear measurements from cone-beam computed tomography-derived surface models of different voxel sizes. Am. J. Orthod. Dentofac. Orthop. **137**(1), 16.e1–16.e6 (2010)
11. Gribel, B.F., Gribel, M.N., Frazao, D.C., McNamara Jr., J.A., Manzi, F.R.: Accuracy and reliability of craniometric measurements on lateral cephalometry and 3D measurements on CBCT scans. Angle Orthod. **81**, 26–35 (2011)
12. Grieshaber, B.M., Osborne, D.L., Doubleday, A.F., Kaestle, F.A.: A pilot study into the effects of X-ray and computed tomography exposure on the amplification of DNA from bone. J. Archaeol. Sci. **35**(3), 681–687 (2008)

13. Harris, E.F., Smith, R.N.: Accounting for measurement error: a critical but often overlooked process. Arch. Oral. Biol. **54**(Suppl. 1), S107–S117 (2009)
14. Hassan, B., Setelt, P., Sanderink, G.: Accuracy of three-dimensional measurements obtained from cone beam computed tomography surface-rendered images for cephalometric analysis: influence of patient scanning position. Eur. J. Orthod. **23**, 1–6 (2009)
15. Kamburoğlu, K., Kolsuz, E., Kurt, H., Kiliç, C., Özen, T., Paksoy, C.S.: Accuracy of CBCT measurements of a human skull. J. Digit. Imaging **24**(5), 787–793 (2011)
16. Kumar, V., Ludlow, J.B., Mol, A., Cevidanes, L.: Comparison of conventional and cone beam CT synthesized cephalograms. Dentomaxillofacial Radiol. **36**, 263–269 (2007)
17. Lascala, C., Panella, J., Marques, M.: Analysis of the accuracy of linear measurements obtained by cone beam computed tomography (CBCT-NewTom). Dentomaxillofac Radiol. **33**, 291–294 (2004)
18. Lorkiewicz-Muszyńska, D., Kociemba, W., Sroka, A., Kulczyk, T., Żaba, C., Paprzycki, W., Przystańska, A.: Accuracy of the anthropometric measurements of skeletonized skulls with corresponding measurements of their 3D reconstructions obtained by CT scanning. Anthropol. Anz. **72**(3), 293–301 (2015)
19. Ludlow, J.B., Laster, W.S., See, M., Bailey, L.J., Hershey, H.G.: Accuracy of measurements of mandibular anatomy in cone beam computed tomography images. Oral Surg. Oral Med. Oral Pathol. Oral Radiol. Endod. **103**, 534–542 (2007)
20. Martin, R., Saller, K.: Kraniometrische Technik. In: Martin, R., Saller, K. (eds.) Lehrbuch der Anthropologie, vol. I. Gustav Fischer, Stuttgart (1957)
21. Nizam, A., Gopal, R.N., Naing, L., Hakim, A.B., Samsudin, A.R.: Dimensional accuracy of the skull models produced by rapid prototyping technology using stereolithography apparatus. Arch. Orofac. Sci. **1**, 60–66 (2006)
22. Park, H.K., Chung, J.W., Kho, H.S.: Use of hand-held laser scanning in the assessment of craniometry. Forensic. Sci. Int. **160**, 200–206 (2006)
23. Richard, A.H., Parks, C.L., Monson, K.L.: Accuracy of standard craniometric measurements using multiple data formats. Forensic. Sci. Int. **242**, 177–185 (2014)
24. Ross, A.H., Slice, D.E., Williams, S.E.: Geometric Morphometric Tools for the Classification of Human Skulls. Department of Justice, Document 231195 (2010)
25. Sholts, S.B., Flores, L., Walker, P.L., Wärmländer, S.K.: Comparison of coordinate measurement precision of different landmark types on human crania using a 3D laser scanner and a 3D digitiser: implications for applications of digital morphometrics. Int. J. Osteoarchaeol. **21**, 535–543 (2011)
26. Sholts, S.B., Wärmländer, S.K., Flores, L.M., Miller, K.W., Walker, P.L.: Variation in the measurement of cranial volume and surface area using 3D laser scanning technology. J. Forensic. Sci. **55**(4), 871–876 (2010)
27. Shrout, P.E., Fleiss, J.L.: Intraclass correlations: uses in assessing rater reliability. Psychol. Bull. **86**(2), 420–428 (1979)
28. Stull, K.E., Tise, M.L., Ali, Z., Fowler, D.R.: Accuracy and reliability of measurements obtained from computed tomography 3D volume rendered images. Forensic. Sci. Int. **238**, 133–140 (2014)

Equivalence of Models of Freeze-Drying

Milena Veneva and William Lee

Abstract Freeze-drying is a preservation process, consisting of two main stages: during the primary drying water is removed by sublimation; during the secondary drying chemically bound water is removed by desorption. Two different models of secondary drying are built. The first one consists of coupled heat and mass balances equations, the second one uses a modified Richards equation. Using scale transformations derived from the PDEs and the BCs, the first model is nondimensionalized. The model is further simplified by asymptotic reduction. It is proven that the reduced model is equivalent to the model that uses the modified Richards equation if the partial pressure of air is negligible compared to that of water vapor in the vials and in the chamber of the freeze-drier. This result shows that there is an opportunity for technology transfer, since solvers developed for modelling groundwater flows using Richards equations can also be used to model the economically important problem of freeze-drying.

1 Introduction

Freeze-drying, also known as lyophilization, is a dehydration process typically used to preserve a perishable material or make the material more convenient for transport. The process is also very popular in the food and pharmaceutical industries as a process which removes water (chemically bound water, or solvents, as a whole) from heat-sensitive products that will be damaged if a standard drying procedure is used instead. Freeze-drying works by freezing the material (usually the product to be dried is in vials) and then reducing the surrounding pressure to allow the frozen water

M. Veneva (✉)
Faculty of Mathematics and Informatics, Sofia University "St. Kliment Ohridski",
5 James Bourchier, 1164 Sofia, Bulgaria
e-mail: milena.p.veneva@gmail.com

W. Lee
MACSI, Department of Mathematics and Statistics, University of Limerick,
Limerick, Ireland
e-mail: william.lee@ul.ie

K. Georgiev et al. (eds.), *Advanced Computing in Industrial Mathematics*,
Studies in Computational Intelligence 681, DOI 10.1007/978-3-319-49544-6_19

in the material to sublimate directly from the solid phase to the gas phase. There are four stages in the complete drying process: pretreatment, freezing, primary drying and secondary drying.

1. Pretreatment—includes any method of treating the product prior to freezing. Usually such methods are used when a theoretical knowledge of freeze-drying of the particular product exists and thus the requirements of the process and the product quality considerations are taken into account;
2. freezing—during this step it is important to cool the material below its triple point—the lowest temperature at which the solid and liquid phases of the material can coexist. This ensures that sublimation rather than melting will occur in the following steps. Usually, the freezing temperatures are between −80 and −50 °C. The freezing stage is the most critical in the whole freeze-drying process, because the product can be spoiled if this is improperly done;
3. primary drying—during this stage the pressure is lowered (to the range of a few millibars), and enough heat is supplied to the material for the water to sublimate. The amount of heat necessary can be calculated using the latent heat of sublimation of water. In this initial drying phase about 95 % of the water in the material is sublimated. This phase may be slow (can be several days in industry), because if too much heat is added, the material's structure could be altered;
4. secondary drying—the aim of this stage of the process is to remove unfrozen water molecules, since the ice was removed in the primary drying phase. This part of the freeze-drying process is governed by the material's adsorption isotherms. In this phase the temperature is raised higher than in the primary drying phase, and can even be above 0 °C, to break any physico-chemical interactions that have formed between the water molecules and the frozen material. Usually the pressure is also lowered in this stage to encourage desorption (typically in the range of microbars, or fractions of a pascal).

1.1 Why Are Mathematical Models Needed?

In the pharmaceutical industry freeze-drying is one of the most important, the most expensive and time-consuming procedures. At the moment the control in the freeze-drier is based on a recipe, which means that the times for primary and secondary drying are set beforehand. This may lead to overdrying, decreasing of the quality of the product or damaging of the product and hence, loss of time and money. This is the reason why a control system based on taking measurements in the freeze-drier during the process and thus determining the endpoints of both the primary and the secondary drying is needed. One should have in mind that measurements inside the vials are not allowed, because they may change or damage the product inside. In order to be able to obtain the correlation between the conditions in the freeze-drier and the desirable conditions in the vials, a mathematical model of freeze-drying is needed. Moreover, this model should be simple enough to be suitable for real-time applications. As a result this model will also help for better understanding of the dynamics and the physics of the process.

2 Secondary Drying

One can see a one-dimensional model of the secondary drying stage in Fig. 1. It shows a vial with height H, radius R and thickness of the wall w. After the end of primary drying the sublimation front (the boundary between the already dried material and the one that is still frozen) does not exist anymore and from a problem with a moving boundary (Stefan problem) a problem for just one phase occurs. Having in mind that the 2D-models are connected with a big amount of numerical computations that require a lot of computer memory and have big computation time, is turns out that one-dimensional models are better for the needs of the real-time applications. For this reason in this paper one-dimensional models are going to be considered.

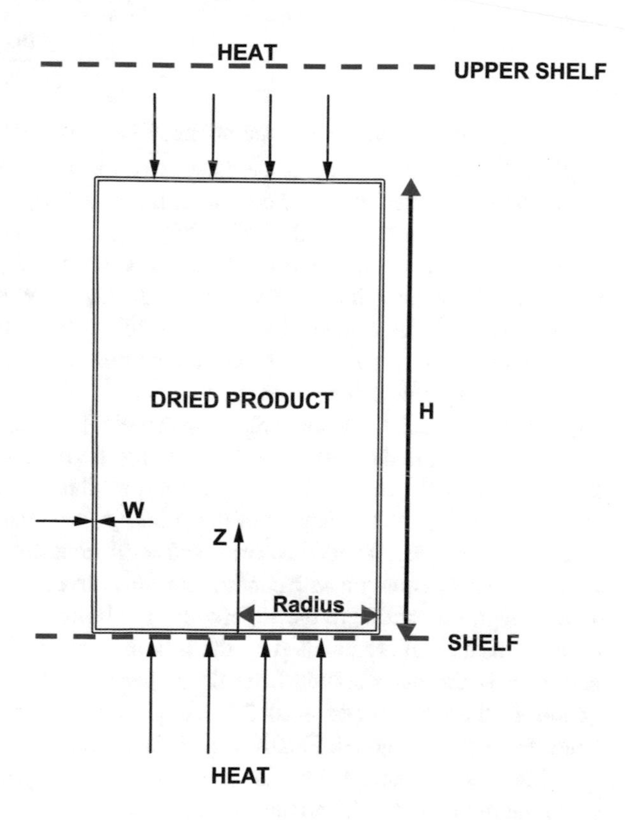

Fig. 1 Secondary drying model

2.1 First Model

In [1] is stated that secondary drying could be modelled from the governing equations used to describe the processes that occur in the dried layer during the primary drying stage of freeze-drying, and the same boundary conditions as during primary drying. The only difference is that some of the coefficients need to be changed. Having in mind that fact and using the model of primary drying described in [2], in the case of lack of air inside the vials and the chamber of the freeze-drier the following model which describes the dynamics of the secondary drying stage of freeze-drying was derived:

$$\rho_e\, c_{p,e}\, \frac{\partial T}{\partial t} + c_{p,w}^{(g)}\, N_w\, \frac{\partial T}{\partial z} = \frac{\partial}{\partial z}\left(k_e \frac{\partial T}{\partial z}\right) + \triangle H_v\, \rho_s\, \frac{\partial c_{sw}}{\partial t}; \tag{1}$$

$$\frac{M_w\, \gamma}{R_{un}}\, \frac{\partial}{\partial t}\left(\frac{p_w}{T}\right) + \frac{\partial N_w}{\partial z} = -\rho_s\, \frac{\partial c_{sw}}{\partial t}; \tag{2}$$

$$N_w = -\frac{M_w}{R_{un}\, T}\,(d_1 + d_2\, p_w)\, \frac{\partial p_w}{\partial z}; \tag{3}$$

$$\frac{\partial c_{sw}}{\partial t} = -r_d\, c_{sw}, \tag{4}$$

where: $c_{p,e}$ is the effective specific heat capacity for the dried layer, $[c_{p,e}] = \mathrm{J\,kg^{-1}\,K^{-1}}$; $c_{p,w}^{(g)}$ is the specific heat capacity for water vapor, $[c_{p,w}^{(g)}] = \mathrm{J\,kg^{-1}\,K^{-1}}$; c_{sw} is the concentration of bound water, $[c_{sw}] = 1$; d_1 is the bulk diffusivity coefficient, $[d_1] = \mathrm{m^2\,s^{-1}}$; d_2 is the bulk diffusivity coefficient, $[d_2] = \mathrm{m^3\,s\,kg^{-1}}$; $\triangle H_v$ is the latent heat of vaporization of bound water, $[\triangle H_v] = \mathrm{J\,kg^{-1}}$; k_e is the effective thermal conductivity for the dried layer, $[k_e] = \mathrm{W\,m^{-1}\,K^{-1}}$; M_w is the molecular weight of water vapor, $[M_w] = \mathrm{kg\,mol^{-1}}$; N_w is the water vapor mass flux, $[N_w] = \mathrm{kg\,m^{-2}\,s^{-1}}$; p_w is the water vapor pressure, $[p_w] = \mathrm{Pa}$; r_d is the rate constant of desorption of bound water, $[r_d] = \mathrm{s^{-1}}$; R_{un} is the universal gas constant, $[R_{un}] = \mathrm{J\,K^{-1}\,mol^{-1}}$; t is the temporal variable, $[t] = \mathrm{s}$; T is the temperature in the vial, $[T] = \mathrm{K}$; z is the spatial variable (along the height of the vial), $[z] = \mathrm{m}$; γ is the porosity of the matrix, $[\gamma] = 1$; ρ_e is the effective density for the dried layer, $[\rho_e] = \mathrm{kg\,m^{-3}}$; ρ_s is the density of the solid, $[\rho_s] = \mathrm{kg\,m^{-3}}$.

Equation (1) represents the conservation of heat inside the vial. It is a standard transient energy equation with conduction and convection heat transfer. Additionally, a latent heat source due to the vaporization of bound water is introduced. The mass transfer in the vial is modelled by the modified continuity equation (2). There, the last term is the mass source from the vaporization of bound water. Equation (3) expresses the water vapor mass flux using the sorption-sublimation model derived from the dusty gas model. Finally, Eq. (4) is a model for the removal of bound water.

Initial conditions: at the beginning of secondary drying the temperature T_{init}, the water vapor pressure p_w^0, and the concentration of bound water c_{sw}^0 are known, where: T_{init} is the initial temperature in the vial, $[T_{init}] = \mathrm{K}$; p_w^0 is the initial water vapor pressure, $[p_w^0] = \mathrm{Pa}$; c_{sw}^0 is the initial concentration of bound water, $[c_{sw}^0] = 1$.

Boundary conditions on the top of the vial:

- radiation heat flux (Stefan-Boltzmann law):

$$- k_e \frac{\partial T}{\partial z} = \sigma F_{up} (T^4 - T_{up}^4), \ z = H, \tag{5}$$

where: F_{up} is the view factor for radiative heat transfer between the upper surface of the dried product and the upper shelf, $[F_{up}] = 1$; H is the height of the vial, $[H] = \mathrm{m}$; T_{up} is the temperature of the upper shelf, $[T_{up}] = \mathrm{K}$; σ is the Stefan-Boltzmann constant, $[\sigma] = \mathrm{J\,s^{-1}\,m^{-2}\,K^{-4}}$;
- the water vapor pressure is known:

$$p_w = constant, \ z = H. \tag{6}$$

Boundary conditions at the bottom of the vial:

- Newton's law of cooling:

$$k_e \frac{\partial T}{\partial z} = Q_v (T - T_{sh}), \ z = 0, \tag{7}$$

where: Q_v is the shelf $\leftrightarrow$ vial heat transfer coefficient, $[Q_v] = \mathrm{kg\,K^{-1}\,s^{-3}}$; T_{sh} is the temperature of the shelf under the vial, $[T_{sh}] = \mathrm{K}$;
- zero water vapor mass flux:

$$N_w = 0, \ z = 0. \tag{8}$$

The model was nondimensionalized and asymptotically reduced. For the needs of the nondimensionalization procedure scale transformations derived from the PDEs and the BCs of the model were applied. Taking into account the nature of the secondary drying process, a timescale for desorption was used. In the course of the asymptotical reduction the terms which contain time derivatives in (1) and (2), and also the convective term in (1) turned out to be small and were dropped from the reduced model. As a result from these two procedures the following system was obtained (no new notation for the dimensionless variables is going to be used):

$$\frac{\partial^2 T}{\partial z^2} = -\frac{\partial c_{sw}}{\partial t}; \tag{9}$$

$$\frac{\partial N_w}{\partial z} = -\frac{\partial c_{sw}}{\partial t}; \tag{10}$$

$$N_w = -\left(\frac{1 + \gamma_w + \eta_w \, p_w}{\gamma_w}\right) \frac{\partial p_w}{\partial z}; \tag{11}$$

$$\frac{\partial c_{sw}}{\partial t} = -c_{sw}, \tag{12}$$

where
$\gamma_w = \frac{d_1}{p_w(z=H)\,d_2}$;
$\eta_w = \frac{p_w(z=0)-p_w(z=H)}{p_w(z=H)}$.

2.2 Second Model—Modified Richards Equation

Another way of modelling secondary drying is using the Richards equation. It is a non-linear partial differential equation which models the movement of liquid water in unsaturated soils. It has three different forms: H-form (head-based), θ-form (saturation-based) and a mixed one. Because of the fact that in the process vapor water was considered, a modified version of the Richards equation was used. Also, according to [3] the mixed form of the equation produces consistently superior numerical results compared to analogous solutions based on the other two forms. Because of this, the mixed formulation of the equation was used. Thus, the model which was considered is the following (the boundary conditions imposed are similar to the ones used in the first model):

$$\frac{\partial \theta_v}{\partial t} = -\frac{\partial q_v}{\partial z} + S(H); \tag{13}$$

$$c_{p,e}\frac{\partial T}{\partial t} + \Delta H_v \frac{\partial \theta_v}{\partial t} = \frac{\partial}{\partial z}\left(\frac{\lambda}{\rho_e}\frac{\partial T}{\partial z}\right) - c_{p,w}^{(g)}\frac{\partial q_v T}{\partial z} - \Delta H_v \frac{\partial q_v}{\partial z}; \tag{14}$$

$$q_v = -K_{vh}\frac{\partial H}{\partial z} - K_{vt}\frac{\partial T}{\partial z}, \tag{15}$$

where: $c_{p,e}$ is the effective specific heat capacity for the dried layer; $c_{p,w}^{(g)}$ is the specific heat capacity for water vapor; H is the hydraulic head (in this model) and the height of the vial (in the first model); ΔH_v is the latent heat of vaporization of bound water; K_{vh} is the isothermal vapor hydraulic conductivity, $[K_{vh}] = \mathrm{m\,s^{-1}}$; K_{vt} is the thermal vapor hydraulic conductivity, $[K_{vt}] = \mathrm{m^2\,K^{-1}\,s^{-1}}$; q_v is the water vapor flux, $[q_v] = \mathrm{m\,s^{-1}}$; $S(H)$ is the source term; t is the temporal variable; T is the temperature in the vial; z is the spatial variable (along the height of the vial); θ_v is the water vapor content; λ is the coefficient of the apparent thermal conductivity, $[\lambda] = \mathrm{W\,m^{-1}\,K^{-1}}$; ρ_e is the effective density for the dried layer.

Equation (13) gives the water vapor flow as the sum of isothermal vapor flow (a flow whose temperature remains constant) and thermal vapor flow (a flow whose temperature is not constant). The total heat flux is defined as the sum of the conduction of sensible heat, convection of sensible heat and convection of latent heat of vaporization, as shown in (14). Equation (15) represents the water vapor flux.

3 Equivalence of the Models

It is going to be proven that the model which contains the modified Richards equation (the second model shown in this paper) is equivalent to the asymptotically reduced first model which consists of coupled heat and mass transfer equations. For that purpose the terms of the second model are going to be expressed as functions of the terms of the first one. The water vapor content θ_v in the vial is equal to the change of the water vapor mass flux. The source term $S(H)$ is equal to the difference between the amount of gasses which goes out from the vial and the ones which goes in plus the water vapor flux. In the terms of the first model both of them are equal to the first derivative of the water vapor mass flux with respect to the spatial variable $= \frac{\partial N_w}{\partial z}$. The hydraulic head is just another way to measure pressure, so H is equivalent to p_w. Thus, Eq. (13) could be rewritten in the following form:

$$\begin{aligned} \frac{\partial}{\partial t}\left(\frac{\partial N_w}{\partial z}\right) &= \frac{\partial}{\partial z}\left(K_{vh}\frac{\partial p_w}{\partial z} + K_{vt}\frac{\partial T}{\partial z}\right) + \frac{\partial N_w}{\partial z} \\ &= \frac{\partial}{\partial z}\left(K_{vh}\frac{\partial p_w}{\partial z}\right) + K_{vt}\frac{\partial^2 T}{\partial z^2} + \frac{\partial N_w}{\partial z}. \end{aligned} \tag{16}$$

Substituting with (9), (10) and (12) into (16), it follows that:

$$\begin{aligned} \frac{\partial c_{sw}}{\partial t} &= \frac{\partial}{\partial z}\left(K_{vh}\frac{\partial p_w}{\partial z}\right) - K_{vt}\frac{\partial c_{sw}}{\partial t} - \frac{\partial c_{sw}}{\partial t} \quad \Leftrightarrow \\ \frac{\partial c_{sw}}{\partial t} &= \frac{1}{(2+K_{vt})}\frac{\partial}{\partial z}\left(K_{vh}\frac{\partial p_w}{\partial z}\right). \end{aligned} \tag{17}$$

Differentiating (11) once with respect to z and using (10), the following equation was obtained:

$$\frac{\partial c_{sw}}{\partial t} = \frac{\partial}{\partial z}\left[\left(\frac{1+\gamma_w+\eta_w\, p_w}{\gamma_w}\right)\frac{\partial p_w}{\partial z}\right]. \tag{18}$$

Since the process is governed by the material's adsorption isotherms, the thermal vapor hydraulic conductivity in (17) could be taken to be zero. Comparing (17) to (18), it yields:

$$K_{vh} = 2\left(\frac{1+\gamma_w+\eta_w\, p_w}{\gamma_w}\right). \tag{19}$$

During the asymptotic reduction of the first model it was found that the time derivative of the temperature and the convection do not play sufficient role and so these terms in (14) (the first and the fourth) could be neglected. Hence (using (9), (10), (12) and (13)):

$$\triangle H_v \frac{\partial \theta_v}{\partial t} = \frac{\lambda}{\rho_e} \frac{\partial^2 T}{\partial z^2} + \triangle H_v \left(\frac{\partial \theta_v}{\partial t} - S(H) \right) \Leftrightarrow$$
$$\frac{\lambda}{\rho_e} \frac{\partial^2 T}{\partial z^2} = \triangle H_v \, S(H). \tag{20}$$

Having in mind that the source term is equal to the first derivative of the water vapor mass flux with respect to the spatial variable and using (10), it follows that (14) is exactly the same as (9). Thus, the equivalence of the models was proven.

4 Discussions and Conclusions

Two different models of secondary drying were built. Using some numerical approaches the first one was simplified. Despite the different motivations of the two models, it was shown that they are equivalent. A very good agreement with the results on the matter of secondary drying that exist in the literature was achieved. Since there are numerical methods and scientific software for solving the Richards equation, e.g. *HYDRUS* [4], the equivalence means that they can be used for obtaining a solution of secondary drying problems. On the other hand, the very simple nature of the two models makes them absolutely convenient to be introduced in the industry manufacturing for the purposes of real-time applications, e.g. soft sensors PAT (Process Analytical Technology) [5]. The models will improve the production, replacing the recipe controlled process with a process controlled in real time.

References

1. Song, C., Nam, G., Kim, C., Ro, S.: A finite volume analysis of vacuuum freeze-drying processes of skim milk solution in trays and vials. Drying Tech. **20**(7), 283–305 (2002)
2. Vynnycky, M.: An asymptotic model for the primary drying stage of vial liophilization. J. Eng. Math. (2015)
3. Celia, M., Bouloutas, E.: A general mass-conservative numerical solution for the unsaturated flow equation. Water Resour. Res. **26**(7), 1483–1496 (1990)
4. Simunek, J., Van Genuchten, M.T., Sejna, M.: Development and applications of the HYDRUS and STANMOD software packages and related Codes. Vadose Zone J. **7**(2), 587–600 (2008)
5. de Beer, T.R.M., Vercruysse, P., Burggraeve, A., et al.: In-line and real-time process monitoring of a freeze drying process using Raman and NIR spectroscopy as complementary process analytical technology (PAT) tools. J. Pharm. Sci. **98**(9), 3430–3446 (2009)

Comparing Bézier Curves and Surfaces for Coincidence

Krassimira Vlachkova

Abstract It is known that Bézier curves and surfaces may have multiple representations by different control polygons. The polygons may have different number of control points and may even be disjoint. This phenomenon causes difficulties in variety of applications where it is important to recognize cases where different representations define same curve (surface) or partially coincident curves (surfaces). The problem of finding whether two arbitrary parametric polynomial curves are the same has been addressed in Pekerman et al. (Are two curves the same? *Comput.-Aided Geom. Des. Appl.* 2(1–4):85–94, 2005). There the curves are reduced into canonical irreducible forms using the monomial basis, then they are compared and their shared domains, if any, are identified. Here we present an alternative geometric algorithm based on subdivision that compares two input control polygons and reports the coincidences between the corresponding Bézier curves if they are present. We generalize the algorithm for tensor product Bézier surfaces. The algorithms are implemented and tested using Mathematica package. The experimental results are presented.

1 Introduction

Comparing Bézier curves and surfaces for coincidence is an important problem in computer-aided design (CAD) which arises in various applications. Suppose we are given the control polygons of two curves (surfaces) which are obtained by different sources, e.g. they can be generated by different software packages. The curves (surfaces) have to be stitched together to obtain a new curve (surface) which is continuous. It is possible that the two control polygons may have different number of control points and may even be disjoint but nevertheless they represent same curve (surface), see Fig. 2. Therefore it is important to find out whether the control polygons represent same curve (surface) and in this case to determine the coincident part of the two curves (surfaces). By *coincident* curves (surfaces) we mean that they

K. Vlachkova (✉)
Faculty of Mathematics and Informatics, Sofia University "St. Kliment Ohridski",
Blvd. James Bourchier 5, 1164 Sofia, Bulgaria
e-mail: krassivl@fmi.uni-sofia.bg

K. Georgiev et al. (eds.), *Advanced Computing in Industrial Mathematics*,
Studies in Computational Intelligence 681, DOI 10.1007/978-3-319-49544-6_20

occupy the same locus of points in $\mathbb{R}^3$, i.e. they are *geometrically equivalent* as defined by Denker and Heron [4]. A *shared domain* of two curves (surfaces) with same parametrization is the intersection of their domains of definition.

Different representations of a parametric polynomial curve occur if it has been degree elevated and/or reparameterized by a composition with a polynomial. While the degree elevation is comparatively easy to be detected, it is usually hard to find out whether the curve has undergone polynomial composition. Decomposition of polynomials is a classical problem in computer algebra. The first algorithm for polynomial decomposition was proposed by Barton and Zippel [1, 2]. An alternative decomposition algorithm was proposed by Kozen and Landau [6]. Later, it has been improved by von zur Gathen [5]. Today polynomial decomposition is a standard built-in function in the computer algebra systems as Maple (the function compoly), Matlab (the function polylib:decompose), Mathematica (the function Decompose). The applied algorithms are known for some of these systems. For example, Matlab applies Barton and Zippel's algorithm [2]. Currently, Mathematica does not disclose such information to the public.

A curve is *irreducible* if it is not a result of a polynomial composition and has not been degree elevated. Sánchez-Reyes [8] has showed uniqueness of the control points (up to reverse order) of an irreducible Bézier curve of arbitrary degree. He pointed out that this result is a straightforward consequence of a previous and more general result by Berry and Patterson [3] for rational Bézier curves. The uniqueness of the control points for cubic curves has also been studied by Wang et al. [10].

The problem of finding whether two arbitrary polynomial curves are the same has been considered by Pekerman et al. [7] where an algorithm for a polynomial decomposition is proposed and used. The authors do not compare the algorithm to any of the previously known algorithms as [1, 2, 5, 6]. They point out that the computations can be done efficiently since the degree of the polynomial to be decomposed is usually low for practical purposes. In [7] the curves are reduced into canonical irreducible forms, then they are compared and their shared domain, if any, is identified.

Here we present an alternative geometric algorithm for comparing Bézier curves for coincidence. Our algorithm works in two phases. In the first phase the control polygons are tested for reducibility. We adopt a set of routines from Mathematica package and transform the control polygons to an irreducible form which is unique as shown in [8]. In the second phase the obtained irreducible control polygons are checked for coincidences. We use a new geometric approach based on subdivision. In this approach the usage of a monomial basis and consequent conversion into canonical form are avoided. In addition, simple geometric conditions for checking whether two control polygons define different irreducible curves are obtained. Given two irreducible Bézier curves our Algorithm 1 reports whether they are different, disjoint, or coincident and in that case reports the control points of the coincident part.

We generalize Algorithm 1 for the case of tensor product Bézier surfaces. Since computer algebra systems currently do not support bivariate polynomial decomposition, our aim was to reduce the problem for surfaces to the problem for curves.

We prove in Theorem 1 that two irreducible tensor product Bézier surfaces of same degree coincide if and only if their control polygons coincide (up to different enumeration of the control points). Given two irreducible tensor product Bézier curves of same degree (m, n) our Algorithm 2 reports whether they are different, disjoint, or coincident and in that case reports the control points of the coincident part. Coincidences also can occur in the case where the degrees of the two irreducible surfaces are (n, m) and $(n + m, n + m)$, respectively. Our algorithm doesn't cover this case.

We implemented and tested our algorithms using Mathematica package. Real examples illustrating our presentation are included.

The paper is organized as follows. In Sect. 2 we consider the problem for Bézier curves and propose Algorithm 1 based on subdivision that reports two irreducible Bézier curves as different, disjoint, or coincident. In case of coincidence Algorithm 1 reports the control points of the coincident part. The problem for tensor product Bézier surfaces is considered in Sect. 3 and Algorithm 2 that compares two irreducible tensor product Bézier surfaces of same degree for coincidence is proposed. Algorithm 2 reports these surfaces as different, disjoint, or coincident and in that case reports the control points of the coincident part.

2 Coincidence of Bézier Curves

In this section we present an algorithm based on subdivision for comparing Bézier curves for coincidence. Our algorithm takes as input two irreducible Bézier curves of same degree $\mathbf{b}^1(t)$, $t \in [0, 1]$, and $\mathbf{b}^2(u)$, $u \in [0, 1]$, and reports if they are different. If not the algorithm reports them as disjoint or coincident. In the case of coincidence the algorithm reports the control points of their coincident part.

Let $\mathbf{b}(t) = \sum_{i=0}^{m} \mathbf{b}_i B_i^m(t)$, where $\mathbf{b}_i$, $i = 0, \ldots, m$, are points in $\mathbb{R}^3$ and $B_i^m(t)$ are the Bernstein polynomials defined for $0 \le t \le 1$ by $B_i^m(t) := \binom{m}{i} t^i (1-t)^{m-i}$, where $\binom{m}{i}$ for $0 \le i \le m$ are the binomial coefficients. We assume that $\binom{m}{i} = 0$ if $i < 0$ or $i > m$.

The curve $\mathbf{b}$ is degree elevated if and only if $\Delta^m \mathbf{b}_0 = \mathbf{0}$, where Δ^r is the forward finite difference of order r, $r \ge 1$, defined recursively by $\Delta^0 \mathbf{b}_i = \mathbf{b}_i$, $\Delta^r \mathbf{b}_i = \Delta^{r-1}\mathbf{b}_{i+1} - \Delta^{r-1}\mathbf{b}_i$. We note that $\Delta^m \mathbf{b}_0$ is the coefficient of t^m in the canonical form of $\mathbf{b}(t)$. If $\mathbf{b}$ has been degree elevated then the control points $\hat{\mathbf{b}}_i$ of the degree reduced curve are computed recursively as

$$\hat{\mathbf{b}}_i = \frac{m}{m-i}\mathbf{b}_i - \frac{i}{m-i}\hat{\mathbf{b}}_{i-1}, \qquad i = 0, \ldots, m-1.$$

In order to check $\mathbf{b}$ for a composition by a polynomial we use the built-in function Decompose in Mathematica. This function returns the decomposed curve $\tilde{\mathbf{b}}(t)$ in its canonical form $\tilde{\mathbf{b}}(t) = \sum_{i=0}^{m_1} \mathbf{a}_i t^i$, where $m_1 \le m$, $\mathbf{a}_i \in \mathbb{R}^3$. Then the control points of $\tilde{\mathbf{b}}(t)$ are

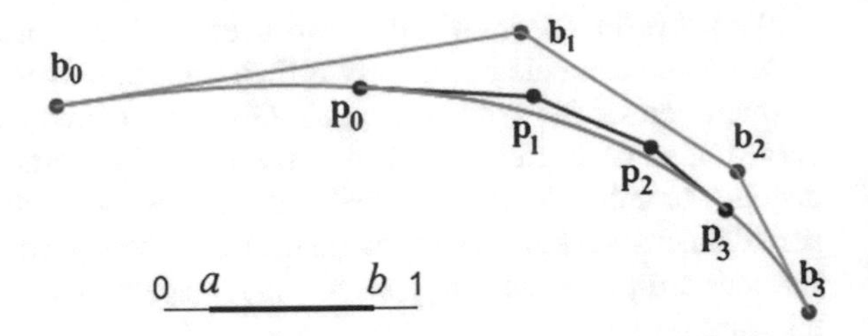

Fig. 1 Curve $\mathbf{b}^2$ with control points $\mathbf{p}_i$, $i = 1, \ldots, n$ is obtained from curve $\mathbf{b}^1$ with control points $\mathbf{b}_i$, $i = 1, \ldots, n$, by subdivision at a and b, where $a,\ b \in \mathbb{R}$

$$\tilde{\mathbf{b}}_i = \sum_{j=0}^{i} \frac{\binom{i}{j}}{\binom{m_1}{j}} \mathbf{a}_j, \qquad i = 0, \ldots, m_1.$$

Degree elevation and decomposition are non-commutative operations, as pointed out in [7]. Hence, in order to check the Bézier curve $\mathbf{b}$ for irreducibility we have to perform alternate check for degree elevation and decomposition while both attempts fail. Finally, we obtain the control points $\mathbf{p}_i$, $i = 0, \ldots, n$ of the irreducible form of $\mathbf{b}$ and we store the last two finite differences $\Delta^{n-1}\mathbf{b}_0$ and $\Delta^n \mathbf{b}_0$. Note that $\Delta^n \mathbf{b}_0 \neq \mathbf{0}$.

Let $\mathbf{b}^1(t) = \sum_{i=0}^{n} \mathbf{b}_i B_i^n(t), t \in [0, 1]$, and $\mathbf{b}^2(u) = \sum_{i=0}^{n} \mathbf{p}_i B_i^n(u), u \in [0, 1], n \in \mathbb{N}$, $n \geq 2$, be two irreducible Bézier curves such that their control polygons do not coincide. Our aim is to check out whether $\mathbf{b}^1$ and $\mathbf{b}^2$ represent same curve and in this case to find their coincident part, if any. Suppose that $\mathbf{b}^1$ and $\mathbf{b}^2$ represent same curve. Then we can consider one of them, say $\mathbf{b}^2$, as obtained from $\mathbf{b}^1$ by subdivision at two parameters a and b, $a, b \in \mathbb{R}$, see Fig. 1. Hence, $\mathbf{b}^2$ is defined in the interval with endpoints a and b which are uniquely defined. This interval is an image of the interval $[0, 1]$ by the affine map $t = (b - a)u + a$, $0 \leq u \leq 1$. We have

$$\sum_{i=0}^{n} \mathbf{p}_i B_i^n(u) = \sum_{i=0}^{n} \mathbf{b}_i B_i^n((b - a)u + a), \qquad 0 \leq u \leq 1. \tag{1}$$

Next, we express a and b in terms of the control points $\mathbf{p}_i, \mathbf{b}_i$, $i = 0, \ldots, n$. We differentiate $n - 1$ times both sides of (1). The derivative of order $n - 1$ is obtained by applying consecutively $n - 1$ finite differences and one evaluation by de Casteljau algorithm. We have

$$n! \sum_{i=0}^{n-(n-1)} \Delta^{n-1}\mathbf{p}_i B_i^{n-(n-1)}(u) = n!(b - a)^{n-1} \sum_{i=0}^{n-(n-1)} \Delta^{n-1}\mathbf{b}_i B_i^{n-(n-1)}((b - a)u + a). \tag{2}$$

Hence, from (2) we have

$$\begin{aligned}(1 - u)\Delta^{n-1}\mathbf{p}_0 + u\Delta^{n-1}\mathbf{p}_1 = (b - a)^{n-1}\big((1 - a - (b - a)u)\Delta^{n-1}\mathbf{b}_0 \\ +(a + (b - a)u)\Delta^{n-1}\mathbf{b}_1\big).\end{aligned} \tag{3}$$

For the derivatives of order n in both sides of (1) we obtain

$$\Delta^n \mathbf{p}_0 = (b-a)^n \Delta^n \mathbf{b}_0. \tag{4}$$

From (4) we have $b - a = \sqrt[n]{|\Delta^n \mathbf{p}_0| / |\Delta^n \mathbf{b}_0|}$ for n odd, and $b - a = \pm\sqrt[n]{|\Delta^n \mathbf{p}_0| / |\Delta^n \mathbf{b}_0|}$ for n even. Note that we have stored in advance $\Delta^{n-1}\mathbf{p}_0$, $\Delta^n \mathbf{p}_0$, $\Delta^{n-1}\mathbf{b}_0$, $\Delta^n \mathbf{b}_0$ while performing check for degree elevation and we have $\Delta^n \mathbf{p}_0 \neq \mathbf{0}$ and $\Delta^n \mathbf{b}_0 \neq \mathbf{0}$.

We have from (3) for $u = 0$ and $u = 1$, respectively

$$\Delta^{n-1}\mathbf{p}_0 = (b-a)^{n-1}\big((1-a)\Delta^{n-1}\mathbf{b}_0 + a\Delta^{n-1}\mathbf{b}_1\big), \tag{5}$$

$$\Delta^{n-1}\mathbf{p}_1 = (b-a)^{n-1}\big((1-b)\Delta^{n-1}\mathbf{b}_0 + b\Delta^{n-1}\mathbf{b}_1\big). \tag{6}$$

Then we compute

$$A = \frac{\Delta^{n-1}\mathbf{p}_0}{(b-a)^{n-1}} - \Delta^{n-1}\mathbf{b}_0, \quad B = \frac{\Delta^{n-1}\mathbf{p}_1}{(b-a)^{n-1}} - \Delta^{n-1}\mathbf{b}_0 = A + \frac{\Delta^n \mathbf{p}_0}{(b-a)^{n-1}}. \tag{7}$$

From (5) and (6) we obtain

$$a\Delta^n \mathbf{b}_0 = A, \quad b\Delta^n \mathbf{b}_0 = B. \tag{8}$$

In the case where n is even there are two possibilities for A and B in (8) but only one of them is correct. For the other possibility there exist no a and b that satisfy (8).

We note that in the case where $a, b \notin [0, 1]$ it is better to subdivide $\mathbf{b}^2$ instead of $\mathbf{b}^1$ to avoid extrapolation which is not numerically stable for large parameter values.

The geometric meaning of (4) and (8) is that the four vectors $\Delta^n \mathbf{b}_0$, $\Delta^n \mathbf{p}_0$, A, and B are collinear. The corresponding coefficients of proportion are $(b-a)^n$ in (4), a and b in (8). Another interpretation of the geometric meaning of the parameters a and b was proposed by Sánchez-Reyes in [9].

The next lemma provides simple sufficient geometric conditions for two irreducible Bézier curves to be different.

Lemma 1 *The irreducible Bézier curves* $\mathbf{b}^1$ *and* $\mathbf{b}^2$ *are different if any of the next three statements is true.*

(i) $\Delta^n \mathbf{b}_0$ *and* $\Delta^n \mathbf{p}_0$ *are not collinear;*
(ii) n *is even and* $\Delta^n \mathbf{b}_0$ *and* $\Delta^n \mathbf{p}_0$ *have opposite directions;*
(ii) $\Delta^n \mathbf{b}_0$ *is collinear with at most one of* A *and* B *defined by (7).*

Proof Statements (i) and (ii) follows from (4). Statement (iii) follows from (8). □

Next, we propose Algorithm 1 for comparing two irreducible Bézier curves for coincidence.

Algorithm 1 Comparison for Coincidence of two Irreducible Bézier Curves

Input: Irreducible Bézier curves $\mathbf{b}^1$ and $\mathbf{b}^2$ represented by their control points $\{\mathbf{b}_i\}_{i=0}^n$, $\{\mathbf{p}_i\}_{i=0}^n$, respectively; $\Delta^{n-1}\mathbf{b}_0$, $\Delta^n\mathbf{b}_0$; $\Delta^{n-1}\mathbf{p}_0$, $\Delta^n\mathbf{p}_0$.

Output: (i) $\mathbf{b}^1$ and $\mathbf{b}^2$ are different;
(ii) $\mathbf{b}^1$ and $\mathbf{b}^2$ are disjoint;
(iii) $\mathbf{b}^1$ and $\mathbf{b}^2$ have coincident part $\bar{\mathbf{b}}$. Report the control points of $\bar{\mathbf{b}}$.

Step 1. **if** $\Delta^n\mathbf{b}_0$ and $\Delta^n\mathbf{p}_0$ are not collinear
  then output (i) and **stop**;
   else if n is even and $\Delta^n\mathbf{b}_0$ and $\Delta^n\mathbf{p}_0$ have opposite directions
     then output (i) and **stop**;
      else compute $l = \sqrt[n]{|\Delta^n\mathbf{p}_0|/|\Delta^n\mathbf{b}_0|}$, $A = \Delta^{n-1}\mathbf{p}_0/l^{n-1} - \Delta^{n-1}\mathbf{b}_0$,
      $B = A + \Delta^n\mathbf{p}_0/l^{n-1}$
    end if
end if

Step 2. **if** both A and B are collinear to $\Delta^n\mathbf{b}_0$
  then go to Step 4;
   else if n is odd
     then output (i) and **stop**;
      else compute $A_1 = -A - 2\Delta^{n-1}\mathbf{b}_0$ and $B_1 = -B - 2\Delta^{n-1}\mathbf{b}_0$
    end if
end if

Step 3. **if** $\Delta^n\mathbf{b}_0$ is collinear with at most one of A_1 and B_1
  then output (i) and **stop**;
   else go to Step 4
end if

Step 4. Compute a and b such that $a\Delta^n\mathbf{b}_0 = A$, $b\Delta^n\mathbf{b}_0 = B$.

Step 5. Subdivide $\mathbf{b}^1$ at a and b and compare the obtained control points to $\mathbf{p}_i$, $i = 0, \ldots, n$.
if there are at least two corresponding non-coincident control points
  then output (i) and **stop**;
   else compute the intersection I of $[0, 1]$ with the interval with endpoints a and b.
end if
if $I = \emptyset$ **then** output (ii) and **stop**;
    else output (iii) with the control points computed in Step 5 and **stop**
end if

In Example 1 the curves $\mathbf{b}^1$ and $\mathbf{b}^2$ are compared for coincidence using Algorithm 1.

Example 1 The control points of the curve $\mathbf{b}^1$ of degree 4 and $\mathbf{b}^2$ of degree 8 are shown in Table 1. The curve $\mathbf{b}^1$ is irreducible. After decomposition we obtain that $\mathbf{b}^2$ is a result of a composition by the polynomial $t^2 + t$ of the curve

$$\mathbf{b}(t) = \big(-1 + 12t - 30t^2 + 24.8t^3 - 4.6t^4, -1 + 4.8t - 4.2t^2 - 0.4t^3 - 0.1t^4 \big).$$

The irreducible form of $\mathbf{b}^2$ has degree $n = 4$, its control points are shown in Table 1 and the control polygons of both forms of $\mathbf{b}^2$ are shown in Fig. 2. The control polygons of the irreducible curves $\mathbf{b}^1$ and $\mathbf{b}^2$ are shown in Fig. 3a. By applying Algorithm 1 we obtain that $\mathbf{b}^2$ is a result of subdivision of $\mathbf{b}^1$ at $a = -0.05$ and $b = 0.4$. The control points of the coincident part $\bar{\mathbf{b}}$ of $\mathbf{b}^1$ and $\mathbf{b}^2$ are shown in Table 1. The curve $\bar{\mathbf{b}}$ and its control polygon are shown in Fig. 3b.

Table 1 Comparison of Bézier curves $\mathbf{b}^1$ and $\mathbf{b}^2$ for coincidence

$\mathbf{b}^1$	$\mathbf{b}^2$	Irreducible $\mathbf{b}^2$	Coincident part $\bar{\mathbf{b}}$
(−1, −1)	(−1.67813, −1.25045)	(−1.67813, −1.25045)	(−1, −1)
(2, 0.2)	(−1.06849, −1.04105)	(0.030555, −0.663532)	(0.2, −0.52)
(0, 0.7)	(−0.513367, −0.822876)	(0.59886, −0.21639)	(0.6, −0.152)
(−0.8, 0.4)	(−0.0474957, −0.602289)	(0.61272, 0.08232)	(0.5968, 0.0976)
(1.2, −0.9)	(0.300317, −0.386845)	(0.46944, 0.21984)	(0.46944, 0.21984)
	(0.513146, −0.18546)		
	(0.59217, −0.0086048)		
	(0.5616, 0.131384)		
	(0.469442, 0.219838)		

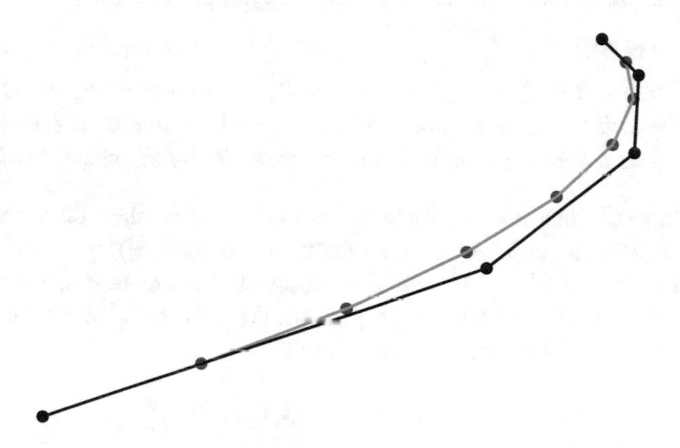

Fig. 2 The control polygons of curve $\mathbf{b}^2$ and its irreducible form. The corresponding control points are shown in Table 1

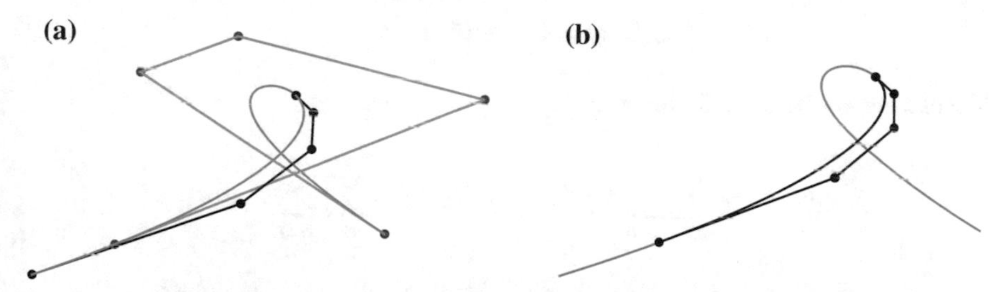

Fig. 3 **a** Curve $\mathbf{b}$ and the control polygons of the irreducible form of $\mathbf{b}^1$ and $\mathbf{b}^2$; **b** Curve $\mathbf{b}$ and the control polygon of the coincident part $\bar{\mathbf{b}}$ of $\mathbf{b}^1$ and $\mathbf{b}^2$

3 Coincidence of Tensor Product Bézier Surfaces

In this section we present an algorithm for comparing tensor product Bézier surfaces for coincidence. Our algorithm takes as input two irreducible tensor product Bézier surfaces of same degree $\mathbf{b}^1(s, t)$, $(s, t) \in [0, 1] \times [0, 1]$, and $\mathbf{b}^2(u, v)$,

$(u, v) \in [0, 1] \times [0, 1]$, and reports if they are different. If not then the algorithm reports them as disjoint or coincident. In the case of coincidence the algorithm reports the control points of their coincident part.

We start with a definition of irreducibility of tensor product Bézier surface which is consistent with the analogous definition for Bézier curve.

Definition 1 The tensor product Bézier surface $\mathbf{b}(u, v) = \sum_{i=0}^{n} \sum_{j=0}^{m} \mathbf{b}_{ij} B_i^n(u) B_j^m(v)$, $0 \le u \le 1$, $0 \le v \le 1$, $m, n \in \mathbb{N}$, is *irreducible* if the curves $\mathbf{c}_j^1(u)$ with control points $\{\mathbf{b}_{ij}\}_{i=0}^{n}$, $j = 0, \dots, m$, and $\mathbf{c}_i^2(v)$ with control points $\{\mathbf{b}_{ij}\}_{j=0}^{m}$, $i = 0, \dots, n$, are irreducible.

Next, we prove a theorem that provides necessary and sufficient condition for coincidence of two irreducible tensor product Bézier surfaces.

Theorem 1 *Let* $\mathbf{b}^1(s, t) = \sum_{i=0}^{n} \sum_{j=0}^{m} \mathbf{b}_{ij} B_i^n(s) B_j^m(t)$ *defined for* $(s, t) \in [0, 1] \times [0, 1]$, *and* $\mathbf{b}^2(u, v) = \sum_{i=0}^{n} \sum_{j=0}^{m} \mathbf{p}_{ij} B_i^n(u) B_j^m(v)$ *defined for* $(u, v) \in [0, 1] \times [0, 1]$, *be irreducible Bézier surfaces. Then* $\mathbf{b}^1$ *and* $\mathbf{b}^2$ *coincide if and only if their control polygons coincide (up to different enumeration of the control points).*

Proof $\Rightarrow$ Suppose that $m \ne n$. Since $\mathbf{b}^1$ and $\mathbf{b}^2$ coincide then there exist smooth functions $s = \varphi(u, v)$ and $t = \psi(u, v)$ such that $\mathbf{b}^1(\varphi(u, v), \psi(u, v)) \equiv \mathbf{b}^2(u, v)$ and the boundaries of $\mathbf{b}^1$ and $\mathbf{b}^2$ coincide. Suppose that $\mathbf{b}_{00} \equiv \mathbf{p}_{00}$. Hence, $\mathbf{b}^1(s, 0)$ coincides with $\mathbf{b}^2(u, 0)$. Therefore $\mathbf{b}^1(\varphi(u, 0), \psi(u, 0)) \equiv \mathbf{b}^2(u, 0)$ for $0 \le u \le 1$ and since both curves are irreducible it follows

$$\varphi(u, 0) = u, \quad \psi(u, 0) = 0. \tag{9}$$

Moreover, $\mathbf{b}^1(\varphi(0, v), \psi(0, v)) \equiv \mathbf{b}^2(0, v)$ for $0 \le v \le 1$, thus

$$\varphi(0, v) = 0, \quad \psi(0, v) = v. \tag{10}$$

We expand φ and ψ as Taylor series,

$$\varphi(u, v) = \sum_{i=0}^{\infty} \sum_{j=0}^{\infty} \alpha_{ij} u^i v^j, \quad \psi(u, v) = \sum_{i=0}^{\infty} \sum_{j=0}^{\infty} \beta_{ij} u^i v^j.$$

From (9) and (10) it follows $\varphi(u, v) = u + \sum_{i=1}^{\infty} \sum_{j=1}^{\infty} \alpha_{ij} u^i v^j$ and $\psi(u, v) = v + \sum_{i=1}^{\infty} \sum_{j=1}^{\infty} \beta_{ij} u^i v^j$. Therefore,

$$\mathbf{b}^1(\varphi(u, v), \psi(u, v)) = \sum_{i=0}^{n} \sum_{j=0}^{m} \mathbf{b}_{ij} B_i^n\big(u + \sum_{i=1}^{\infty} \sum_{j=1}^{\infty} \alpha_{ij} u^i v^j\big) B_j^m\big(v + \sum_{i=1}^{\infty} \sum_{j=1}^{\infty} \beta_{ij} u^i v^j\big). \tag{11}$$

Suppose that $\varphi(u, v) \ne u$. Then there exists $\alpha_{ij} \ne 0$ for some i, j, $1 \le i, j < \infty$. Hence

$$\deg\Big(\sum_{i=1}^{\infty}\sum_{j=1}^{\infty}\alpha_{ij}u^i v^j\Big) \geq 2 \;\Rightarrow\; \deg\big(B_i^n(u+\cdots)\big) \geq 2n.$$

We obtain that the degree of the polynomial in the right hand side in (11) is $\geq 2n+m$ which is not possible since $\mathbf{b}^1$ has degree $n+m$. Therefore $\varphi(u,v)=u$. Similarly we obtain $\psi(u,v)=v$ and hence $\mathbf{b}^1(u,v)\equiv \mathbf{b}^2(u,v)$.

The other possibilities for the boundaries are treated analogously. In the case where $\mathbf{b}_{00}=\mathbf{p}_{n0}$ we obtain $\varphi(u,v)=1-u$, $\psi(u,v)=v$, and $\mathbf{b}^1(1-u,v)=\mathbf{b}^2(u,v)$. If $\mathbf{b}_{00}=\mathbf{p}_{0m}$ then $\mathbf{b}^1(u,1-v)=\mathbf{b}^2(u,v)$, and if $\mathbf{b}_{00}=\mathbf{p}_{nm}$ then $\mathbf{b}^1(1-u,1-v)=\mathbf{b}^2(u,v)$. In the case where $m=n$ there are four more possibilities:

$$\begin{aligned}
\mathbf{b}_{00}=\mathbf{p}_{00},\ \mathbf{b}_{n0}=\mathbf{p}_{0m} &\Rightarrow \mathbf{b}^1(v,u)=\mathbf{b}^2(u,v),\\
\mathbf{b}_{00}=\mathbf{p}_{n0},\ \mathbf{b}_{n0}=\mathbf{p}_{nm} &\Rightarrow \mathbf{b}^1(1-v,u)=\mathbf{b}^2(u,v),\\
\mathbf{b}_{00}=\mathbf{p}_{0m},\ \mathbf{b}_{n0}=\mathbf{p}_{00} &\Rightarrow \mathbf{b}^1(v,1-u)=\mathbf{b}^2(u,v),\\
\mathbf{b}_{00}=\mathbf{p}_{nm},\ \mathbf{b}_{n0}=\mathbf{p}_{n0} &\Rightarrow \mathbf{b}^1(1-v,1-u)=\mathbf{b}^2(u,v).
\end{aligned}$$

In all cases the control polygons of $\mathbf{b}^1$ and $\mathbf{b}^2$ coincide up to different enumeration of the control points.

$\Leftarrow$ Straightforward □

Next, we propose Algorithm 2 that compares for coincidence two irreducible tensor product Bézier surfaces of degree (n,m). In Example 2 tensor product Bézier surfaces $\mathbf{b}^1$ and $\mathbf{b}^2$ are compared for coincidence using Algorithm 2.

Example 2 The control points of the irreducible tensor product Bézier surfaces $\mathbf{b}^1$ and $\mathbf{b}^2$ of degree (2, 3) are shown in Table 2. The two surfaces and their control polygons are shown in Fig. 4a, b, respectively. In Fig. 5a, $\mathbf{b}^1$ and $\mathbf{b}^2$ are shown together.

Table 2 Comparison of the irreducible tensor product Bézier surfaces $\mathbf{b}^1$ and $\mathbf{b}^2$ for coincidence

$\mathbf{b}^1$	(0, 0, 0)	(0, 0.7, −0.5)	(0, 2, −0.8)
	(1, 0, 1)	(1, 1, 0)	(1, 1, 1)
	(2, 0, 0.75)	(2, 1, 1.2)	(2, 2, 0.75)
	(3, −1, 0.15)	(3, 1, 0.4)	(3, 2, 0.2)
$\mathbf{b}^2$	(−1/2, 0.223122, −0.663304)	(−1/2, 0.685108, −0.814381)	(−1/2, 1.70446, −1.34961)
	(1/6, 0.253472, 0.0865162)	(1/6, 0.716898, −0.317969)	(1/6, 1.14913, −0.292144)
	(5/6, 0.322917, 0.517303)	(5/6, 0.799306, 0.259462)	(5/6, 1.1651, 0.332682)
	(3/2, 0.225694, 0.606424)	(3/2, 0.81875, 0.504948)	(3/2, 1.26719, 0.533203)
$\bar{\mathbf{b}}$	(0, 0.25, −0.161111)	(0, 0.716667, −0.433333)	(0, 1.3875, −0.6375)
	(1/2, 0.277778, 0.280556)	(1/2, 0.754167, −0.05)	(1/2, 1.1625, −0.00625)
	(1, 0.298611, 0.539583)	(1, 0.804167, 0.320833)	(1, 1.19062, 0.382812)
	(3/2, 0.225694, 0.606424)	(3/2, 0.81875, 0.504948)	(3/2, 1.26719, 0.533203)

Algorithm 2 Comparison for Coincidence of two Irreducible Tensor Product Bézier Surfaces

Input: Irreducible Bézier surfaces $\mathbf{b}^1$ and $\mathbf{b}^2$ represented by their control points $\{\mathbf{b}_{ij}\}_{i=0,j=0}^{n,m}$, $\{\mathbf{p}_{ij}\}_{i=0,j=0}^{n,m}$, respectively

Output: (i) $\mathbf{b}^1$ and $\mathbf{b}^2$ are different;
(ii) $\mathbf{b}^1$ and $\mathbf{b}^2$ are disjoint;
(iii) $\mathbf{b}^1$ and $\mathbf{b}^2$ have coincident part $\bar{\mathbf{b}}$. Report the control points of $\bar{\mathbf{b}}$.

Step 1. Set $M_0 := [0, 1]$.
for $j = 0, \ldots, m$ apply Algorithm 1 to curves $\mathbf{c}_j^1$ and $\mathbf{c}_j^2$ with control points $\{\mathbf{b}_{ij}\}_{i=0}^n$ and $\{\mathbf{p}_{ij}\}_{i=0}^n$, respectively.
  if $\mathbf{c}_j^1$ and $\mathbf{c}_j^2$ are different
    then output (i) and **stop**;
      else find the corresponding coincidence interval I_j and set $M_{j+1} = M_j \cap I_j$
  end if
end for

Step 2. **if** $M_{m+1} := [a, b] \neq \emptyset$
  then for $j = 0, \ldots, m$ subdivide $\mathbf{c}_j^1$ and $\mathbf{c}_j^2$ at a and b. Use the same notation for the new control points **end for**;
    else go to Step 3
end if

Step 3. Set $N_0 := [0, 1]$.
for $i = 0, \ldots, n$ apply Algorithm 1 to curves $\bar{\mathbf{c}}_i^1$ and $\bar{\mathbf{c}}_i^2$ with control points $\{\mathbf{b}_{ij}\}_{j=0}^m$ and $\{\mathbf{p}_{ij}\}_{j=0}^m$, respectively.
  if $\bar{\mathbf{c}}_i^1$ and $\bar{\mathbf{c}}_i^2$ are different
    then output (i) and **stop**;
      else find the corresponding coincidence interval J_i and set $N_{i+1} = N_i \cap J_i$
  end if
end for

Step 4. **if** $N_{n+1} := [c, d] \neq \emptyset$
  then for $i = 0, \ldots, n$ subdivide $\bar{\mathbf{c}}_i^1$ at c and d
    end for
  output (iii) and **stop**;
    else output (ii) and **stop**
end if

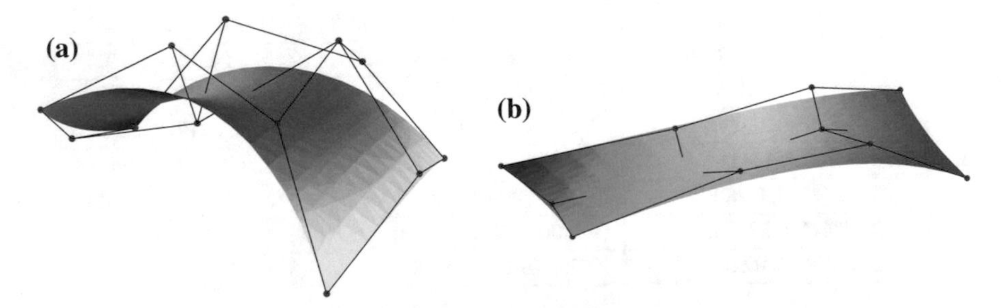

Fig. 4 The irreducible tensor product Bézier surfaces $\mathbf{b}^1$ and $\mathbf{b}^2$ of degree (2, 3) with their control polygons: **a** surface $\mathbf{b}^1$; **b** surface $\mathbf{b}^2$. The corresponding control points are shown in Table 2

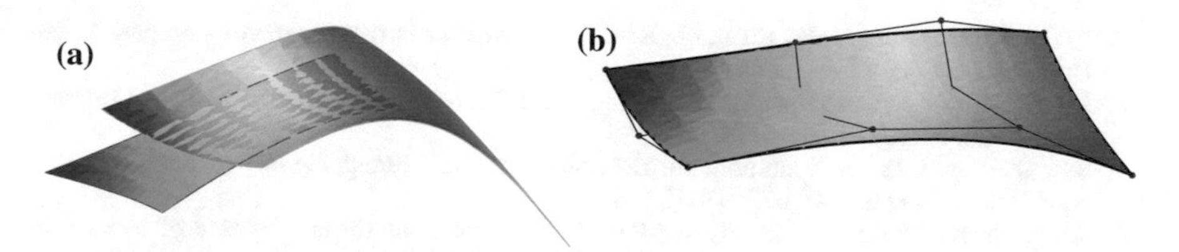

Fig. 5 **a** The irreducible tensor product Bézier surfaces $\mathbf{b}^1$ and $\mathbf{b}^2$ from Example 2; **b** The coincident part $\bar{\mathbf{b}}$ of $\mathbf{b}^1$ and $\mathbf{b}^2$ and its control polygon. The corresponding control points are shown in Table 2

Their coincident part $\bar{\mathbf{b}}$ with its control polygon is shown in Fig. 5b. This surface is obtained by subdivision of $\mathbf{b}^1$ in direction u at $a = 0,\ b = 1/2$ and direction v at $c = 1/6,\ d = 3/4$. The corresponding control points are shown in Table 2.

4 Conclusion

In this paper we present a new geometric algorithm based on subdivision that compares irreducible Bézier curves for coincidence and reports their coincident part if it is present. We generalize the algorithm to pairs of irreducible tensor product Bézier surfaces of degree (n, m), $m, n \in \mathbb{N}$. We believe that our approach can be successfully applied to the open problem for comparing pairs of irreducible tensor product Bézier surfaces of degree (n, m) and $(n + m, n + m)$, respectively. Another task for future research is to develop and implement an algorithm for comparing triangular Bézier surfaces for coincidence.

Acknowledgements This work was partially supported by the Bulgarian National Science Fund under Grant No. DFNI-T01/0001 and Sofia University Science Fund Grant No. 48/2016.

References

1. Barton, D.R., Zippel, R.: Polynomial decomposition. In: Proceedings of SYMSAC, vol. 76, pp. 356–358 (1976)
2. Barton, D.R., Zippel, R.: Polynomial decomposition algorithms. J. Symb. Comput. **1**, 159–168 (1985)
3. Berry, T.G., Peterson, R.R.: The uniqueness of Bézier control points. Comput. Aided Geom. Des. **14**, 877–879 (1997)
4. Denker, W.A., Heron, G.J.: Generalizing rational degree elevation. Comput. Aided Geom. Des. **14**, 399–406 (1997)
5. von zur Gathen, J.: Functional decomposition of polynomials: the tame case. J. Symb. Comput. **9**, 281–299 (1990)
6. Kozen, D., Landau, S.: Polynomial decomposition algorithms. J. Symb. Comput. **7**, 445–456 (1989)

7. Pekerman, D., Seong, J.-K., Elber, G., Kim, M.-S.: Are two curves the same? Comput.-Aided Des. Appl. **2**(1–4), 85–94 (2005)
8. Sánchez-Reyes, J.: On the conditions for the coincidence of two cubic Bézier curves. J. Comput. Appl. Math. **236**(6), 1675–1677 (2011)
9. Sánchez-Reyes, J.: The conditions for the coincidence or overlapping of two Bézier curves. Appl. Math. Comput. **248**, 625–630 (2014)
10. Wang, W.-K., Zhang, H., Liu, X.-M., Paul, J.-C.: Conditions for coincidence of two cubic Bézier curves. J. Comput. Appl. Math. **235**, 5198–5202 (2011)

Spectral Theory of sl(3, $\mathbb{C}$) Auxiliary Linear Problem with $\mathbb{Z}_2 \times \mathbb{Z}_2 \times \mathbb{Z}_2$ Reduction of Mikhailov Type

A.B. Yanovski

Abstract We consider an auxiliary system $L_{S_{\pm 1}}$, see (1) below, used recently as L operator in a Lax pair for some soliton equations. $L_{S_{\pm 1}}$ could be regarded as a generalization of a pole gauge Generalized Zakharov-Shabat system on sl(3, $\mathbb{C}$) on the whole real axis involving rational dependence on the spectral parameter. We consider the system on the condition that its 'potentials' $u(x)$ and $v(x)$ tend sufficiently fast to constant values u_0, v_0 when $x \to \pm\infty$ in the general situation when both $u_0, v_0 \neq 0$. We show that in this case the spectral theory for $L_{S_{\pm 1}}$ should be considered on a suitable Riemann surface and discuss the symmetry properties of the fundamental analytic solutions to $L_{S_{\pm 1}}\psi = 0$.

1 Introduction

Let us introduce the auxiliary system, which attracted attention recently [7, 8]:

$$L_{S_{\pm 1}}\psi = (i\partial_x + \lambda S_1(x) + \lambda^{-1} S_{-1}(x))\psi = 0. \tag{1}$$

In the above $S_1(x)$ and $S_{-1}(x)$ are 3×3 traceless matrix functions. In addition, it is assumed that the set of the fundamental solutions of $L_{S_{\pm 1}}\psi = 0$ is invariant under a group generated by the following transformations:

$$\begin{aligned} g_0(\psi)(x, \rho) &= \left[\psi(x, \rho^*)^\dagger\right]^{-1} & & \\ g_1(\psi)(x, \rho) &= H_1 \psi(x, -\lambda) H_1, & H_1 &= \mathrm{diag}\,(-1, 1, 1) \\ g_2(\psi)(x, \rho) &= H_2 \psi(x, \tfrac{1}{\lambda}) H_2, & H_2 &= \mathrm{diag}\,(1, -1, 1). \end{aligned} \tag{2}$$

where diag (a_1, a_2, a_3) denotes a diagonal matrix with diagonal elements a_1, a_2, a_3, * denotes complex conjugation and † stands for Hermitian conjugation. The above elements define a Mikhailov Reduction Group, [11, 12], and since g_i^2 is equal to identity the group is isomorphic to $\mathbb{Z}_2 \times \mathbb{Z}_2 \times \mathbb{Z}_2$. Imposing reduction group forces S_{-1} to be equal to $H_2 S_1 H_2$ and S_1 to be of the form

A.B. Yanovski (✉)
University of Cape Town, Cape Town, Rondebosch 7700, South Africa
e-mail: Alexandar.Ianovsky@uct.ac.za

K. Georgiev et al. (eds.), *Advanced Computing in Industrial Mathematics*, Studies in Computational Intelligence 681, DOI 10.1007/978-3-319-49544-6_21

$$S_1 = \begin{pmatrix} 0 & u(x) & v(x) \\ u^*(x) & 0 & 0 \\ v^*(x) & 0 & 0 \end{pmatrix} \tag{3}$$

where $u(x)$ and $v(x)$ are complex-valued functions ('potentials') defined on the real line. Also, it is assumed that $|u(x)|^2 + |v(x)|^2 = 1$. One can show that this condition ensures that the values of $S_1(x)$ and S_{-1} will be in the orbit of the element $J_0 = \mathrm{diag}\,(1, 0, -1)$ with respect to the adjoint action of the group $SU(3)$ in $i\mathfrak{su}(3)$. The requirement is motivated by the fact that in this case the L-operator $L_{S_1} = i\partial_x + \lambda S_1(x)$ could be regarded as a generalization of a pole gauge Generalized Zakharov-Shabat system on $\mathrm{sl}(3, \mathbb{C})$ which permits to develop the whole theory for L_{S_1} using cases already studied [14]. We consider the system $L_{S_{\pm 1}}$ on the condition that $u(x)$ and $v(x)$ tend sufficiently fast to constant values u_0, v_0 when $x \to \pm\infty$ in the non-degenerate case, that is, when both $u_0, v_0 \neq 0$. The study of the spectral theory of this system (that is the study of its fundamental solutions and how the reductions (2) affect them) has been started in [9] where the degenerate cases $u_0 = 0$ and $v_0 = 0$ have been considered. In order to study the non-degenerate case one must consider some analytic continuation of the function μ defining the exponent of the asymptotic of the fundamental solutions (FAS), see below. We exploited the idea of making cuts and select a regular branch of μ but it seems that this leads to some difficulties later, so now we explore another route. We are trying to formulate everything—fundamental solutions, scattering data, etc. in the terms of the Riemann surface that corresponds to the analytic continuation of μ. We believe that such formulation is free from the flaws that involve selecting regular branches and will be more useful in the future.

2 Rational GMV. Spectral Theory

We would like to now to develop the spectral theory of the problem $L_{S_{\pm 1}}\psi = 0$ that we introduced in (1) with boundary conditions

$$\lim_{x\to\pm\infty} u(x) = u_0, \qquad \lim_{x\to\pm\infty} v(x) = v_0. \tag{4}$$

Naturally, $|u_0|^2 + |v_0|^2 = 1$.

2.1 *Asymptotic Behavior of the Fundamental Solutions*

For the purposes of the Inverse Scattering Method it is essential to study the Fundamental Analytic Solutions (FAS) of the system that we are interested in. We shall write down later the integral equations that will give us the FAS but now we discuss their asymptotic behavior. We expect that when $x \to \pm\infty$ the solutions of $L_{S_{\pm 1}}\psi = 0$ will behave as $(\exp \mathrm{i}J(\lambda)x)A$ where $A = A(\lambda)$ is a matrix that does not depend on x and

$$J(\lambda) = (\lambda S_1 + \lambda^{-1}S_{-1})|_{u=u_0, v=v_0}. \tag{5}$$

It is not hard to find that $J(\lambda)$ has eigenvalues $\mu_0 = 0,\ \mu_\pm = \pm\mu$ where

$$\mu = \sqrt{2(|v_0|^2 - |u_0|^2) + (\lambda^2 + \lambda^{-2})}. \tag{6}$$

Of course, since $|u_0|^2 + |v_0|^2 = 1$ one can cast μ in the following equivalent forms:

$$\mu = \sqrt{4|v_0|^2 + (\lambda - \lambda^{-1})^2} = \sqrt{-4|u_0|^2 + (\lambda + \lambda^{-1})^2}. \tag{7}$$

Hence $J(\lambda)$ is diagonalizable and there is a constant matrix C (depending of course on u_0 and v_0 and λ and μ) such that

$$\begin{aligned} C^{-1}J(\lambda)C &= \mu\mathrm{diag}\,(1, 0, -1) = \mu J_0 \\ J_0 &= \mathrm{diag}\,(1, 0, -1). \end{aligned} \tag{8}$$

Denote

$$F(\lambda) = \sqrt{(|v_0|^2 - |u_0|^2) + \frac{1}{2}(\lambda^2 + \lambda^{-2})}. \tag{9}$$

Naturally, $\mu = \sqrt{2}F(\lambda)$ where by F is denoted one of the branches of the square root of the function

$$f(\lambda) = (|v_0|^2 - |u_0|^2) + \frac{1}{2}(\lambda^2 + \lambda^{-2}).$$

$f(\lambda)$ is defined in the punctured plane $\mathbb{C}^* = \mathbb{C} \setminus \{0\}$ and is a composition of the functions $w = g(\lambda) = \lambda^2$, $h(w) = \frac{1}{2}(w + w^{-1})$ and the shift $s(z) = z + a$, $a = |v_0|^2 - |u_0|^2$. In case of boundary conditions in general position it has poles of order 2 at $\lambda = 0$ and $\lambda = \infty$ and simple zeros at the four points z_1, z_2, z_3, z_4:

$$z_1 = |u_0| + \mathrm{i}|v_0|,\ z_2 = -|u_0| + \mathrm{i}|v_0|,\ z_3 = -|u_0| - \mathrm{i}|v_0|,\ z_2 = |u_0| - \mathrm{i}|v_0|. \tag{10}$$

The four zeros degenerate into two in case either u_0 or v_0 equals zero. (Into ± 1 in case $v_0 = 0$ and into $\pm\mathrm{i}$ in case $u_0 = 0$). All the zeros lie on the unit circle $\mathbb{S}^1 = \{\lambda : |\lambda| = 1\}$. Since the function $f(\lambda)$ is invariant under the involutions mapping the Riemann sphere into itself:

$$\varphi_1(\lambda) = \lambda^*, \quad \varphi_2(\lambda) = -\lambda, \quad \varphi_3(\lambda) = \lambda^{-1}, \tag{11}$$

the set of zeros is also invariant under these involutions. As mentioned, in this article we consider the general position case, $u_0 \neq 0$, $v_0 \neq 0$. If one separates the plane into simply connected regions such that in each of them the function $f(\lambda)$ does not have zeros, then in each of them there will be exactly two branches of the square root of $f(\lambda) = s \circ h \circ g(\lambda)$ and hence of μ. Alternatively, one can consider the analytic continuation of the square root, making some cuts in order to obtain regular branches, as it has been done in our work [15]. Here however we shall try to adopt another view, we shall try to formulate everything on the Riemann surface Y related to the above analytic continuation. Let

$$r(\lambda) = 2(|v_0|^2 - |u_0|^2) + (\lambda^2 + \lambda^{-2})$$

which is analytic everywhere with the exception of $\lambda = 0$ where it has a pole of order 2. It is equal to zero at the points z_i, $i = 1, 2, 3, 4$ (the branch points) so it can be written as

$$r(\lambda) = \lambda^{-2}(\lambda - z_1)(\lambda - z_2)(\lambda - z_3)(\lambda - z_4).$$

The function $r(\lambda)$ could be considered as an analytic function on the Riemann sphere $\mathbb{P}^1$ with values in $\mathbb{P}^1$ ($r(0) = \infty$ and $r(\infty) = \infty$). Denote by $\mathcal{M}(\mathbb{P}^1)$ the field of meromorphic functions over $\mathbb{P}^1$ and as usual by $\mathcal{M}(\mathbb{P}^1)[T]$ the ring of polynomials in T with coefficients in $\mathcal{M}(\mathbb{P}^1)$. Consider the polynomial $P(T) = T^2 - r(\lambda)$. The function $r(\lambda)$ is written as $\lambda^{-2}p_4(\lambda)$ where $p_4(\lambda)$ is polynomial of degree 4 with simple zeros so $P(T)$ is irreducible over $\mathcal{M}(\mathbb{P}^1)$. Then according to [3], Proposition 8.9, the polynomial $P(T)$ defines a two-fold covering $(Y, p, \mathbb{P}^1)$ of $\mathbb{P}^1$ and a meromorphic function F on Y such that $(p^*P)(F) = 0$. Here Y is a compact Riemann surface, p is proper holomorphic 2-fold covering map. The covering is unique up to fiber preserving bi-holomorphic maps, in other words if $(Y_1, p_1, \mathbb{P}^1)$ and F_1 are a covering and a meromorphic function with the same properties as $(Y, p, \mathbb{P}^1)$ and F then there exists unique fiber-preserving map $\sigma : Y_1 \mapsto Y$ such that $\sigma^* F = F_1$. Thus the covering is unique if one requires that F obeys a normalization $F(\zeta_0) = r(\lambda_0)$ for a fixed $\zeta_0 \in Y$, $\lambda_0 \in \mathbb{P}^1$ such that $p(\zeta_0) = \lambda_0$.

Of course, the covering is defined by the analytic continuation of $\sqrt{r(\lambda)}$ which has two branches μ and $-\mu$, so the function F is in fact $\pm\mu$ (the sign 'chooses' the corresponding sheet) and the choice of the initial germ defined by (λ_0, ζ_0) chooses the point from which one starts the analytic continuation and the value of μ at that point, that is $\zeta_0 = \mu(\lambda_0)$.

Since the poles of $r(\lambda)$ are of second order the covering space Y has two points over $\lambda = 0$ and $\lambda = \infty$ which we denote by ∞_1 and ∞_2.

In the degenerate cases occurs the following. The polynomial $P(T)$ is reducible, it is a square of a polynomial of first order $Q(T)$ (in the case $v_0 = 0$ we have $Q(T) = T - (\lambda - \lambda^{-1})$ and in the case $u_0 = 0$ we have $Q(T) = T - (\lambda + \lambda^{-1})$). Then the two-fold covering we have in the above transforms into two one-fold coverings of $\mathbb{P}^1 \mapsto \mathbb{P}^1$, the covering maps being simply the identity so all the above is redundant. That is why the degenerate cases are very simple. We are not going to discuss them in this article.

The points on the Riemann surface $(Y, p, \mathbb{P}^1)$ will be denoted by $\rho = (\lambda, \mu)$. (Of course $\mu^2 = r(\lambda)$). We have a natural involution $W(\lambda, \mu) = (\lambda, -\mu)$ and taking into account (11) it is also natural to introduce the following involutions:

$$R_0(\lambda, \mu) = (\lambda^*, \mu^*), \; R_1(\lambda, \mu) = (-\lambda, \mu), \; R_2(\lambda, \mu) = (\lambda^{-1}, \mu). \qquad (12)$$

It is clear that we have $p \circ W = W$ and that $R_s^2 = \mathrm{id}$ for $s = 0, 1, 2$.
Let us first discuss the matrix $C = C(\lambda, \mu)$ that diagonalizes $J(\lambda)$ (it is of course an object on the Riemann surface Y, so we shall write $C = C(\rho)$. We set

$$C = \frac{1}{\sqrt{2}} \begin{pmatrix} 1 & 0 & 1 \\ \mu^{-1}(\lambda - \lambda^{-1})u_0^* & -\sqrt{2}\mu^{-1}(\lambda + \lambda^{-1})v_0 & -\mu^{-1}(\lambda - \lambda^{-1})u_0^* \\ \mu^{-1}(\lambda + \lambda^{-1})v_0^* & \sqrt{2}\mu^{-1}(\lambda - \lambda^{-1})u_0 & -\mu^{-1}(\lambda + \lambda^{-1})v_0^* \end{pmatrix}. \qquad (13)$$

It is easy to check that $C^{-1}J(\lambda)C = \mu J_0 = \mathrm{diag}\,(\mu, 0, -\mu)$. The matrix C that diagonalizes $J(\lambda)$ is not unique. We have chosen it to be unitary for real λ since in this case $J(\lambda)$ is Hermitian. Changing μ to $-\mu$, that is, passing from $C(\lambda, \mu)$ to $C(\lambda, -\mu)$, amounts to multiplying $C(\lambda, \mu)$ to the left by $\mathrm{diag}\,(1, -1, -1) = -H_1$ so we obtain $C(W(\rho)) = -H_1 C(\rho)$.

2.2 The Fundamental Solutions. Integral Equations

Now, let us assume that ϕ is a solution to the equation

$$L_{S_{\pm 1}}\phi = (\mathrm{i}\partial_x + \lambda S_1(x) + \lambda^{-1} S_{-1}(x))\phi = 0. \qquad (14)$$

In order to investigate the fundamental solutions of (14) it is useful to introduce the functions

$$\Phi_Y(x, \rho) = C^{-1}(\rho)\phi(x, \rho)\exp\,(-\mathrm{i}\mu J_0 x), \quad \rho = (\lambda, \mu) \qquad (15)$$

which satisfies the equation

$$\mathrm{i}\partial_x \Phi_Y + [\lambda(C^{-1}S_1C) + \lambda^{-1}(C^{-1}S_{-1}C)]\Phi_Y - \mu\Phi_Y J_0 = 0. \qquad (16)$$

Conversely, if $\Phi_Y(x, \rho)$ satisfies (16) then $\psi = C\Phi_Y \exp\,(\mathrm{i}\mu J_0 x)$ satisfy the Eq. (14). For the sake of brevity let is put

$$S(x, \rho) = \lambda(C^{-1}(\rho)S_1(\lambda, x)C(\rho)) + \lambda^{-1}(C^{-1}(\rho)S_{-1}(\lambda, x)C(\rho)). \qquad (17)$$

We shall omit frequently the argument x writing simply $S(x, \rho) = S(\rho)$ and then (16) is written as

$$\mathrm{i}\partial_x \Phi_Y + S(\rho)\Phi_Y - \mu\Phi_Y J_0 = 0. \tag{18}$$

For the purposes of the spectral theory one need functions $\zeta_Y^{n(p)}$ that satisfy the above equation and in addition have asymptotic $\lim_{x\to-\infty}\zeta^n(x) = \mathbf{1}$, $\lim_{x\to+\infty}\zeta^p(x) = \mathbf{1}$. Here and below, for the sake of brevity we shall not write the subscript Y.

Now, from the experience with the Beals-Coifman system and its gauge-equivalent, [2, 6] we know that one needs to consider the cases (a) $\mathrm{Im}(\mu) > 0$ and (b) $\mathrm{Im}(\mu) < 0$ separately. The systems will be written for the functions $\zeta^n(x,\rho)$ and $\zeta^p(x,\rho)$, as already mentioned, we omit the index Y for brevity. Let us introduce the following projections: π_+, π_- and π_0, acting in the set of the 3×3 matrices:

1. If A is a 3×3 matrix, then $\pi_+ A$ is the part of A that lies above the main diagonal, that is $(\pi_+ A)_{ij} = A_{ij}$ if $i < j$ and $(\pi_+ A)_{ij} = 0$ otherwise.
2. If A is a 3×3 matrix, then $\pi_- A$ is the part of A that lies below the main diagonal, that is $(\pi_+ A)_{ij} = A_{ij}$ if $i > j$ and $(\pi_+ A)_{ij} = 0$ otherwise.
3. If A is a 3×3 matrix, then $\pi_0 A$ is the part of A that lies on the main diagonal, that is $(\pi_+ A)_{ij} = A_{ij}$ if $i = j$ and $(\pi_+ A)_{ij} = 0$ otherwise.

With the above notation we write:
Case (a). $\mathrm{Im}(\mu) > 0$, asymptotic at $-\infty$. Solutions are denoted by $\zeta^n(x,\rho)$ and the system of the integral equations runs as:

$$\begin{gathered}(\pi_+ + \pi_0)\zeta^n(x,\rho) = \\ \mathbf{1} + \mathrm{i}\int_{-\infty}^{x} \mathrm{d}y(\pi_+ + \pi_0)\left\{e^{\mathrm{i}\mu(x-y)J_0}(S(y,\rho) - \mu J_0)\zeta^n(y,\rho)e^{-\mathrm{i}(\mu(x-y)J_0}\right\}\end{gathered} \tag{19}$$

and

$$\begin{gathered}(\pi_-)\zeta^n(x,\rho) = \\ \mathrm{i}\int_{-\infty}^{x} \mathrm{d}y(\pi_-)\left\{e^{\mathrm{i}\mu(x-y)J_0}(S(y,\rho) - \mu J_0)\zeta^n(y,\rho)e^{-\mathrm{i}(\mu(x-y)J_0}\right\}.\end{gathered} \tag{20}$$

Case (b). $\mathrm{Im}(\mu) < 0$, asymptotic at $-\infty$. Then we must consider the following system:

$$\begin{gathered}(\pi_- + \pi_0)\zeta^n(x,\rho) = \\ \mathbf{1} + \mathrm{i}\int_{-\infty}^{x} \mathrm{d}y(\pi_- + \pi_0)\left\{e^{\mathrm{i}\mu(x-y)J_0}(S(y,\rho) - \mu J_0)\zeta^n(y,\rho)e^{-\mathrm{i}(\mu(x-y)J_0}\right\}\end{gathered} \tag{21}$$

and

$$\begin{gathered}(\pi_+)\zeta^n(x,\rho) = \\ \mathrm{i}\int_{-\infty}^{x} \mathrm{d}y(\pi_+)\left\{e^{\mathrm{i}\mu(x-y)J_0}(S(y,\rho) - \mu J_0)\zeta^n(y,\rho)e^{-\mathrm{i}(\mu(x-y)J_0}\right\}.\end{gathered} \tag{22}$$

The system are constructed in such a way that in the integrands the jk components of matrices $(S(y,\rho) - \mu J_0)\zeta^n(y,\rho)$ are always multiplied by terms of the type $\exp\left[ia_{jk}(x-y)\right]$ where $a_{jk} = \mu_{jk}$ in case $x > y$, $\mathrm{Im}(\mu) > 0$ and $a_{jk} = -\mu_{jk}$ in case $x < y$, $\mathrm{Im}(\mu) < 0$. From its side $\mu_{kk} = 0$, $\mu_{12} = -\mu_{21} = \mu$, $\mu_{13} = -\mu_{31} = 2\mu$, $\mu_{23} = -\mu_{32} = \mu$. Thus in the integrands of the above integral equations we always have falling exponents ensuring that the kernels of the above integral operators fall exponentially when $x \to \pm\infty$ and $\mathrm{Im}(\mu) \neq 0$. The above systems have been written for the first time in [9], where authors considered degenerate cases.

Then, provided that the function $(S[x,\rho] - \mu J_0)$ has a sufficiently small $L^1(\mathbb{R})$ norm, the above equations have bounded solutions, but the general question about the existence of bounded solutions of the above equations is complicated and according to our knowledge is not discussed anywhere up to now. In this paper we shall assume that such solutions exist. Quite in a similar way are constructed integral equations for solutions $\zeta^p(x,\rho)$ that have normalization at $+\infty$, that is, $\zeta^p(x,\rho) \to \mathbf{1}$ when $x \to +\infty$. In all the cases it is readily checked that if $\zeta^n(x,\rho)$, $\zeta^p(x,\rho)$ are bounded, satisfy the above systems of integral equations, and one can differentiate under the sign of the integrals, then $\zeta^n(x,\rho)$, $\zeta^p(x,\rho)$ are fundamental solutions to (18).

Now what we have when $\mathrm{Im}(\mu) = 0$. This happens on some curves in Y. Assuming that such a curve has orientation, we can consider (if they exist) the extensions of the above solutions from the left and from the right. In this case we shall write to them superscripts '+' and '−' becoming $\zeta^{n,+}(x,\rho)$, $\zeta^{n,-}(x,\rho)$ or $\zeta^{p,+}(x,\rho)$, $\zeta^{p,-}(x,\rho)$.

As we have seen, is important to know the sign of $\mathrm{Im}(\mu)$ for some fixed ρ, so we find first find when we have $\mu = 0$. This is not very hard, one finds after some calculations, that $\mathrm{Im}(\mu) = 0$ either if λ is real ($\lambda \neq 0$) or if λ belongs to the arcs $a_\pm$ which are the closed arcs joining z_2 and z_3 and z_4 and z_1 on the unit circle $\mathbb{S}^1$. We assume them oriented with the orientation induced from the canonical orientation of $\mathbb{S}^1$. So we see that $\mathrm{Im}(\mu)$ must change sign when the projection λ of ρ crosses the real line or one of the arcs $\bar{a}_\pm$. The curves on Y that project into $\mathbb{R}_\pm \cup \infty$, where $\mathbb{R}_\pm$ are the sets of positive and negative real numbers respectively consist of two closed curves γ_1 and γ_2 on the two sheet of Y passing through the points ∞_1 and ∞_2 that lie above $\infty \in \mathbb{P}^1$. We can assume them oriented in such a way that the orientation of the projections are from $-\infty$ to 0 and from 0 to $+\infty$ (on the real line $\pm\infty$ makes sense). The inverse images of each the arcs $\bar{a}_\pm$ are also closed curves $\Gamma_{1,2}$ 'half' of each of them belonging to one of the sheets and 'half' to the other, joining the points ρ_2 and ρ_3 and ρ_4 and ρ_1, where $\rho_i = (z_i, 0)$ and ρ_i belong to both sheets. We can assume that on $\Gamma_{1,2}$ there is the orientation but this time the projection p takes 'half' of each of them to the corresponding arc with positive orientation and the other half (on the other sheet) with the opposite orientation.

As mentioned, the existence of bounded solutions is a very hard problem, which we are not going to address now, so for the future we shall simply assume the following: (a) bounded solutions (except for some points where we have poles and these poles are not on the curves $\gamma_{1,2}$ and $\Gamma_{1,2}$) exist. (b) The limits of the fundamental solutions from right and left moving along the curves $\gamma_{1,2}$ and $\Gamma_{1,2}$ exist and are fundamental solutions to (18). In this case one can make almost immediately some observations.

Proposition 1 *Suppose for given potentials $u(x)$, $v(x)$ and the bounded fundamental solutions $\zeta^n(x,\rho)$, $\zeta^p(x,\rho)$ exist. Then they are unique.*

The proof follows the ideas developed for the Zakharov-Shabat and Beals and Coifman system (CBC system) and is quite straightforward, see [15].

Quite similar to the CBC system some of the reasoning in the proof of the above Proposition could be applied to investigate what is the relation between the solutions $\zeta^n(x,\rho)$ and $\zeta^p(x,\rho)$. We get that

Proposition 2 *Suppose for given potentials $u(x)$, $v(x)$ and the bounded fundamental solutions $\zeta^n(x,\rho)$, $\zeta^p(x,\rho)$ exist and have no singularities in some open set X in Y' where*

$$Y' = Y \setminus (\{\gamma_1\} \cup \{\gamma_2\} \cup \{\Gamma_1\} \cup \{\Gamma_2\})$$

(for a given curve γ by $\{\gamma\}$ is denoted its trace). Then there exists a diagonal matrix $K(\rho)$ analytic in X such that

$$\zeta^n(x,\rho)K(\rho) = \zeta^p(x,\rho), \quad \rho \in X.$$

The proof is based on the observation that since both $\zeta^p(x,\rho)e^{\mathrm{i}\mu x J_0}$ and $\zeta^n(x,\rho)e^{\mathrm{i}\mu x J_0}$ are fundamental solutions of a system (14) one must have a non-degenerate matrix $K(\rho)$ such that

$$\zeta^n(x,\rho) = \zeta^p(x,\rho)e^{\mathrm{i}\mu x J_0}K(\rho)e^{-\mathrm{i}\mu x J_0}$$

and because $\zeta^n(x,\rho)$, $\zeta^p(x,\rho)$ are bounded this can happen only if K is diagonal. Since $\lim_{x\to+\infty}\zeta^n(x,\rho) = K(\rho)$ one can recover for example $\zeta^p(x,\rho$ from $\zeta^n(x,\rho)$ so it is enough to consider only one of them. For this reason *we shall consider only the solutions $\zeta^n(x,\rho)$ and for shortness we shall omit the superscript 'n'*. The above relations are also true for the extensions of the solutions $\zeta(x,\rho)$ on the set $\Sigma = \{\gamma_1\} \cup \{\gamma_2\} \cup \{\Gamma_1\} \cup \{\Gamma_2\}$, from the left (denoted by superscript '+') and from the right (denoted by superscript '−'):

$$\zeta^{n,\pm}(x,\rho)K^{\pm}(\rho) = \zeta^{p,\pm}(x,\rho), \quad \rho \in \Sigma. \tag{23}$$

It is more interesting however to consider the relation between $\zeta^+(x,\rho)$ and $\zeta^-(x,\rho)$. Of course, we are speaking here about ρ such that $\mathrm{Im}(\mu) = 0$ which means that we are on some of the curves $\gamma_{1,2}$, $\Gamma_{1,2}$. In this case the exponential factors are always bounded and we have

$$\zeta^+(x,\rho) = \zeta^-(x,\rho)e^{\mathrm{i}\mu x J_0}G(\rho)e^{-\mathrm{i}\mu x J_0}$$

for some non-degenerate matrix $G(\rho)$ defined on $\Sigma = \{\gamma_1\} \cup \{\gamma_2\} \cup \{\Gamma_1\} \cup \{\Gamma_2\}$, that is this will be the Riemann-Hilbert problem associated with our spectral problem (of course, if we can define it properly). Please note that the set Σ is invariant under the involutions W and R_i, $i = 0, 1, 2$.

2.3 The Effect of the Symmetries

Let us now find how the symmetries we had for our linear problem (1) affect the solutions we introduced. As mentioned, we assume that the fundamental solutions $\zeta(x, \rho)$ exist.

Lemma 1 *Suppose we have the general position boundary conditions. Then we get*

$$\begin{aligned} C(W(\rho)) = -H_1 C(\rho), \quad [C^\dagger(R_0(\rho))]^{-1} = C(\rho) \\ C(R_1(\rho)) = -H_1 C(\rho), \quad H_2 C(R_2(\rho)) H_2 = C(\rho). \end{aligned} \tag{24}$$

As a consequence we obtain that

Corollary 1 *In the case of general position boundary conditions the function* $S(x, \rho)$ *satisfy:*

$$\begin{aligned} S^\dagger(x, R_0(\rho)) = S(x, \rho), \quad (R_1^* S)(x, R_1(\rho)) = S(x, \rho), \\ H_2 S(x, R_2(\rho)) H_2 = S(x, \rho). \end{aligned} \tag{25}$$

Then, using the uniqueness of $\zeta(x, \rho)$, from the integral equations they satisfy one gets

Proposition 3 *In the case of general position boundary conditions the solutions* $\zeta(x, \rho)$ *have the following properties:*

$$\begin{aligned} [(\zeta(x, R_0(\rho))^\dagger]^{-1} = \zeta(x, \rho) \\ \zeta(x, R_1(\rho)) = \zeta(x, \rho) \\ H_2(\zeta(x, R_2(\rho))) H_2 = \zeta(x, \rho). \end{aligned} \tag{26}$$

Finally, in terms of the solutions $\chi(x, \rho) = C(\rho)\zeta(x, \rho) \exp(\mathrm{i}\mu x J_0)$ of the linear problem (1) the above symmetries take the form:

Proposition 4 *In the case of general position boundary conditions the solutions* $\chi(x, \rho)$ *have the following properties:*

$$\begin{aligned} [(\chi(x, R_0(\rho))^\dagger]^{-1} = \chi(x, \rho) \\ H_1 \chi(x, R_1(\rho)) H_1 = \chi(x, \rho)(-H_1) \\ H_2(\chi(x, R_2(\rho)) H_2 = \chi(x, \rho). \end{aligned} \tag{27}$$

3 Elements of Inverse Scattering Method

Let us discuss now briefly how the Inverse Scattering method for our system could be developed in case we will be able to overcome all the difficulties related with the existence of bounded fundamental solutions. We have seen that if ρ is such that $\mathrm{Im}(\mu) = 0$ (ρ is on the trace of one of the curves $\gamma_{1,2}$ or $\Gamma_{1,2}$) we have

$$\zeta^{+}(x,\rho) = \zeta^{-}(x,\rho)e^{\mathrm{i}\mu x J_0}G(\rho)e^{-\mathrm{i}\mu x J_0} \tag{28}$$

where this time we cannot claim that $G(\rho)$ is a diagonal matrix. Strictly speaking we should add here an index to refer on what curve we have these relations but for brevity we shall assume that writing ρ already gives this information, though this is not completely true since the curves intersect. The solutions $\zeta^{+}(x,\rho)$, $\zeta^{-}(x,\rho)$ (in fact these are the solutions $\zeta^{n,+}(x,\rho)$,$\zeta^{n,-}(x,\rho)$) still satisfy the integral equations through which they were defined. That is why the limits when $x \to \pm\infty$

$$e^{-\mathrm{i}\mu x J_0}\zeta^{n,+}(x,\rho)e^{\mathrm{i}\mu x J_0}, \quad e^{-\mathrm{i}\mu x J_0}\zeta^{n,-}(x,\rho)e^{\mathrm{i}\mu x J_0}$$

exist. Let us assume that from the left side of the corresponding curve $\mathrm{Im}(\mu) > 0$ and from the right side $\mathrm{Im}(\mu) > 0$. (If this is not the case one can invert the orientation). Since the solutions are bounded one sees that the limits of $e^{-\mathrm{i}\mu x J_0}\zeta^{n,+}(x,\rho)e^{\mathrm{i}\mu x J_0}$ when $x \to -\infty$ should be upper triangular matrices with units on the diagonal and the limits of $e^{-\mathrm{i}\mu x J_0}\zeta^{n,-}(x,\rho)e^{\mathrm{i}\mu x J_0}$ should be lower triangular with units on the diagonal. Using upper index '+' for upper triangular matrix and '−' for lower triangular matrix we denote:

$$S^{\pm}(\rho) = \lim_{x\to-\infty}\left(e^{-\mathrm{i}\mu x J_0}\zeta^{n,\pm}(x,\rho)e^{\mathrm{i}\mu x J_0}\right). \tag{29}$$

Further, quite analogously, when $x \to +\infty$ the limits of $\zeta^{p,+}(x,\rho)$, $\zeta^{p,-}(x,\rho)$ should be lower (upper) triangular matrices with units on the diagonal.

$$T^{\mp}(\rho) = \lim_{x\to+\infty}\left(e^{-\mathrm{i}\mu x J_0}\zeta^{p,\pm}(x,\rho)e^{\mathrm{i}\mu x J_0}\right). \tag{30}$$

Taking into account that $\zeta^{p,+}(x,\rho) = \zeta^{n,+}(x,\rho)D^{+}$ and $\zeta^{p,-}(x,\rho) = \zeta^{n,-}(x,\rho)D^{-}$ one can simply write

$$\begin{aligned} S^{\pm}(\rho) &= \lim_{x\to-\infty}\left(e^{-\mathrm{i}\mu x J_0}\zeta^{n,\pm}(x,\rho)e^{\mathrm{i}\mu x J_0}\right) \\ T^{\mp}(\rho)D^{\pm}(\rho) &= \lim_{x\to+\infty}\left(e^{-\mathrm{i}\mu x J_0}\zeta^{n,\pm}(x,\rho)e^{\mathrm{i}\mu x J_0}\right) \end{aligned} \tag{31}$$

where we have denoted by 'hat' the inverse because the superscripts become complicated. Finally, passing to the limits $x \to +\infty$ and $x \to -\infty$ in (28) we get

$$G(\rho) = \hat{S}^{-}(\rho)S^{+}(\rho) = \hat{D}^{-}(\rho)\hat{T}^{+}(\rho)T^{-}(\rho)D^{+}(\rho). \tag{32}$$

Thus as usual we obtain for G two Gauss decompositions. The factors in these decompositions are unique and are candidates for the scattering data. Let us see how they are affected by the involutions. Taking into account Proposition 3 we get that

Proposition 5 *In case we have the general position boundary conditions the Mikhailov reduction symmetries affect the Gauss decomposition factors in the following way:*

$$
\begin{aligned}
&((S^{\pm}(\rho))^{\dagger})^{-1} = S^{\pm}(\rho), \quad S^{\pm}(R_1(\rho)) = S^{\pm}(\rho), \quad H_2 S^{\pm}(R_2(\rho)) H_2 = S^{\pm}(\rho) \\
&((T^{\pm}(\rho))^{\dagger})^{-1} = T^{\pm}(\rho), \quad T^{\pm}(R_1(\rho)) = T^{\pm}(\rho), \quad H_2 T^{\pm}(R_2(\rho)) H_2 = T^{\pm}(\rho) \\
&((D^{\pm}(\rho))^{\dagger})^{-1} = D^{\pm}_{+}(\rho), \quad D^{\pm}(R_1(\rho)) = D^{\pm}(\rho), \quad H_2 D^{\pm}(R_2(\rho)) H_2 = D^{\pm}(\rho).
\end{aligned}
\tag{33}
$$

4 Conclusions

From the above we can conclude that the questions about the symmetries of the fundamental solutions to the system $L_{S_{\pm 1}}\psi = 0$ indeed could be formulated on a certain Riemann surface and this makes these issues more transparent than before. However, the most essential part of the spectral theory for $L_{S_{\pm 1}}$ is still to be developed. On the first place, a key question is that of the existence of bounded fundamental solutions and their analytic properties and the work on it has not started yet. After some progress in that issue is achieved, of course there will come the question about the completeness of the adjoint solutions and the application of the corresponding completeness relations, a task that is needed for the proper formulation of the Inverse Scattering Method for $L_{S_{\pm 1}}$ along the lines of the so called AKNS method, [1, 4–6, 10]. The hopes that this could be achieved are strong, since in fact the so called Generating Operators Λ_{S_1} related to the adjoint solutions of $L_{S_{\pm 1}}$ has been found [9, 13].

Acknowledgements The author is grateful to the NRF of South Africa Incentive Grant 2015 for the financial support.

References

1. Ablowitz, M.J., Kaup, D.J., Newell, A.C., Segur, H.: The inverse scattering transform—fourier analysis for nonlinear problems. Stud. Appl. Math. **53**, 249–315 (1974)
2. Beals, R., Coifman, R.R.: Scattering and Inverse scattering for First Order Systems. Commun. Pure Appl. Math. **37**, 39–89 (1984)
3. Foster, O.: Riemannsche Flächen. Springer, Berlin, Heidelberg, New York (1977)
4. Gerdjikov, V.S.: Generalized Fourier transforms for the soliton equations. Gauge-covariant formulation. Inverse Probl. **2**, 51–74 (1986)
5. Gerdjikov, V.S., Yanovski, A.B.: Completeness of the eigenfunctions for the Caudrey-Beals-Coifman system. J. Math. Phys. **35**, 3687–3721 (1994)
6. Gerdjikov, V.S., Vilasi, G., Yanovski, A.B.: Integrable hamiltonian hierarchies—spectral and geometric methods. Lecture Notes in Physics, vol. 748. Springer, Berlin-Heidelberg (2008)
7. Gerdjikov, V.S., Mikhailov, A.V., Valchev, T.I.: Reductions of integrable equations on **A.III**-symmetric spaces. J. Phys. A: Math. Gen. **43**, 434015 (2010)

8. Gerdjikov, V.S., Mikhailov, A.V., Valchev, T.I.: Recursion operators and reductions of integrable equations on symmetric spaces. J. Geom. Symm. Phys. **20**, 1–34 (2010)
9. Gerdjikov, V.S., Mikhailov, A.V., Grahovski, G.G., Valchev, T.I.: Rational bundles and recursion operators for integrable equations on A. III-type symmetric spaces. Theor. Math. Phys. **167**, 740–750 (2011)
10. Iliev, I.D., Khristov, E.Kh., Kirchev, K.P.: Spectral methods in soliton equations. Pitman Monographs and Surveys in Pure and Applied Mathematics, vol. 73. Wiley, New-York (1994)
11. Lombardo, S., Mikhailov, A.V.: Reductions of integrable equations. Dihedral Group J. Phys. A **37**, 7727–7742 (2004)
12. Mikhailov, A.V.: The reduction problem and inverse scattering method. Physica **3D**, 73–117 (1981)
13. Yanovski, A.: Recursion operators for rational bundle on sl$(3, \mathbb{C})$ system with $\mathbb{Z}_2 \times \mathbb{Z}_2 \times \mathbb{Z}_2$ reduction of Mikhailov type. In: Mladenov, I., Ludu, A., Yoshioka, A. (eds.) Proceedings of the Sixteenth International Conference on Geometry, Integrability and Quantization, pp. 301–311 (2015). doi:10.7546/giq-16-2015-301-311, https://projecteuclid.org/euclid.pgiq/1436815752
14. Yanovski, A.: Gauge equivalent integrable equations on sl(3) with reductions. J. Phys. Conf. Ser. **621**, 012017 (2015). http://iopscience.iop.org/article/10.1088/1742-6596/621/1/012017
15. Yanovski, A.: Some aspects of the spectral theory for sl$(3, \mathbb{C})$ system with $\mathbb{Z}_2 \times \mathbb{Z}_2 \times \mathbb{Z}_2$ reduction of Mikhailov type with general position boundary conditions. In: Mladenov, I., Meng, G., Yoshioka, A. (eds.) Proceedings of the Seventeenth International Conference Geometry, Integrability and Quantization, pp. 379–391 (2015). ISSN 1314-3247, https://projecteuclid.org/euclid.pgiq/1450194170

Printed in the United States
By Bookmasters